PLUTONIUM AND THE RIO GRANDE

PLUTONIUM AND THE RIO GRANDE

Environmental Change and Contamination in the Nuclear Age

WILLIAM L. GRAF

New York Oxford
Oxford University Press 1994

Oxford University Press

Oxford New York Toronto
Delhi Bombay Calcutta Madras Karachi
Kuala Lumpur Singapore Hong Kong Tokyo
Melbourne Auckland Madrid

and associated companies in
Berlin Ibadan

Published by Oxford University Press, Inc.,
200 Madison Avenue, New York, New York 10016

Oxford is a registered trademark of Oxford University Press

Library of Congress Cataloging-in-Publication Data
Graf, William L., 1947–
Plutonium and the Rio Grande : environmental change and
contamination in the nuclear age / William L. Graf.
p. cm. Includes bibliographical references and index.
ISBN 0-19-508933-2
1. Plutonium–Environmental aspects.
2. Radioactive pollution of water–Rio Grande Watershed.
3. Radioactive pollution of sediments–Rio Grande Watershed.
4. Los Alamos National Laboratory.
I. Title.
TD427.P63G73 1994
628.1′685′097896–dc20 93-40028

2 4 6 8 9 7 5 3 1

Printed in the United States of America
on acid-free paper

For Kelly, and a future better than the past

The Rio Grande near Otowi, New Mexico. (*a*) The river and the Chili Line in about 1920. (Museum of New Mexico, photo 11886) (*b*) The same view in 1991, when large cottonwood trees had grown up to obscure Black Mesa in the background. The superstructure of the rail bridge has disappeared, but the bridge pilings remain. (W. L. Graf, photo 104-22/23)

Preface

This is a special place on a special river: In the warm September sunlight, the Rio Grande shimmers with bright gloss as it curls and whirls past the old wooden pilings of the abandoned railroad bridge at Otowi. The rail route here was officially known as the Denver and Rio Grande Western Santa Fe Branch, but residents, even fifty years after its demise, still refer to it affectionately as the Chili Line, because of its autumn freight of red chili peppers. There are no chilies now, or trains. Along the river and the old rail route, yellow flowers of chamisa dazzle against the soft gray-green of the plants' leaves and the background of dark volcanic rocks lining the canyon walls and cliffs. The water swirls around the dark wood of the old bridge posts and then drifts downstream to tumble through boulder-studded rapids made by debris brought down to the main river by the stream in Los Alamos Canyon. The white water sings in the rapids with a hissing sigh, punctuated with occasional lapping splashes. The only other sounds are the breeze through the wavering cottonwood leaves, and the scratchy noise of small, curious lizards as they scamper through the paperlike dry, brittle undergrowth. Peaceful, quiet.

Beneath the peace and quiet, however, there is another current in addition to that of the river, and other sounds, other visions. The spirits of this place are strong, and sitting on these old bridge pilings, you can feel, hear, and envision them. The pueblo dwellers who lived nearby and fished from this bank, the Spaniards who came and left, the Mexicans who came and stayed, all left subliminal parts of themselves in this landscape. There are Anglo-American spirits here too: the boys from the ranch on the mesa who used to come here to swim; the engineers who coaxed a little more steam out of their narrow-gauge engines on the Chili Line through this canyon; Aldo Leopold, who rode in the tiny narrow-gauge passenger cars on his way from his Forest Service job to visit his future wife; General Leslie Groves, who came here to build America's atomic bomb factory at Los Alamos; Edith Warner, who served dinner and chocolate cake to her infrequent guests in a little house just up the river; and Robert Oppenheimer who occasionally came down from Los Alamos to have dinner with his wife and friends at Edith's house.

They all are gone now, but their legacy remains. The boys' ranch is now part of the town of Los Alamos. The rails of the Chili Line disappeared for the war effort in the 1940s, but the roadbed still winds through the boulders and chamisa. Edith's house has been replaced, and an ugly highway now covers her beloved garden. The work of Groves and Oppenheimer changed the landscape in subtle but important ways, and their legacy is physical and chemical as well as spiritual. The gray sand and gravel brought down to the Rio Grande from Los Alamos during the spring snowmelt and summer flash floods carries with it plutonium that was released during the early nuclear weapons programs at Los Alamos. In the 1940s and early 1950s, the country was concerned with national survival, and few were worried about the plutonium that made its way onto the sandy canyon floors. But now the nation's agenda includes a deep-seated concern for public health and environmental quality, and so the plutonium in the river system has become a major issue. How much plutonium is in the river system? Does it pose a hazard to life-forms, including people? Where is it? How does it move? Where will it go? How long will it take to get there?

These questions are part of my life's work because I am a geomorphologist, an earth scientist who studies earth-surface processes. I am a geographer who specializes in the analysis of river mechanics. For me, there is a deep sense of pleasure in using science to analyze questions such the issue of plutonium in the Rio Grande. It is much like the sense of pleasure a carpenter feels in using a finely made tool. And there is a sense of satisfaction in the scientific resolution of these questions, much like the satisfaction of the carpenter with a pleasing piece of woodwork. The science notwithstanding, I always return to this place, this landscape, these spirits, and I know again why this work is more than science. America has accomplished many great things, but preserving a quality of life as well as a quality environment is our greatest challenge for the twenty-first century. Efforts to understand places like the Northern Rio Grande in their entirety, with their human and natural histories, are the surest path to achieving this ambition of a quality national life.

The following pages contain a complete account of my effort to answer the questions about plutonium and the Rio Grande and to develop an approach useful to address similar questions in other areas. I have attempted to devise a method of analysis that can be used by workers in other environments faced by similar problems. Because of the sensitive nature of these issues and because there seems to be a great deal of mistrust and misunderstanding about them, I have described in great, and sometimes excruciating, detail how I arrived at my conclusions. My objective has been to produce research that is absolutely transparent, so that its results can be fairly evaluated and duplicated by anyone.

The results of my work show that the plutonium from Los Alamos National Laboratory and atmospheric fallout has been deposited along the Rio Grande in small, though detectable, quantities in certain predictable places. Whether the quantities are hazardous is for other workers to determine—my self-appointed task was to find the element, to explain its "natural" history, and to create a model for other studies in places where similar questions have been raised. I hope that the general approach used for the Rio Grande study can be used, with some modification, for many other parts of the world. In addition, I hope that the results of my work will provide for a safer, more confident future for all those who love and enjoy this particular landscape. In the final analysis, perhaps it will allow all our spirits to rest more easily with the chamisa, the canyon, and the river.

Otowi, N.M.
November 1993

W. L. G.

Acknowledgments

The research into plutonium in the Northern Rio Grande reported on the following pages is the combined product of the efforts of many supportive individuals. I am deeply indebted to Leonard J. Lane, a hydrologic engineer with the U.S. Agricultural Research Service in Tucson, who first brought to my attention the issue of plutonium and the Rio Grande. He supplied important data and ideas during the research, but more important, his encouragement during the several-year effort made the rough spots smoother. Thomas E. Hakonson, a radioecologist in the Environmental Studies Group at Los Alamos National Laboratory (and now at Colorado State University), provided information, encouragement, and an enthusiasm for the research when one or more of those critical ingredients was in short supply. Thomas E. Buhl, a health physicist in the Environmental Surveillance Group at Los Alamos, coordinated my early contacts with the laboratory, managed the bureaucratic aspects early in the project, and provided professional and personal support without which the work simply could not have been done. Steven McLin, a hydrologist in the Environmental Surveillance Group, continued that support in an exemplary fashion. McLin; William D. Purtymun, a geologist; and Alan Stoker, a radiation and materials specialist, all in the Environmental Surveillance Group at Los Alamos, shared their knowledge and experience, making substantial improvements in the final product.

The Department of Geography, University College London, generously provided the venue for the final calculations and revision of the manuscript during my appointment there as Visiting Professor. I am grateful to my hosts there, Professors Richard Munton and Ronald Cooke, for their support. Stanley W. Trimble of the Department of Geography, University of California at Los Angeles, provided valuable historical documents that offered an important perspective on environmental changes along the Rio Grande. John Grubbs, Chair of the Department of

Geography and Environmental Engineering at the U.S. Military Academy, provided useful and thoughtful reflections on the project. Pat D'Andrea of Santa Fe, New Mexico, kindly spent considerable effort to provide helpful details for the final draft of the manuscript. Several anonymous reviewers produced what seemed to be reems of pages filled with useful comment and advice that improved the final product. Joyce Berry and Irene Pavitt, editors at Oxford University Press, provided the endless patience, sound judgment, and perceptive professionalism needed to convert the manuscript into a book.

My graduate students in the Department of Geography at Arizona State University made substantial contributions to this research effort. Judith K. Haschenburger, Martin T. Kammerer, Ted Lehman, and Sandra Clark struggled with the data and tracked down elusive but important pieces of information. Barbara Trapido, the cartographer of the department, provided helpful advice and unmatched skills in generating the maps and diagrams that tell part of the story in the following pages.

Like fine wine and good art, scientific research is not cheap. Through the Environmental Surveillance Group, Los Alamos National Laboratory generously supported my exploration of plutonium in the Northern Rio Grande (Contract 9-X38-2886P-1). The laboratory provided unrestricted access to field areas and data and gave me a completely free hand in pursuing my research. Prepublication copies of the manuscript on which this book was based were circulated at the laboratory as Los Alamos Unclassified Informal Report LA-UR-93-1963. Through its Geography and Regional Science Program, the National Science Foundation supported some of the theoretical development (Grant SES 85-1842601). The Gladys Cole Memorial Award, Geological Society of America, supported the initial work on this and a related project. I am grateful to these organizations for making possible a challenging and exciting adventure in the little-known borderland that is the conjunction of geomorphology, radioecology, botany, hydrology, and human history on the Northern Rio Grande.

Contents

PLUTONIUM AND THE RIO GRANDE

1

Introduction and Related Research

The Basic Issue

Plutonium occurs throughout the earth's environmental systems, though usually in quantities so small that they are barely detectable. Because this artificial element is so toxic, it is necessary to identify those few locations where the concentrations are likely to be the highest. Because almost all plutonium released into the environment is ultimately attached to soil and sediment particles,[1] the behavior of constantly changing natural transport systems such as water and sediment flows (Figure 1.1) provide the key to understanding the ultimate geographic disposition of the element.[2] The general purpose of the work discussed in this book is to explain the distribution of plutonium in the Northern Rio Grande system of northern New Mexico and southwestern Colorado (Figure 1.2) by forging a link among the available data and general principles of environmental sciences such as hydrology, geomorphology, and radioecology.

Between 1945 and 1952, Los Alamos National Laboratory handled large amounts of plutonium as part of the Manhattan Project (the effort to construct the first atomic weapons) and as part of the weapons programs related to the early years of the cold war. During this time, the laboratory emptied untreated plutonium waste into the alluvium of Los Alamos Canyon.[3] After 1952, the laboratory released relatively small amounts of treated plutonium waste. Although the vertical movement of plutonium through the alluvial materials has been largely limited to the upper 10 m,[4] the horizontal movement of the contaminants has had much larger dimensions. The plutonium was adsorbed onto sedimentary particles, and so the fate of those sediments is also the fate of the plutonium. Natural processes of erosion have resulted in substantial movement of contaminated sediments through the canyons. Research during the 1960s and early 1970s showed that since the war years, surface flows within the laboratory's boundaries

Figure 1.1. The changing environment of the Rio Grande is seen along the stream immediately north of Otowi Bridge, New Mexico. (*a*) Looking north from the east bank of the river in about 1910. (Museum of New Mexico, photo 3700) (*b*) The same view in 1991, showing a much narrower channel, filling of the meander on the right with sediment, and much denser growth of cottonwood trees. (W. L. Graf, photo 104–17)

had redistributed at least some of plutonium.[5] Laboratory researchers later estimated that fluvial (river-related) processes in Los Alamos Canyon had probably removed significant quantities from the laboratory area by carrying the plutonium into the Rio Grande.[6] They predicted that early in the twenty-first century almost all of the plutonium would have been emptied from Los Alamos Canyon into the Rio Grande.

Figure 1.2. The general location of the Rio Grande in New Mexico and Colorado.

During the 1990s, increasing national concern for environmental quality and a more stringent regulatory system require the design and maintenance of effective monitoring and surveillance programs to account for environmental contaminants such as plutonium. Los Alamos National Laboratory has an extensive store of data collected from a variety of environments, but the processes in Los Alamos Canyon and the Rio Grande indicate the need for a better understanding of the river systems that move and store the plutonium. The present sampling programs aimed at active river sediments have not detected plutonium in concentrations above those expected from fallout alone.[7] Therefore, either the plutonium from the laboratory has been dispersed by river processes or the sampling effort has not yet identified the plutonium's storage sites.

A major issue in designing an effective contaminant-sampling program for river systems is understanding the context in which the processes operate. Because rivers behave erratically over periods of decades, those contaminants introduced into the system at one time (when diluting uncontaminated sediments are available in large quantities) may be less significant than they are at other times (when only small quantities of diluting sediments are available). The deposition of potentially contaminated sediments is geographically discontinuous, a fact that influences the choice of sample sites in an effective sampling program. Although the Northern Rio Grande system is relatively rich in data describing river processes, these data have not heretofore been used to explain contaminant movement and storage. Thus we have little information about the geographic distribution of dated sediments and landforms that might be plutonium storage sites.

The geographic components of hydrology (the science of water) and geomorphology (the science of earth-surface processes) offer useful principles for explaining the dynamics of river systems that transport and store radionuclides. Unfortunately, there has been little specific theoretical and applied work in these sciences to explain the physical mobility of radionuclides. Whereas the biological sciences have employed radioecology as a distinct body of knowledge pertaining to the dynamics of radionuclides in life-forms, hydrology and geomorphology have spawned only an embryonic literature concerning the dynamics of radionuclides in physical environments. Scientifically and legally defensible sampling programs and explanations for their findings therefore must reinterpret hydrogeomorphic principles in light of their radiological implications.

The study area for this work will be referred to in the following pages as the "Northern Rio Grande," a term that includes the main stream and its tributaries from Española (essentially its confluence with the Rio Chama) to San Marcial (the headwaters of Elephant Butte Reservoir south of Socorro). Although there are no formally accepted definitions of the terms, "Upper Rio Grande" as used by residents and writers in New Mexico often refers to the river north of Española, a usage adopted in this work. I do not use, however, the term "Middle Rio Grande" (except when it is part of the proper name of an organization) because it has had a variety of usages in official government parlance, generally referring to some segment of the river between Cochiti and San Marcial.

Specific Objectives

In order to simultaneously develop general principles and contribute to the practical aspects of an informed sampling program for the Northern Rio Grande, this book has the following objectives. In addition, I hope that by using the Rio Grande in northern New Mexico as a representative example, the results of this effort will be general enough for other, similar environments. My specific objectives are to

1. Outline the physical environmental context of the Northern Rio Grande. The river-channel processes affecting the plutonium's mobility operate within a particular matrix of landforms, soils, geology, climate, and vegetation. The geographic variability of these subsystems partly explains the behavior of the entire river system.
2. Specify the mass budgets of surface water and river sediments in the Northern Rio Grande system. Although the water transports little plutonium, it is the primary vehicle for energy in the system, and its variability in terms of total annual flows and flood peaks strongly influences the movement of sediment through the system. The sediment is important because plutonium is strongly adsorbed onto sedimentary particles. No process explanation can be effective unless it is couched in a clear understanding of the magnitudes of mass transfers in the general system.
3. Describe the major changes in the operation of the Northern Rio Grande system from the 1940s to the 1980s. Hydroclimatic changes and the construction of major engineering works in the region significantly altered the channel processes of the river from the 1940s to the 1980s. Channels, near-channel landforms, and deposits reflect these changes and are potential storage sites for plutonium.
4. Develop mapping techniques to identify the locations of river deposits likely to contain plutonium. Methods to date the deposits are also required because releases of plutonium from laboratory and fallout sources vary from year to year. The mapping and dating methods rely on connections between the hydrogeomorphic systems and riparian vegetation.
5. Explore 11 limited but representative reaches of the Northern Rio Grande to discover the nature of sediment storage in the system and to identify useful sampling environments. Because the length of river possibly of interest is more than 300 km, a detailed analysis of a few relatively short reaches can replace a costly and detailed investigation of the entire length.

6. Use the water and sediment mass budgets with previously collected radiological data to construct a regional budget for plutonium in the surface environment. Because of the quality of radiological data, this effort is only a first approximation, but it is useful in outlining the relative roles of laboratory and fallout contributions and provides an order-of-magnitude picture of the plutonium system.
7. Use the geographic analysis of the distribution of sediment storage to estimate the distribution of plutonium in storage in the river system. A distributional analysis is the foundation of an effective, rational sampling program.
8. Design a simple, effective, inexpensive, and defensible sampling program based on the preceding points. Such a program would consist of a series of guidelines useful for radiological professionals but understandable to the educated layperson.
9. Assemble a data bank for the physical system of the Northern Rio Grande that can serve as a supporting document for public discussions about policies for the Northern Rio Grande. Others can also use the data from diverse sources to address their own research questions.

Related Research on Radionuclides

Although most radioecological literature views "soils and sediments" as a general catchall term, from a geomorphologic perspective the terms refer to specific environmental components. Soils are surface materials whose organic and inorganic components host living organisms. Soils may develop on materials that have been transported and deposited in accumulations, or they may develop *in situ* as the result of the weathering of bedrock. In contrast, sediments are unconsolidated surface materials that have been transported and deposited. The difference is significant from a radioecological standpoint because soils may receive contaminants from either atmospheric fallout or direct surface additions. Sediments may also receive such inputs, but they also may contain contaminants collected elsewhere that have been subsequently transported along with the sediments. Because sediments have the additional property of sorting, which occurs during transport (especially by flowing water), the physical separation of small or light-weight particles from large or heavy ones is common. And because heavy metals and radionuclides tend to adsorb to higher concentrations on the finest particles, the natural variability of sediments is reflected in the variability of their contaminant concentrations. Heavy metals and radionuclides also have an affinity for organic materials that may be a significant component of soils but that may be entirely absent from sediments.

The purpose of the following pages is to summarize the generally available scientific literature concerning the relationship between river sediments and the contaminants of radionuclides and heavy metals. Soils are mentioned only in relation to this general theme. This review is not exhaustive but is designed to define the primary threads of research that have emerged. I discuss heavy metals and radionuclides for two reasons. First, the literature on radionuclides is so limited that few well-defined threads have yet emerged. A clearer picture, however, is available if we include the literature on heavy metals. Second, heavy metals and radionuclides are likely to behave in a similar manner in rivers because most of the radionuclides important to long-term environmental quality are heavy metals.

Surface materials, whether soil or sediment, are the principal reservoir of heavy metals and radionuclides in the natural environment. Of all the various organic and inorganic compartments in the natural environment, sediments almost always contain the largest quantities of heavy metals.[8] In river systems, water quality is the most immediate concern because of its use for human consumption, but river sediment generally contains much higher concentrations of heavy-metal pollutants than water does.[9] Radionuclides have a similar distribution, and researchers have often concluded that soils and sediments are the major repository for plutonium.[10] For example, in contaminated forest ecosystems at Oak Ridge, Tennessee, and Los Alamos, New Mexico, the soil and surface sediments contain more than 99 percent of the plutonium; the biotic components of the systems contain less than 1 percent.[11] The compartmentalization of other heavy metals in rivers is similar: a global survey showed that more than 90 percent of most heavy metals in the river systems is carried in the sediment load.[12] Because the variability of plutonium concentrations in soils and sediments is greater than that in the biotic compartments,[13] small sample volumes or numbers are problematic.[14] That is, the radionuclides that find their way into biotic systems usually do so from soils and sediments.[15]

Of all the actinides, plutonium is the most-studied element in regard to its behavior in the natural environment. Much of the previous work has focused on chemical characteristics and has shown that at the pH values of most natural soils and sediments, plutonium may be chemically unstable until it is adsorbed onto soil particles. Thereafter, the element becomes generally stable in chemical terms. Reviews of the chemical properties of plutonium indicate that fallout and industrial plutonium behave similarly, and that the most important soil properties affecting plutonium's chemical mobility are pH, together with clay, calcium carbonate, manganese, iron, and organic content.[16]

In highly acidic conditions ($1 < pH < 2$), most of the soluble plutonium occurs in ionic form.[17] In alkaline conditions ($pH > 8$), plutonium becomes relatively unextractable.[18] The fixation of plutonium in soils and sediments is most strongly developed with clay particles, and high cation-exchange capacity results in rapid adsorption onto soil particles when clay is present.[19] Calcium bentonite rapidly fixes a variety of forms of plutonium,[20] and the fixing properties of calcium are especially pronounced in near-neutral or slightly alkaline pH conditions, those most commonly encountered in the natural environment.[21] Organic material also accelerates the fixation of plutonium, and in soils with more than about 3 percent organic content, the impact of organics on plutonium binding dominates other considerations.[22] In aquatic environments, manganese and occasionally iron behave as scavanging agents for plutonium.[23] In some coastal sediments, more than 85 percent of the plutonium is associated with the iron or manganese coatings on particles.[24] Whatever the process of fixation, however, most chemical studies indicate that the plutonium adsorbed onto soil particles is generally unavailable to plants, that only up to about 10 percent may become involved with organic processes.[25]

The overall effect of these chemical activities resulting in rapid and secure fixation in most natural settings is a distinct vertical distribution of plutonium in soils and sediments. Fallout plutonium and plutonium released onto soil surfaces by means of industrial processes are highly concentrated near the surface. In the Savannah River area, for example, 84 percent of the soil plutonium is within 5 cm of the surface, and 90 percent is within 15 cm.[26] The concentration of fallout plutonium in Japanese soils declines exponentially with depth, and more than 80 percent of plutonium 239 and plutonium 240 is within 10 cm of the surface.[27]

In soils some plutonium moves vertically, partly because of the physical movement of small particles with adsorbed plutonium. The rate of movement depends on the distribution of particle sizes in the soil. In German soils, the vertical movement is 0.8 to 1.0 cm per year.[28] In dryland conditions near Trinity Site, New Mexico (site of the first nuclear detonation), 20 years after the initial release, about 50 percent of the soil plutonium had moved from the 0- to 5-cm layer to the 5- to 20-cm layer.[29] At a geomorphological scale, such movement is inconsequential because it is slow and has little effect on the horizontal mobility of the materials.

Process studies (rather than static mapping approaches) are a key to the development of sampling or monitoring programs, the explanation of observed distributions of contaminants, and the predictions of future distributions because the sediments that contain most of the contaminants are highly mobile, especially on time scales of several decades. In just one storm at Los Alamos, surface water runoff transported 1 to 2 percent of the entire sediment-bound inventory of plutonium.[30] The geomorphic literature on radionuclides and heavy metals has identified five distinct avenues of inquiry and explanation for these processes: (1) slope processes, (2) drainage basin processes, (3) channel processes, (4) floodplain processes, and (5) general mass-budget approaches.

Geomorphic studies of contaminants in slope processes have focused on the transport and storage of cesium 137, by using the radionuclide as a tracer for erosion studies. Cesium 137, produced primarily during atmospheric detonations of nuclear weapons, occurred as global atmospheric fallout, with the peak rates of deposition in the Northern Hemisphere in 1957–1959, 1962–1964 (major peak), and 1974.[31] Because the cesium strongly adsorbed onto soil particles (especially the clay)[32] and because it was rarely taken up by vegetation, the element is concentrated in the upper soil layer. Overland water flows that erode the soil also erode its associated cesium burden, so that in limited areas the amount of cesium remaining is an indication of the amount of erosion since it was added to the soil.[33] The difference between the cesium added to the soil (usually directly related to annual precipitation in the United States) and the concentrations measured at a later date (adjusted for radioactive decay) is an indication of the magnitude of erosion.[34]

Numerous case studies using this cesium method have refined our understanding of soil loss rates.[35] Loss rates can be mapped using cesium to define the regions and locations of greatest erosion.[36] The occurrence of cesium in the vertical profile of sedimentary deposits and the destination of soil materials eroded from slopes also help determine sedimentation rates along foot slopes and stream courses.[37] But despite these advances using cesium as an indicator of erosion processes, there are no basinwide studies that use these processes to explain the distribution of cesium. No cesium mass budgets have been published in the readily accessible literature, although that literature may be extensive and detailed enough to permit the development of such budgets. The major value of the cesium-erosion literature for this book, therefore, is that it demonstrates how slope processes deliver fallout radionuclides to rivers for further transport.

Once radionuclides attached to sediment particles enter stream channels, their fate becomes intertwined with channel processes. A large-scale effort to explore the connection between channel processes and radionuclide transport was a research project by the U.S. Geological Survey during the early 1960s. The two primary topics of this project were investigations of the dispersion of radionuclides suspended in flowing water, and processes related to bed materials. Controlled experiments in flumes,[38] in a wide range of natural streams,[39] and in the Clinch River, Tennessee, showed that when the contaminants were released at a particular point, downstream processes dispersed the suspended materials.[40] As the distance from the point of injection increased, the concentrations of contaminants declined exponentially.[41]

The U.S. Geological Survey work also explored bedload processes related to radionuclides. Although the concentration of contaminants per unit weight was highest in the fine particles, significant quantities also adhered to the larger bed sediments, mostly through a process of cation exchange.[42] The bottom sediments of the Clinch River emerged as a significant component of the radionuclide storage in the river system.[43] Bedload transport models therefore became important to predicting the movement of the contaminants, and research in the Clinch

River as well as in the North Loup River of Nebraska (using marked tracer sediment) revealed much about the bedload transport process.[44] The survey also conducted some research using radioactive tracers in sediment in a conveyance channel of the Rio Grande near Bernalillo, but the results apparently never appeared in formal publications.[45]

Studies of radionuclides and heavy metals in transit through river channels languished after the U.S. Geological Survey projects were completed. Radioecologists, however, correctly identified the importance of fluvial sediment transport for contaminants. Research by biologists and geoscientists at Los Alamos has highlighted the importance of runoff in transporting plutonium in the surface environment,[46] and it has demonstrated the importance of physical transport of plutonium in making the element available to life-forms.[47] Evidence collected in investigations of plutonium and cesium 137 in Los Alamos Canyon has shown that stream sediments are the main reservoir of radioactivity from Los Alamos National Laboratory's waste-disposal activities and that runoff processes are moving the sediments downstream.[48] In a more general sense, radioecologists found that "the distribution of transuranic elements from point sources at nuclear facilities typically produces decreasing concentrations with distance from the source."[49]

Although applicable to short-term atmospheric and suspended river-sediment processes in highly controlled conditions, this statement is probably not true for most natural river systems. Because contaminants adhere mainly to the fine sediments that are sorted by fluvial processes,[50] most streams produce unequal geographic distributions of contaminants. L. J. Lane and T. E. Hakonson offered an enrichment ratio, borrowed from agricultural researchers,[51] to account for the resulting variations in concentrations.[52] They and their associates then used engineering-based models of hydraulic processes to illuminate further the probable mechanisms of transport in small to medium channels and predicted the movement of contaminated sediments through Los Alamos Canyon.

A geomorphological approach to the contaminant–sediment transport problem complements the engineering-based interpretations. The concentration of thorium 230 downstream from an accidental spill in the Puerco River, New Mexico, does not decrease with distance from the source, as might be predicted by the standard diffusion models.[53] On a time scale of several weeks, the wavelike distribution of thorium concentrations in the downstream direction may result from the movement of contaminated sediment through the system in pulses, a phenomenon commonly observed in rivers.[54] On a shorter time scale, the wavelike downstream distribution of thorium results from geographic variation in the stream's ability to transport heavy metals and sediment.[55]

Researchers in Australia also investigated radionuclide transport in rivers, mostly in regard to mill tailings.[56] The explanation of the distribution of contaminated sediments through river systems in Australia depended on (1) the characteristics of the sedimentary environments along the channels, (2) the physical properties of the contaminated sediments that influenced their transport, and (3) the mixing of contaminated and uncontaminated materials.[57] The researchers argued that environmental managers and planners could take into account these points in a general, if not a precise, quantitative way.[58]

Related Research on Heavy Metals

Although little published research is available pertaining to the fluvial transport of radionuclides, more information appears in the literature regarding other heavy metals. Using a commonly held definition of heavy metals as those elements with a positive chemical valence and a density equal to or greater than 4.5 g/cu cm,[59] all the elements in the periodic table beyond actinium (that is, the actinide series which includes uranium and plutonium) are "heavy metals." The significance of this definition is that all heavy metals behave similarly from the perspective of their physical transport in rivers. Although the metals may exhibit various characteristics of chemical mobility, they all adsorb onto sedimentary particles,[60] and those particles are likely to be distributed by physical processes into depositional patterns that result from sorting by weight.

The geographic and geomorphologic implications of high density for heavy metals are connected to the shear stress required to move the particles with adsorbed metals. Particles of common quartz, for example, have a density of 2.65 g/cu cm, and those of plutonium dioxide have a density of 11.46 g/cu cm.[61] Standard, widely accepted hydraulic principles provide the means of defining the relationship between flowing water and sedimentary particles,[62] and they indicate that particles of plutonium dioxide require six times more shear stress for their initial motion than do quartz particles of similar size.

On a larger scale, the movement of sediments and attached metals through stream systems creates a particular geography of concentrations downstream from their sources. K. E. Carpenter, a fisheries biologist, first recognized the significance of river-channel processes in metal transport.[63] Her work in Welsh lead-and zinc-mining districts led to later studies in the same region that indicated some materials were temporarily deposited before being remobilized, thereby demonstrating that internal storage along and near channels was a source of pollutants.[64] Further analyses of a variety of streams in Europe, Great Britain, and the United States showed that in channels the downstream decline in concentrations could be modeled using simple distance decay functions.[65] In active sediments, this phenomenon is partly the product of mixing and being diluted with uncontaminated sediments.[66]

Heavy-metal concentrations are usually highest in the finest stream sediments,[67] and in most rivers the finest materials accumulate on flood plains. As a result, much of the research by fluvial geomorphologists interested in metals has focused on flood plains as repositories for the metals, in which each layer of sediments and associated metals is often the product of a flood.[68] As sediments accumulate and bury previous deposits, a vertical stratigraphy develops with varying amounts of metals in each layer. The metal content can indicate the date of emplacement of individual strata if the date of the pollutant's introduction into the system is known.[69] Channel and flood-plain studies have been concerned with a variety of common heavy metals, usually copper, zinc, lead, and cadmium, but the conclusions are probably applicable as well to the chemically stable elements of the actinide series, particularly some forms of plutonium and americium. Unlike humid-region rivers where flood plains are the focus of storage, in some dryland rivers the metal concentrations may be highest in active channels where frequent additions by flows inject metals more often than they do in overbank areas.[70]

Sediments and attached heavy metals and radionuclides that originate from the erosion of hillslopes or waste sites therefore do not simply enter river systems and move directly to seas, oceans, or reservoirs. Instead, many of them are stored internally in channel sediments and on flood plains, thereby reducing the amount output by the system. However, attempts to construct mass budgets of heavy metals or radionuclides for entire river systems have been uncommon, partly because of the lack of data. Such budgets require information about the water and sediment fluxes over large areas, because the movement of these materials determines the movement of the contaminants. Water and sediment data are expensive to collect and are not generally available in the same locations or regions for which reliable metal or radionuclide data are available. There are, however, at least four regional budget studies of heavy metals. In Europe, investigators derived aspatial budgets (those with compartments for various components of the environment without geographic identity) for lead, copper, zinc, nickel, chromium, and cadmium for the 4,488-sq-km basin of the Ruhr River.[71] A more general geographically correct budget for mercury is available for the Upper Colorado River Basin in the 279,500-sq-km drainage area above Lake Powell in the western United States.[72] In the Netherlands, sediment and soil sampling have produced aspatial budgets for lead, copper, zinc, and cadmium for the River Geul, a 350-sq-km watershed.[73] An investigation of lead and arsenic along a 121-km study reach of the Belle Fourche River in South Dakota revealed that flood plains have stored one-third to one-half of the metals entering the system from mine tailings.[74]

Regional budgets for radionuclides are not available for river systems, although budget studies for parts of the environment have used limited compartmental approaches. Estimates of the yield of fallout plutonium from hillslope erosion into rivers suggest that probably less than 10 percent traveled from the rivers to the oceans.[75] Lakes, reservoirs, channel sediments, and flood plains must therefore store the remaining 90 percent. In studies that bordered on budgetary approaches, small-scale, probabilistic, agricultural-based models have predicted rates of removal of plutonium from a 33-sq-km watershed at the Nevada Test Site.[76] Efforts to model the movement of plutonium in Los Alamos Canyon have also contained some mass-budget perspectives.[77] However, none of the previous work on radionuclides has attempted a detailed, regional, geographically specific budget for a large watershed.

Summary and Generalizations

This brief review indicates that the formal literature on the physical mobility of heavy metals and radionuclides in river systems is sparse. The few studies in print indicate that monitoring and surveillance efforts, construction of theoretical explanations, and practical prediction of the distribution of contaminants in the river-channel environment must take into account the following established generalizations for radioactive, heavy-metal contaminants:

1. Contaminants adhere to sedimentary particles and do so in the greatest concentrations in the finest material.
2. Once adsorbed onto sedimentary particles, most metal contaminants (including plutonium) remain relatively stable, especially in environments with a high pH.
3. Contaminants are denser than are most natural sedimentary particles, and therefore fluvial processes are likely to preferentially sort and deposit them.
4. River systems store large quantities of sediments and associated contaminants in flood plains and along channels, especially in those systems undergoing aggradation.
5. Contaminants exist in most river systems because of either the erosion of waste disposal sites or the addition of global atmospheric fallout.
6. There is considerable geographic variation in the concentration of contaminants in river systems as a result of geographically and/or temporally discontinuous sedimentation and variable transport rates.

Important unresolved issues include the following:

1. It is not clear whether channel sediments or flood-plain sediments contain higher concentrations of contaminants (the flood-plain materials are generally finer, but the channel materials may contain more contaminants).
2. The distribution of contaminants in flood plains is not clear for those rivers undergoing radical channel change, especially channel shrinkage with lateral accretion rather than simple vertical flood-plain accumulation.
3. For most systems, the relative importance of industrial versus atmospheric inputs for radionuclides over distances greater than a few kilometers is not known.

4. Except for a few relatively small and isolated systems, the rates of change for transport and storage for contaminants is not known.
5. The connections among rates of energy expenditure, geographic distribution of energy, and the resulting distribution of contaminants are not clear.

Because soils and sediments are the major reservoir for heavy metals and radionuclides and because they are mobile, geographic and geomorphologic analyses of surface processes are critical to understanding the distribution of the contaminants. Radioecologists recognize the need for surface process studies: "Water flow . . . can and does act as an important agent of radionuclide concentration and redistribution."[78] Environmental geochemists recognize the same need: "Studies of geomorphology . . . are basic to a deep understanding of the geochemistry of landscapes."[79] For plutonium, an international review of research showed that "physical transport mechanisms become significant when compared with chemical transport mechanisms."[80] A summary review of heavy metals in general concluded that "a proper understanding of the dynamics (erosion, transportation, sedimentation) is not only essential within most sedimentological contexts but additionally for the prediction of the fate of sediment-bound contaminants."[81] Geomorphologists, however, have yet to explore to any significant degree the subject of heavy-metal or radionuclide dynamics. Although geochemists have devised techniques using stream sediments to locate ore bodies,[82] process specialists have made only a few studies of the subject.

The distribution of plutonium in the Northern Rio Grande system thus is analyzed against a background of some general knowledge, but with significant theoretical gaps. The information from previous work on the contaminants provides clues to geomorphic change, as in the case of the cesium 137 approach to quantifying hillslope erosion. In the case of rivers such as the Rio Grande, however, theoretical structures and empirical data are stronger for the geomorphic system than for the plutonium system. Therefore, my research attempts first to define and explain geomorphic processes and then to use that understanding to unravel the likely fate of the contaminant.

2

Plutonium and Los Alamos

The History of Plutonium

The plutonium in the Northern Rio Grande is entirely artificial. Small amounts of plutonium may have formed in exceptionally rich uranium deposits in south-central Africa, but for practical purposes, until its manufacture in 1939, the element did not occur in the earth's environment. Although the detailed story of the origins of plutonium are beyond the scope of this book, a summary of that history does clarify the issues regarding plutonium in the Northern Rio Grande in the late twentieth century. The purposes of this chapter are to review the origins of plutonium and to examine briefly the nature of that element.[1]

Modern nuclear physics, which ultimately led to the production of plutonium, began with the publication of the discovery of X-rays by Wilhelm Conrad Röntgen in 1896. His work showed that the physical world was much more complicated than previously thought and that energy could be emitted from substances. In the same year, Henri Becquerel of Paris showed that uranium emitted radiation, and soon thereafter Marie and Pierre Curie coined the term *radioactivity* to describe the emissions they recorded from two newly discovered elements, radium (named after its radiative properties) and polonium (named after Marie Curie's home country of Poland).[2] Between 1898 and 1902, Ernest Rutherford of Cambridge University and, later, McGill University explored processes of radioactive decay that generated free electrons (beta radiation) and bursts of energy (gamma radiation) and discovered that some elements changed their basic properties during the emission. Rutherford termed these changes *transmutation* and laid the philosophical foundations for understanding atomic structure.[3]

The transmutation of elements was a significant addition to the rapidly expanding knowledge about the number and types of elements in the natural world. Between 1894 and 1900, William Ramsey enlarged the periodic table with an entire family of inert gases, and by 1903 more than a dozen radioactive elements were known. By 1903, it was obvious that the decay process explained many observed elemental changes: Americans Bertram B. Boltwood and Herbert N. McCoy showed that radium descended from uranium, and Otto Hahn connected several types of thorium. Rutherford moved his research to the University of Manchester in 1907, and there, along with Hans Geiger, he began to examine the rates of emission and decay processes. Geiger later developed, to count emissions, the now-familiar instrument that bears his name.[4]

Niels Bohr joined Rutherford, and together they developed a comprehensive theory to describe and explain the structure of atoms.[5] Between 1908 and 1920, the theory evolved to include both alpha radiation (consisting of helium atoms) and protons and neutrons. An associate of the group, Frederick Soddy, introduced the term *isotope* to describe the different atomic varieties of the same element that had the same chemistry but different atomic weights.[6]

It was a short step from description and explanation to manipulation. Once the general nature of the atom was known, several workers set about the task of changing it artificially. The idea of bombarding atoms with protons to change their structure was current during the 1920s. This bombardment was most effective if the protons could be accelerated to high energy levels, leading to the invention of several schemes to increase the efficiency of the "atom smashers." One of the more successful efforts was by Ernest O. Lawrence and M. Stanley Livingston of the University of California at Berkeley. Their system, based on the precept of accelerating ions around a curved track using magnets, resulted in the production of several cyclotrons, or particle accelerators, during the 1930s.

After about 1932, several research groups experimented with bombarding elemental targets with alpha particles, protons, positrons (positively charged electrons), and neutrons. The result was a prolific industry in the production of new isotopes, previously unseen varieties of many common elements. The most active research groups were those in Paris (Irène Curie and Frédéric Joliot), Rome (directed by Enrico Fermi), and Berlin (Otto Hahn, Lisa Meitner, and Fritz Strassmann). The Rome group systematically explored the periodic table and found that the heavy elements captured and retained neutrons most readily while emitting a beta particle. As a result, the element moved one step up in the periodic table. This discovery was significant because it meant that transmutation could occur not only down the table with a loss of energy, as discovered earlier, but also occur up the table with an addition of energy.[7]

The Berlin group worked with the heaviest elements and found that they could create an entirely new family of elements. They bombarded uranium with neutrons and created several transuranic elements (elements higher than uranium in the periodic table). The chemical mixture resulting from these experiments was so complex, however, that they were unable to sort out the various components. In 1935, Hans Bethe, R. F. Backer, and M. Stanley Livingston undertook an extensive review and summary of the research to that point. They developed the theory of "fusion" to explain the creation of the new elements in which the nuclei of various atoms fuse together to form the new element. In 1938, the Paris and Berlin groups explained some of their new products as the result of "fission," in which the nuclei divide or split to create two new daughter products.[8]

Because the processes of fusion and fission release huge quantities of energy, the researchers began to consider the military implications of their work. By the late 1930s, a global war of proportions not earlier contemplated appeared inevitable, and accordingly, the possibility of controlling energy through fusion and fission became a strategic consideration. The power of weapons manufactured to take advantage of this energy became even more awesome when the possibility of a chain reaction appeared in 1939. Because fission might cause a uranium atom to release a neutron, that neutron might trigger fission in another nearby uranium atom, making the explosive release of energy self-sustaining. Leo Szilard, an associate of Fermi, along with Albert Einstein, petitioned President Franklin D. Roosevelt to support the exploration, for military applications, of the new developments in nuclear physics. They were worried that German research might produce similar weapons capable of destroying American cities. Roosevelt approved, and in 1940 Fermi (now at Columbia University) received military funds to continue his work attempting to generate a chain reaction. After two years of work and a move to the University of Chicago, Fermi and his group successfully generated a chain reaction on December 2, 1942.[9]

Fissionable materials therefore suddenly became very important in the early 1940s because they would be the fundamental substance of an atomic weapon. Uranium 235 is fissionable but is found in only tiny amounts in deposits in which it is associated with the much more abundant uranium 238. Separating the two isotopes is difficult and requires large quantities of ore, which were in short supply. Alternative fissionable materials were therefore highly desirable, and transuranic elements offered possible substitutes. In 1940, Philip Abelson and Edwin McMillan of the University of California at Berkeley conceived of elements 93 (eventually named neptunium) and 94 (eventually named plutonium) as the beginning of a series of transuranic elements in the periodic table similar to the rare earths. They obtained evidence that element 93 emitted beta particles and therefore that it must transmute into element 94. Louis Turner of Princeton University deduced that element 94 must have a fissionable isotope and that this isotope could be created by adding neutrons to uranium 238.[10]

Element 94 was the special interest of Glenn T. Seaborg of the University of California at Berkeley. In 1939, chagrined at missing the discovery of fission, he focused his energy on exploring elements 93 and 94, especially the production and isolation of element 94. During the summer and fall of 1940, he, Edward McMillan, and Arthur Nahl tried bombarding uranium with neutrons, and they produced what appeared to be an isotope of a new element with the atomic number of 94. On February 23, 1941, Seaborg and his team of chemists isolated element 94 combined with thorium, but they could not yet produce it in pure form. On March 28, 1941, Seaborg demonstrated that plutonium 239 was fissionable. The chemical isolation of plutonium was a formidable task that required the development of special tools for handling microscopic quantities of material. It was not until August 20, 1942, that Seaborg's group successfully precipitated a particle of pure plutonium. Generated from more than 1 kg of uranium, the plutonium was a microscopic particle of less than 1 μg, but from that time on, plutonium was a physical reality rather than a hypothetical concept.[11] On September 10, 1942, B. B. Cunningham and L. B. Lerner of the Metallurgical Laboratory at the University of Chicago isolated and weighed the first visible quantity of plutonium: 2.77 μg.[12]

Seaborg named element 94 plutonium after the planet Pluto, discovered in 1930. Martin Kaproth had followed a similar logic in 1789 when he named uranium after the then newly discovered planet Uranus. Edwin McMillan suggested that accordingly, element 93 should be named neptunium after the planet Neptune. Seaborg chose Pu as the chemical symbol for the new element rather than Pl, partly to avoid confusion with platinum (Pt) but also partly on a whim to create attention. He thought the element would be nasty to deal with, so he derived the symbol from "P.U.," a slang term for "putrid" or "smelly."[13] The new element 94 therefore came to carry the name of the ancient Greek god Pluto, ruler of the underworld, god of the earth's fertility, and god of the dead.

Plutonium and Los Alamos

Once defined by science, plutonium became a primary object of engineering and industrial activity. The industrial structure needed to build an atomic bomb had begun even before the plutonium was isolated. In April 1941, at the direction of President Roosevelt, Vannevar Bush, director of the Carnegie Institution of Washington, created a special section (S-1) of the Office of Scientific Research and Development. James B. Conant, president of Harvard University, became his deputy in charge of the theoretical and industrial aspects of bomb development. Arthur H. Compton, a physicist at the University of Chicago, was the major managing committee leader. Within a year, laboratories at several institutions across the nation were connected into an interlocking network, with each unit addressing a specific part of the problem.[14]

The acquisition of restricted industrial materials and plants during World War II, however, made a civilian production effort impossible. It was clear to Bush and Conant that the military would have to be directly involved, and it was at their behest that the U.S. Army became a working partner in the bomb project. The army established a special office in New York for the project in August 1942, labeling it the Manhattan Engineer District and appointing General Leslie R.

Groves as its commanding officer. Groves quickly orchestrated the construction of an industrial effort larger and more complex than any previously attempted.[15]

The major installations in the Manhattan Project were at Oak Ridge, Tennessee; Hanford, Washington; and Los Alamos, New Mexico. Oak Ridge was the site of four major industrial plants: Y-12 to generate uranium 235 by electromagnetic processes, Y-25 to generate uranium 235 by gaseous diffusion, S-50 to augment the first two by thermal diffusion, and the Clinton Laboratories to conduct research on locally generated plutonium. By late 1942, it became clear to the project managers that the Clinton Laboratories could be used for only experimental and pilot purposes because there was not enough space available at the site to construct plutonium production facilities.[16]

Groves selected Hanford, Washington, as the site for the huge plutonium production plants. Begun in January 1943, the facilities on the 500,000-acre site included three water-cooled reactors and four separation plants. The construction was completed, and the production of plutonium began in September 1944. Hanford became the primary source of plutonium, and except for small experimental quantities, it produced all the American-made plutonium that eventually reached the Northern Rio Grande system in New Mexico.

The culminating activity of the Manhattan Project was the refinement of the plutonium and the construction of weapons. By early 1942, the project's far-flung laboratory and industrial elements had become so diverse and used such a wide variety of equipment and techniques that Conant and Groves decided to include in the system a final research design and assembly facility. In December 1942, they formally appointed J. Robert Oppenheimer of the University of California at Berkeley to oversee the operation of this final aspect of the project, although he already was essentially acting in that capacity. Oppenheimer suggested to Groves that the site of the final laboratory in the system be in northern New Mexico, an area he knew from earlier vacation visits. They finally settled on the Los Alamos Ranch School, a boy's boarding school on the Pajarito Plateau northwest of Santa Fe (Figures 2.1 and 2.2). The location had some housing already available, was easily acquired by the government, had large uninhabited spaces, and could be made militarily secure. Condemnation proceedings began in November 1942; in February 1943, the school closed, the Manhattan Project arrived, and Los Alamos Scientific Laboratory became a reality.[17]

Although Oppenheimer and Groves originally anticipated that the site would house about 30 scientists, the complexities and problems of design and bomb construction using plutonium were so great that the laboratory quickly expanded. By 1945, Los Alamos had become a city of 5,000 inhabitants supported by dozens of buildings and laboratories (Figures 2.3 and 2.4). Much of the plutonium was handled in Building D in Technical Area 1 (TA-1).[18] The first plutonium delivery to Los Alamos, in February 1944, was a small experimental quantity (probably less than 1 mg) from the Clinton Laboratories at Oak Ridge.[19] The first shipment of

Figure 2.1. Ashley Pond, the focal point of the Los Alamos Ranch School and the city of Los Alamos, which replaced the school. (*a*) Looking north across the pond to the buildings of the school in about 1938. (Los Alamos Historical Society, photo R-4689) (*b*) The same view in 1992 with only a few of the original buildings remaining, mostly hidden by the trees. The pond now has a slightly different shape. (W. L. Graf, photo 107–10)

small quantities of plutonium from Hanford arrived at Los Alamos on February 2, 1945.[20] Large quantities of uranium 235 and plutonium 239 arrived at Los Alamos in the early summer of 1945 and made up the cores of the first nuclear weapons: the plutonium for the experimental Trinity blast east of Socorro, New Mexico, on July 16, and for the bomb dropped on Nagasaki on August 9. The bomb dropped on Hiroshima, on August 6, which was also manufactured at Los Alamos, was fueled by uranium. The significance of this history of plutonium shipments is that there was no plutonium at Los Alamos until mid-1944, and it was not present in large

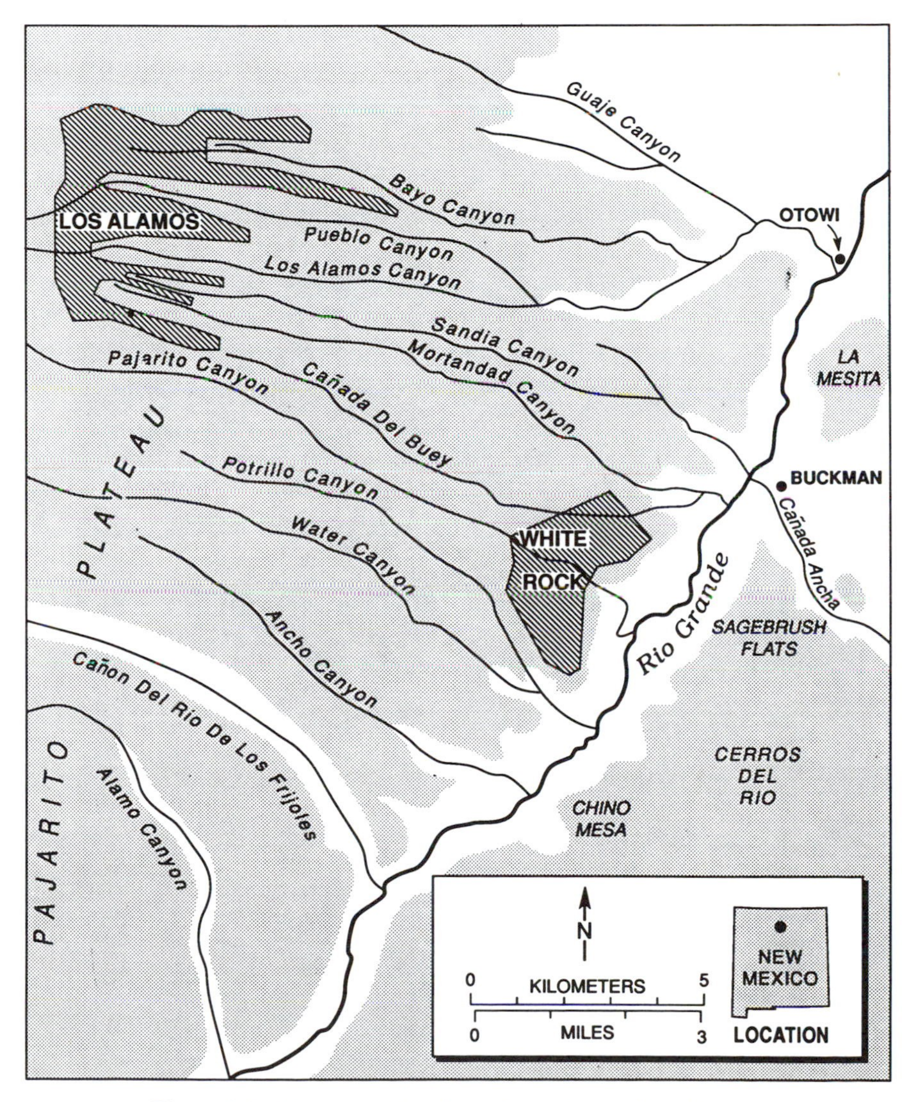

Figure 2.2. The general location of Los Alamos, New Mexico.

quantities there until mid-1945. Significant amounts of plutonium were therefore unlikely to have been released from the laboratory into the surrounding environment until mid- to late 1945.

During and after World War II, two factors drove the management at Los Alamos to produce atomic weapons. First was the perceived need for the United States to develop the ultimate weapon before its adversary could do so, and thereby win the war.[21] Many scientists at the laboratory were refugees from countries conquered earlier in the war, and their most common after-dinner toast was "Death to Tyrants." Their sense of political and scientific history imparted an uncommon devotion to their work. The second driving force was an inertia inherent in the process.[22] After spending their lives on the project, these scientists could not stop before its completion, even after the surrender of their primary enemy, Germany. After four years of building the first bomb, the process was larger than any individual or one group of individuals, and extensions of the original project continued after the war despite disagreements about how to control the weapons and their materials. During the early days of the Cold War the Soviet Union became the major international military competitor of the United States. The result was further experimentation with plutonium at Los Alamos, with continued attendant releases of plutonium into the environment.

These considerations are important to explaining why the potential pollution hazard of plutonium received less attention in the 1940s than it does in the 1990s.

Figure 2.3. The Main Technological Area (TA-1) at Los Alamos in 1945, with its temporary buildings and wartime security.

Figure 2.4. The Main Technological Area (TA-1) at Los Alamos in 1955, with its extensive development and permanent buildings. Ashley Pond is at the left.

In the midst of a global war followed by global competition between the remaining superpowers, environmental quality was a minor issue. It has been only in recent years, especially since 1970, that American society has embraced environmental quality as a national goal.[23] Nonetheless, even early in its development, researchers recognized the hazardous nature of plutonium, and in the design of laboratories and industrial plants Groves was greatly concerned about the workers' safety. But at that time, there was no scientific expertise to deal with radiation and toxic hazards from the material.

To address this lack of knowledge, in July 1942 Compton established a health division at the Metallurgical Laboratory at the University of Chicago. From then on, biophysicists, physicians, and biologists began to explore the implications of plutonium and radiation for human health. Many of the early developments in the health physics of plutonium were at Los Alamos. Joseph Kennedy, director of the Chemistry and Metallurgy Division, and Louis Hempelmann, medical director of the laboratory, worked to devise methods to detect plutonium in human tissues and to find ways to prevent plutonium poisoning.[24] Hempelmann designed safety rules for handling plutonium at the laboratory that were similar to rules previously established for the radium-processing industry.[25]

During World War II and the early days of the Cold War, an understanding of the dynamics of plutonium in the natural environment was several years in the future. The term *radioecology* did not appear in the scientific literature until

1956.[26] Reports by British researchers in 1941 and by Los Alamos chemist George Kistiakowsky in 1944 mentioned potential environmental contamination by fallout, but the issue was not taken seriously until efforts by physical chemists Joseph Hirschfelder and John Magee at Los Alamos in April 1945.[27] By the time of the Trinity blast, there was considerable effort to assess the intensity and distribution of fallout from nuclear detonations. But the problem was too complex for solution at that time, as was the issue of waste disposal from experiments and bomb manufacturing. General Groves adequately summarized the perspective of the project participants in the 1940s and 1950s regarding environmental pollution by plutonium: "We had always thought that it would be possible by intensive research to eliminate much of this radioactive problem in the future."[28]

From 1945 to 1949, health officials at Los Alamos made periodic surveys of the radiological characteristics of water and sediment in the Los Alamos system. The results of these investigations appeared in internal laboratory reports that remained classified until the late 1950s. From 1949 to 1971, the Water Resources Division of the U.S. Geological Survey studied the effects of plutonium releases from the laboratory, with the results appearing occasionally in publications of the Survey (Water-Supply Papers, Professional Papers, and the general scientific literature). After 1970, laboratory staff in the Environmental Studies Group and the Environmental Surveillance Group undertook a series of investigations, in many cases reporting the results in the general scientific literature. Significant data and conclusions appeared in annual "surveillance reports," laboratory publications that were publicly available but not widely distributed. Alan Stoker and his associates collated most of the important data and published a summary of the work done from 1945 to 1975.[29]

Despite more than four decades of research, General Groves's optimistic view of a future when the problem would be "solved," a clear understanding of the dynamics of plutonium in the environment remains elusive. Although physicists and engineers were able to surmount the problems of nuclear fission and bomb building, environmental scientists have yet to unravel completely the fate of plutonium in the environment. The explanation for the difference between the expected and actual outcomes relates to money and complexity. The amount of money invested in environmental research is minuscule in comparison with the amount invested in weapons development. Also, the natural environment is far more complex than the relatively simplistic conditions of the physical or chemical laboratory. Environmental measurements are difficult and inaccurate compared with laboratory efforts; many variables are often not assessed; and control cases are difficult to identify.

The Nature of Plutonium

The plutonium that was the focal point of the Los Alamos industrial activity and that now is an environmental pollutant is a metallic element formed by neutron

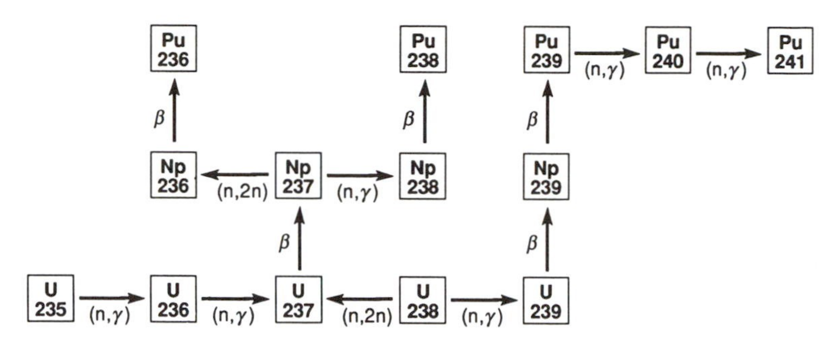

Figure 2.5. The formation scheme for plutonium. (Redrawn from R. L. Watters, T. E. Hakonson, and L. J. Lane, "The Behavior of Actinides in the Environment," *Radiochimica Acta* 32 [1983]: 89)

capture in uranium 238 (Figure 2.5). Although 14 isotopes of plutonium are known, only four occur in quantities great enough to be of concern as contaminants: plutonium 241 (half-life of 13.2 years), plutonium 240 (6,580 years), plutonium 239 (24,400 years), and plutonium 238 (86.4 years). Of the four, plutonium 239 and 240 are the most common isotopes found in the natural environment, being about 21 times more often found in sediment than is plutonium 238. Plutonium 241 emits beta radiation, whereas the other three emit alpha radiation during their decay. As plutonium decays to more stable isotopes and elements, it passes through a series of predictable stages, eventually becoming stable as lead 207 (Figure 2.6).

In nearly pure form, plutonium can be separated chemically from its precursor, uranium 238, but its chemical characteristics and behavior are complex. In aqueous systems, for example, plutonium can exist in four oxidation states simultaneously.[30] The most common, stable plutonium compound at earth environment temperatures and pH ranges is plutonium dioxide (Figure 2.7). Most plutonium in the natural environment is likely to be plutonium dioxide, although it also occurs in six allotropic metal forms. Plutonium in natural sedimentary environments in the form of a metal or plutonium dioxide is relatively insoluble. And this insolubility explains why in terrestrial ecosystems more than 99 percent of the plutonium is found in soils and sediments. It is not chemically mobile enough to pass easily through organic membranes into plant roots. Concentrations in plants are usually 0.001 to 0.000001 times the concentration in underlying soils and sediments, and concentrations usually decline by a mean factor of 0.0001 through each of the transitions from soil to plant to animal systems.[31]

The human health hazard from plutonium is its radiotoxicity. Because the most common forms of the element are alpha emitters, radiation from decaying plutonium does not readily pass through physical barriers, including the human skin. However, if particles containing plutonium are inhaled or ingested and become internally lodged in the human body, the persistent emission of alpha radiation leads to the destruction of cells and the development of cancers. Like most actinides, internally mobile plutonium is often deposited in bone tissue, where its residence in humans is longer than a century.[32] This concentration process is nonuniform, so that more plutonium can be present in some parts of the skeleton than in others.[33] Plutonium ingested in animal systems often concentrates in the liver, whereas that inhaled concentrates in the lungs, in each case leading to the development of cancers. Small amounts of plutonium taken into animal systems are absorbed under normal conditions: about 0.0001 of that ingested is absorbed by the intestinal tract.[34] Larger amounts enter typical animal systems through airways. About 0.05 percent of that inhaled enters the bloodstream, and about 0.15 percent enters the lymphatic system.[35] For these reasons, plutonium-contaminated sediments that might become airborne and then inhaled are probably the most important from the standpoint of human contamination.

The degree of plutonium's chemical toxicity for humans is relatively well known, probably because "perhaps no single element has ever been so intensively studied."[36] Although plutonium is hazardous to human health, arsenic and some biological toxins are probably more poisonous.[37] Indeed, among radioactive isotopes, cesium 137 and strontium 90 are more hazardous because they are more common and/or more mobile in the earth-surface environment.[38] Although there are no public records of physical damage to humans through exposure to plu-

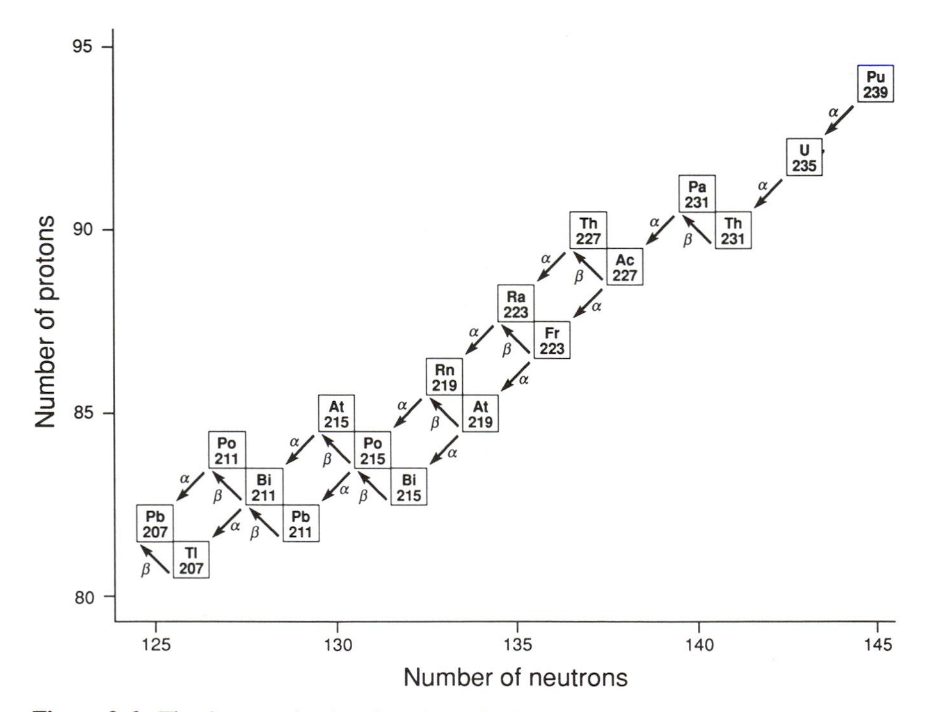

Figure 2.6. The decay series for plutonium. (Redrawn from J. Dennis, ed., *The Nuclear Alamanac: Confronting the Atom in War and Peace* [Boston: Addison-Wesley, 1984], p. 462)

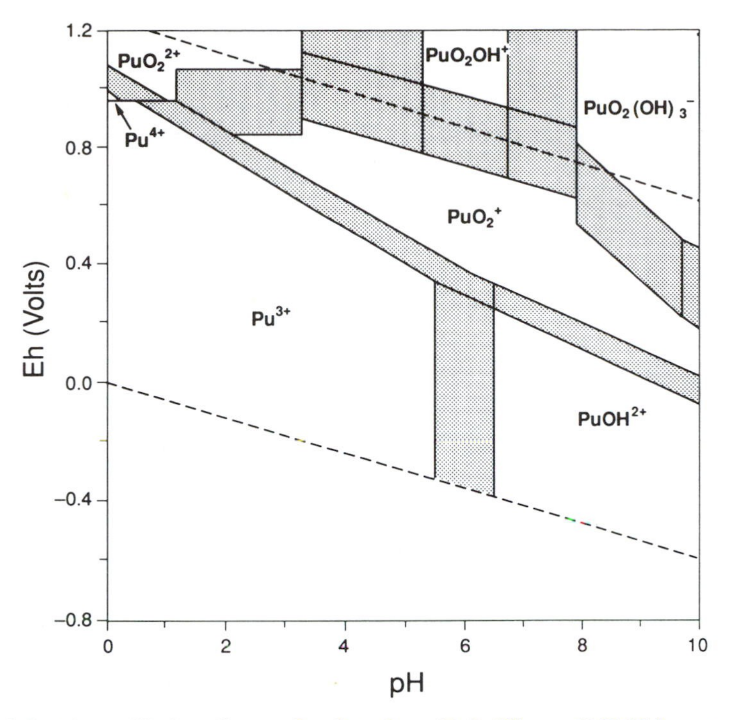

Figure 2.7. The equilibrium diagram for plutonium. (R. L. Watters, T. E. Hakonson, and L. J. Lane, "The Behavior of Actinides in the Environments," *Radiochimica Acta* 32 [1983]: 93)

tonium, the potential for poisoning and radioactivity damage to humans is great enough to warrant the careful monitoring, measurement, and assessment of concentrations. A responsible program to protect human health thus includes an accurate evaluation of the location and magnitudes of plutonium in those environments used by the general population.

Measurements and Safe Limits

The amount of plutonium in samples of environmental materials is most often measured as a concentration, a certain amount of plutonium per unit mass for sediment (per unit volume for air and per unit liquid measure for water). For nonradioactive metal isotopes, the units are usually milligrams per kilogram or micrograms per kilogram (mg/kg or μg/kg; that is, parts per million or parts per billion), but for radioactive isotopes in sediments, the units are atomic disintegrations per minute per unit mass.[39] In the United States, the most common unit for the rate of atomic disintegrations is a curie, which is equal to about 37 billion atomic disintegrations per minute (that is, about 37 billion atoms disintegrating by giving up a unit of radiation per second). The standard for determining a curie is the number of atomic disintegrations per second experienced by 1 g of pure radium. Plutonium occurs in the natural environment in such small quantities that a curie is too large to use as a convenient measure, and so fractions of a curie are the usual unit of measure. Common measures are the picocurie (pCi, 1×10^{-12} Ci) and the femtocurie (fCi, 1×10^{-15}) (Appendix A). One picocurie is equal to about 2.2 atomic disintegrations per second.

General amounts of radioactive isotopes as measured in curies include the megacuries produced by a nuclear detonation. Kilocuries represent the amount used in medical treatments to reduce tumors. Microcurie amounts serve as tracers in environmental systems research. Because of the worldwide presence of nuclear fallout, most people have picocurie amounts of radioactive isotopes in their tissues. One picocurie per gram of plutonium in sediment represents much less than one part per billion of the metal in natural material and could not be detected by ordinary physical or chemical methods.[40]

The picocurie per gram is an odd combination of metric and nonmetric units, but it is used in this book because the original data and most of the publications cited use this measure. The International System of Units measure of atomic decay is the becquerel (Bq), which is equal to one disintegration per second. The standard measure of concentration in environmental materials is therefore becquerels per gram, and one picocurie per gram equals 0.0367 bequerel per gram. Neither the curie nor the bequerel identifies the isotope creating the decay emissions, a problem that can be resolved only by analyzing the number of kinds of radiation produced during the decay.[41]

Laboratory methods assess plutonium concentrations in environmental samples by counting the number of atomic disintegrations per minute and then comparing that value with the mass of material involved. The laboratory environment and instruments create some emissions of their own that are counted in the process, so that a standardized number of emissions representing the background must be subtracted from the total number of emissions counted. The resulting value represents the number of emissions presumably derived from the sample. Because the background is inconsistent from one time to the next, the correction factor is the mean of many attempts to measure the background values. The reported values for the number of emissions from a sample is a number equaling the measured emissions minus a mean value representing a statistical distribution of values for the background. For some samples, if the actual background at the time of the sample analysis happens to be exceptionally low and far below the mean value used for correction, a negative number may be reported for the sample. Some negative numbers of this type appear in the following pages. For analytic purposes, they are carried forward in further calculations to preserve the integrity of the statistical distributions that created them. But for practical purposes, they indicate that plutonium was either absent or present in quantities too small to be detected.

Because the principal health hazard from environmental plutonium is related to radiation, safe amounts of plutonium concentrations in environmental materials are predicated on the radiation hazard if the material is ingested. As guidelines for water and air, the U.S. Department of Energy uses concentrations above "background" levels, because the purpose of the guidelines is to determine the safety of occupational environments.[42] For water, the standard is 0.000005 mCi/ml, or 5,000 pCi/l. For air, it is 6×10^{-14} mCi/ml, 60,000 aCi/m^3, or 0.06 pCi/m^3.

No agencies in the United States have plutonium standards for sediment quality. This oversight is especially important for the heavy metals that are radioactive isotopes, because most research shows that almost all the metals in the natural environment are associated with sediments or soils. An example of such a standard is in the Netherlands, where governmental guidelines for sediment quality include the evaluation of copper and zinc: 100 to 500 ppm copper and 500 to 3,000 ppm zinc require further evaluation, and more than 500 ppm copper or 3,000 ppm zinc indicate the need for removal and disposal.[43] If the United States decided to establish sediment quality standards for metals, including plutonium, it would undoubtedly require considerable bureaucratic effort, but the legal, monitoring, surveillance, and scientific rewards would be substantial.

3

The Northern Rio Grande Basin

Drainage Network

In northern New Mexico, the environmental plutonium bound to sedimentary particles is the most mobile in river systems, particularly the Rio Grande. This chapter describes the physical characteristics of the drainage basin into which Los Alamos National Laboratory has released plutonium. I review those characteristics of the basin that most strongly influence the movement of sediment and its associated plutonium: landforms, geology and soils, climate, vegetation, and precipitation.

Precipitation and elevation provide the energy that is the primary driving force behind river processes in the Northern Rio Grande Basin. The geographic variation in stream flow and the temporal characteristics of its magnitude and frequency explain how water, sediment, and contaminants such as plutonium move through the system. An accurate accounting of stream flow is therefore essential to the development of a basinwide budget for water, sediment, and contaminants. Calculations for the mechanics of sediment transport (and the transport of associated contaminants) thus depend on measurements of stream flow from a variety of places within the system. In this chapter I examine the basic data for stream flow in the basin and then define and explain the temporal and geographical variation in the system's river flows. The result is a regional stream-flow budget.

The portion of the Northern Rio Grande emphasized in this book consists of the watershed upstream from the U.S. Geological Survey stream gage[1] on the Rio Grande at San Marcial, at the headwaters of Elephant Butte Reservoir. The drainage network in this 71,700-sq-km area is the principal mechanism for the surface transport and storage of plutonium. The Rio Grande begins as a trickle of meltwater from a semipermenant snowbank at Stoney Pass in the San Juan Mountains in southwestern Colorado. Steep mountain tributaries are the primary sources of water, joining the main stem as it trends southeastward to the San Luis Valley and the Alamosa, Colorado, area (Figures 3.1 and 3.2). Additional mountain waters from the Rio Conejos, which drains the southern San Juan Mountains in southern Colorado, join the main stream as it flows southward into New Mexico.

The northern Sangre de Cristo Mountains in Colorado generate surface runoff, but relatively little reaches the main river. About 7,500 sq km of the San Luis Valley constitute a closed basin, with no direct surface contributions to the Rio Grande. The Sangre de Cristo Mountains contribute only about 10 percent as much water as do the San Juan Mountains.[2] The Rio Chama, which drains New Mexico's San Juan Mountains, joins the Rio Grande near Española. Its combined 37,000-sq-km area drains into the main river at the upstream end of the study area. Of the two major basins above Española (and thus above Los Alamos), the Upper Rio Grande above Española produces more water, and the Rio Chama produces more sediment.

The Rio Grande watershed almost doubles its drainage area between the Otowi Bridge (a short distance downstream from Española) and San Marcial (Figure 3.3), but all the large basins added to the system within this reach are on the west bank. The Jemez River, draining the Jemez Mountains, is small compared with the Rio Chama and Upper Rio Grande components. The Rio Puerco drains a large area, but it is mostly plateau rather than mountain terrain, and so it adds much sediment but little water. The Rio Salado also drains nonmountainous terrain and is principally a sediment producer.

General Geomorphology

An appreciation of the landscape through which this channel network flows is important to understanding the transportation and storage of plutonium, because steep slopes shed their debris and associated fallout contaminants more rapidly than does moderately sloping terrain. Topographic variation is also significant because high elevations with their greater amounts of rain are likely to receive more fallout.[3] Altitudes in the Northern Rio Grande range from about 1,400 m on the river at San Marcial to about 4,360 m on Blanca Peak in the Sangre de Cristo Mountains. The Northern Rio Grande Basin includes portions of the Southern Rocky Mountains, the Colorado Plateau, and the Basin and Range Province (Figure 3.4).

The components of the Southern Rocky Mountains included in the basin are the San Juan, Sangre de Cristo, and Jemez mountains as well as the San Luis Valley.[4] The San Juan Mountains dominate the northwestern part of the basin. The range is the first mountain mass exceeding 3,000 m in altitude downwind from the Nevada Test Site. It is therefore a logical locale for the deposition of fallout plutonium that might be in greater concentrations than the latitudinal or global

Figure 3.1. The Rio Grande near Creede, Colorado, is typical of the stream in the upper basin in the area immediately west of the San Luis Valley. This reach contains significant amounts of stored sediment and probably stores radionuclides from fallout in mountain watersheds. The general character of the river has not changed in nearly 90 years, despite the construction of the Rio Grande Reservoir upstream. (*a*) Looking west along the stream at the site of a proposed reservoir in 1911. (O. T. Davis, photo 1312, Denver Public Library, Western History Collection) (*b*) The same view in 1990. (W. L. Graf, photo 81-9)

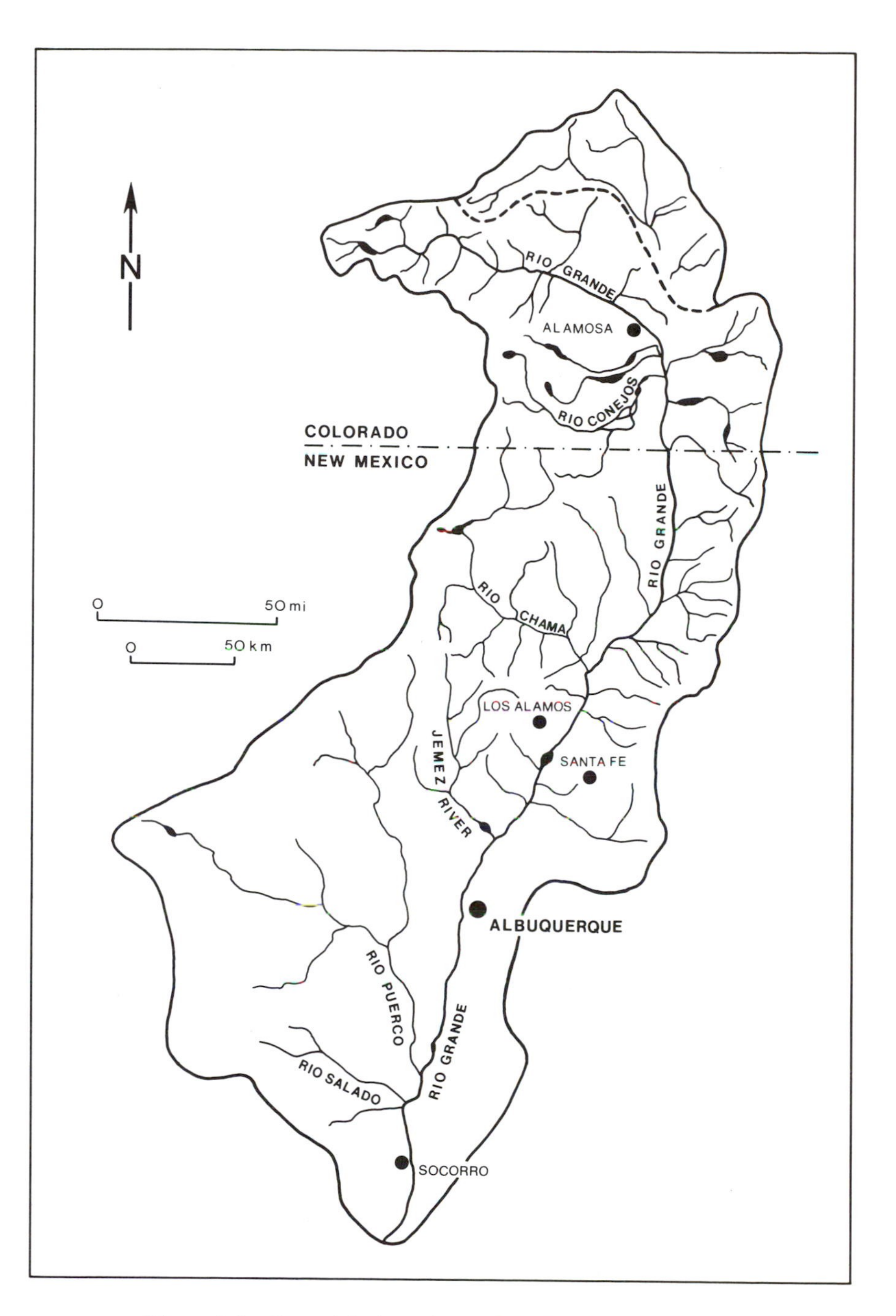

Figure 3.2. General drainage basin of the Northern Rio Grande.

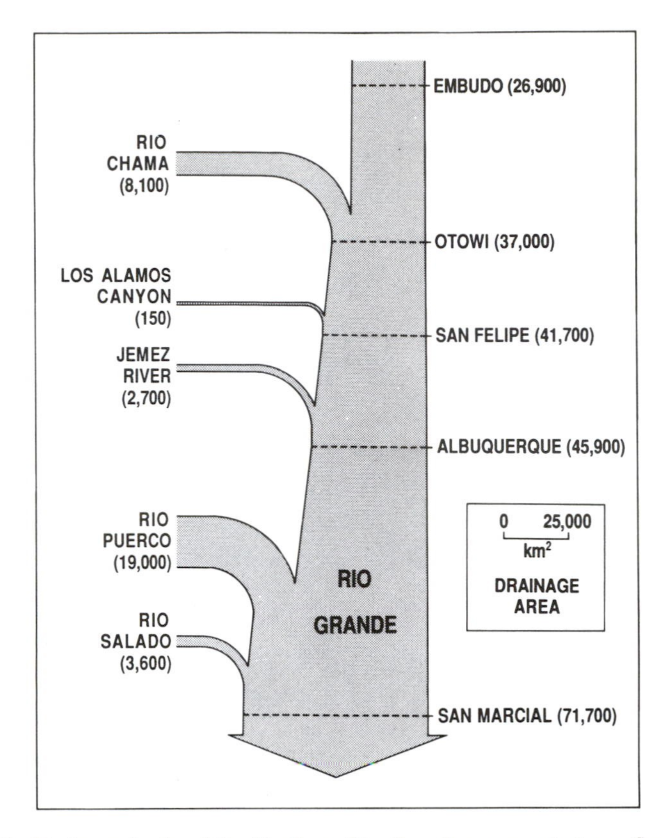

Figure 3.3. Drainage basin of the Northern Rio Grande represented as a flow diagram, with the width of the arrow representing the drainage area upstream or within tributary basins. Parenthetical drainage areas are given in km^2.

average, because of the possible additions from activities at the test site directly upwind. The mountains have steep, glaciated slopes with thin soils,[5] so erosion of fallout is likely. Recent sampling of Rio Grande Reservoir, located in the high-altitude portion of the San Juan Mountains, revealed elevated levels of plutonium 238, plutonium 239, and plutonium 240, reflecting these conditions.[6]

The Sangre de Cristo Mountains are even higher than the San Juan Mountains, but their location downwind with respect to the prevailing westerly winds makes them drier. They also probably contribute some fallout plutonium to the Rio Grande system, but no direct measurements are available. The Jemez Mountains are a relatively small unit west of the Rio Grande. Essentially a massive caldera complex, they give rise to the Jemez River and several smaller streams, but within the region, the mountains produce water, sediment, and probable radionuclide output in much smaller quantities than do the other mountains.

The topographic highs of the mountains are complemented by the topographic lows of several major valleys in the drainage basin of the Northern Rio Grande.

The San Luis Valley is a deep structural basin about 160 km by 80 km and is filled to depths of several thousand meters with alluvium derived from the surrounding mountains. The valley includes broad alluvial plains, isolated volcanic cones and mesas such as the San Luis Hills, and the alluvial fan of the Rio Grande as it exits the San Juan Mountains.[7] The valley is important from the perspective of mass transport in the Rio Grande system because it is a zone of especially low stream

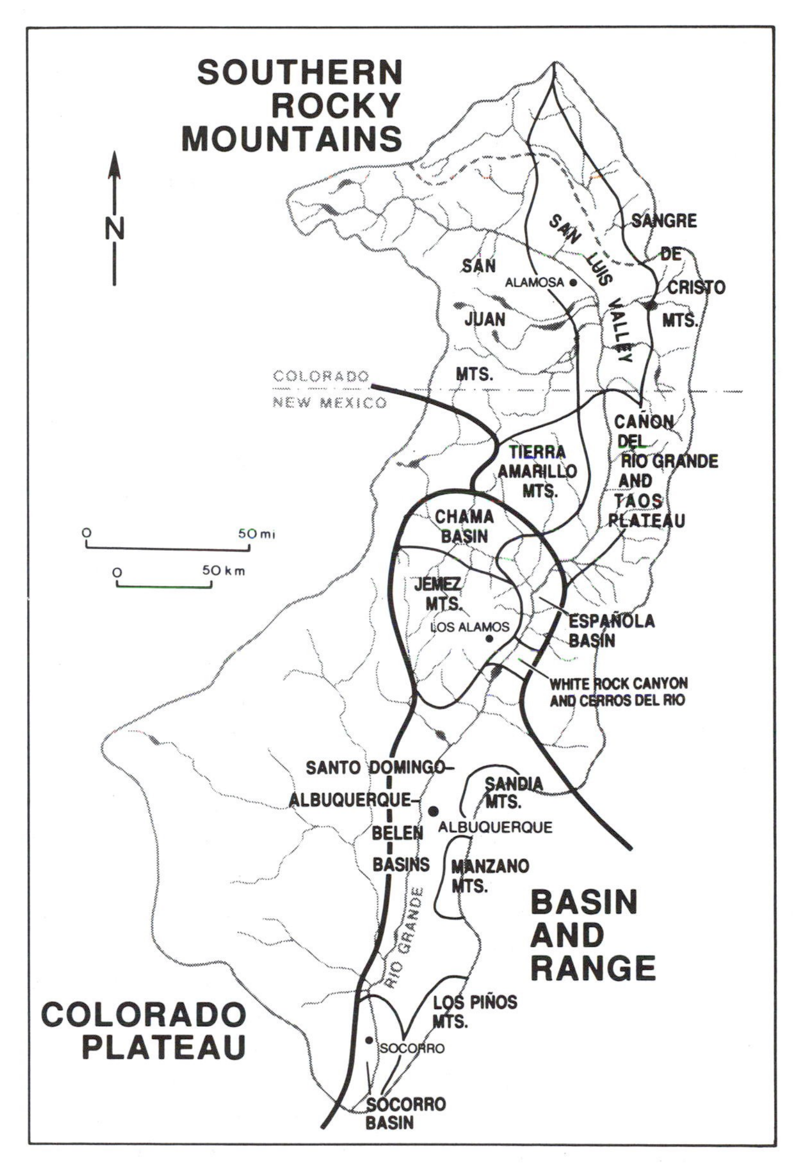

Figure 3.4. Landform divisions of the basin of the Northern Rio Grande.

gradients.[8] The deposition of sediment and associated contaminants is common in the valley, producing meandering channels and extensive zones of channel and flood-plain deposits.[9] Under present conditions, it is likely that the majority of sediments and contaminants eroded from the surrounding mountains move no farther south than the San Luis Valley.

The Cañon del Rio Grande, a deep gorge etched into volcanic rock at the southern end of the San Luis Valley, marks the passage of the river into the Basin and Range Province. Farther downstream, the Española Basin provides a broad alluvial valley for the course of the river, but volcanic rock associated with the Jemez Mountains and the Pajarito Plateau (the location of Los Alamos) confines the river to the narrow White Rock Canyon. Thereafter, the Rio Grande flows southward through a series of structural basins, with fault-block mountains on the east and west sides. Erosion of the fault zones produces abrupt boundaries between the steep mountain slopes and the gently inclined pediments leading down to the flood plain on the valley floor. Between White Rock Canyon and San Marcial, the flood plain varies between 5 and 12 km in width.[10]

Geology and Sediments

A general review of the geologic materials underlying the Northern Rio Grande is necessary to explain the transport of plutonium, because of the strong association between radionuclides and sediment. Weathering of the bedrock and erosion of the resulting soils are primary pathways for movement of the contaminants. The distribution of highly erodible outcrops directly controls the geographic characteristics of the sediment and contaminant budget (Figure 3.5).

At the headwaters of the Rio Grande, the San Juan Mountains are an immense pile of Tertiary volcanic rocks that present formations of variable erodibility at the surface.[11] For general locations of this and other geomorphic areas discussed later, see Figure 3.2. The mountains are a dome structure dissected by glacial and fluvial erosion. In detail, however, they consist of an array of basalt, latite, rhyolite flows, collapse calderas, ash flows, breccia, and reworked volcanic tuff formed during a series of eruptions, probably during the early Tertiary.[12] Considerable erosion altered the surfaces of these deposits. Subsequently, during the Oligocene, Miocene, and Pliocene, new eruptions created intrusions of quartz latite, rhyolite, and some basalt into the older rock.[13] The later eruptions apparently emplaced many of the mineral deposits that led to gold and silver mining in the area. Recent erosion of this vast array of rock types has produced sediment of various particle sizes. The basalt flows create boulders that the streams fail to transport for long distances. The ash flows produce fine particles through the rapid erosion of steep slopes. They may be a major source of fallout plutonium in the Rio Grande Reservoir because some ash occurs in the higher portions of the watershed above the reservoir, combining high erodibility with likely infusions of fallout.

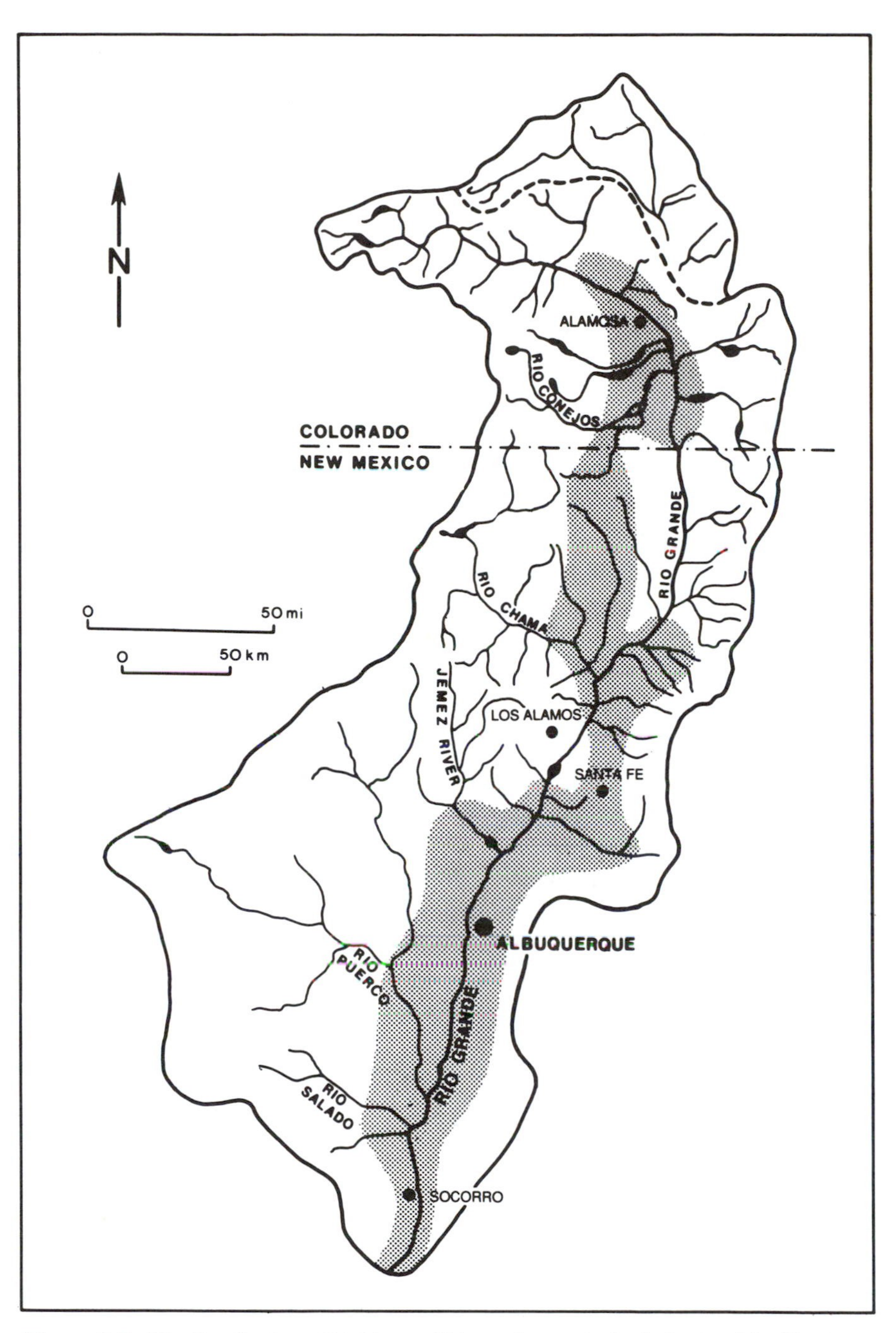

Figure 3.5. The distribution of highly erodible surface rocks in the basin of the Northern Rio Grande.

The Sangre de Cristo Mountains are a fault-block range made up of a core of Precambrian gneiss with deformed Permian and Pennsylvanian sedimentary rocks and some Tertiary volcanics on the eastern side.[14] On the west, the portion that drains into the Northern Rio Grande, sharply defined faults divide the resistant core from the neighboring San Luis Valley.[15] Although the slopes are steep in this area, the crystalline rock is relatively resistant to erosion, and sediment-bound radionuclides are probably not as common here as in the San Juan Mountains.

On a geologic time scale, erosion of the San Juan and Sangre de Cristo mountains has produced huge quantities of debris that fill the intervening San Luis Valley (structurally, the Alamosa Basin) to a depth of up to 9,000 meters.[16] The sediments filling the basin are interbedded gravels, sands, and clays, some resulting from deposition in Tertiary lake beds.[17] Present-day processes continue the general trend toward deposition: The Rio Grande does not incise the basin's sediments more than 1 m, and the modern river aggrades continuously throughout its course in the San Luis Valley.[18] Erosion of materials stored in the basin is unlikely, and the area is a sink for sediment and fallout plutonium.

The Cañon del Rio Grande, immediately south of the San Luis Valley, contains geologic materials radically different from those of the valley. The canyon is a defile eroded by the Rio Grande into the Taos Plateau, a complex sheet of sediments and volcanic materials that extends from the San Juan and Tierra Amarilla mountains on the west to the Sangre de Cristo Mountains on the east.[19] The Quaternary basalts and interbedded sedimentary units tilt gently toward the east and the Rio Grande, and the alluvial fans and coalesced fans slope westward toward the river.[20] The stratigraphy of the area indicates tectonic instability characterized by warping and faulting throughout the Tertiary, accompanied by repeated drainage interruption and fan building.[21] The basalts shed large blocks into the canyon of the river, and the sedimentary beds make the side slopes unstable, but the geologic conditions do not encourage large amounts of sediment to enter the river.

Some radionuclides may enter the main river in the Taos Plateau and the Cañon del Rio Grande reach because uranium deposits occur naturally in some of the volcanic rocks of the area. Mining for uranium has been limited, but uranium is naturally associated with other minerals that can be found in economically viable concentrations, especially in the Red River Valley area.[22] Erosion of mine tailings or of natural country rock may introduce some uranium into the Rio Grande, but it is probably in extremely small quantities, and no plutonium is involved.

The Española Basin, one of several basins along the river through the Rio Grande Rift Zone in New Mexico, is a shallow structural depression at least partly bounded by faults and is located immediately downstream from the Cañon del Rio Grande and the Taos Plateau.[23] Because the confluence of the Rio Grande and its principal tributary, the Rio Chama, is within the Española Basin, it is the site of deposition for large amounts of poorly consolidated sands and gravels with interbedded lenses of finer materials. It is the first reach downstream from the San Luis Valley that contains significant quantities of Quaternary alluvium along the channel.[24] The Santa Fe Group and particularly the Tesuque Formation are geologic units that crop out over large areas of the basin and that provide easily eroded materials for fluvial transport.[25] These geologic formations form the buff and yellow hills so often depicted by local artists in paintings. General erosion of the formations has produced a landscape dominated by ramplike pediments extending at a number of levels from the river to the foot slopes of the mountains.[26] Local erosion of these materials has produced badlands and emptied large quantities of fine-grained debris into the Rio Grande.[27] The Española Basin therefore represents the entry point for a significant portion of the sediment load of the Northern Rio Grande downstream from the San Luis Valley. Although prospectors have found some uranium in the Santa Fe Group in the basin, the amounts have been so small that they are not economical to mine.[28] Natural erosion probably adds some uranium to the Rio Grande from this source, but there is probably little fallout plutonium because of the low amounts of precipitation in the area.

The Rio Chama drains a particularly erodible part of the Española Basin and thus contributes more sediment to the downstream areas than does the Upper Rio Grande. These erodible materials are parts of the Santa Fe Group that extend from the area of the confluence with the Rio Grande upstream to the vicinity of Abiquiu.[29] As is the case elsewhere in the Española Basin, badland terrain is common on the Sante Fe Group near Abiquiu,[30] and it was a common subject of many of Georgia O'Keefe's best-known landscape paintings. The Rio Chama enters the Abiquiu area after draining the southern San Juan Mountains and flowing through the Chama Basin, a broad shallow depression at the eastern edge of the Colorado Plateau.[31] The rocks in the Chama Basin are predominately sandstones and shales,[32] but the erosion rates are moderate and sediment production is less than in the areas downstream from Abiquiu. Fallout plutonium eroded from the San Juan Mountains probably does not reach the lower river because of intervening large reservoirs and dams. Uranium occurs naturally in several locations in the Rio Chama drainage (particularly in the Morrison Formation and Dakota Sandstone), so erosion may introduce some radioactivity into the stream system. There are no known large or highly concentrated deposits.[33]

South of the Española Basin, the Rio Grande flows into White Rock Canyon, a 300-m-deep gorge between the Jemez Mountains on the west and the Cerros del Rio on the east. The Cerros del Rio, a westward extension of the Santa Fe Plateau, is an elevated basalt platform that is exceptionally resistant to erosion.[34] Although generally little sediment enters the river from the east side throughout the canyon, an exception is Cañada del Ancho, a stream draining the exposed Santa Fe Group slopes of arkosic sands in the Buckman area, about 5 km downstream from Otowi Bridge.

The significance of the reach near Otowi in regard to plutonium is that it includes the entry point of sediment and radionuclides from Los Alamos Canyon which drains the only industrial source of plutonium in the northern Rio Grande. Los Alamos National Laboratory is located on the Pajarito Plateau, a broad, dissected apron of Bandelier Tuff on the west side of White Rock Canyon.[35] Erosion by streams has etched deep canyons into the relatively smooth surfaces sloping from the Jemez caldera downward toward the Rio Grande and Rio Chama. Most of the erosion took place during three periods of the Pleistocene,[36] so that the resulting landforms have relatively unstable side slopes. Erosion of all members of the Bandelier Tuff unit except a welded tuff member of limited extent produces mostly sand- to gravel-size particles for transport in the region's river system. Contaminants released into these materials are therefore associated with sediment that is unlike the silt-size particles with which the fallout plutonium is associated.

South of White Rock Canyon lies a structural low in the Rio Grande Rift Zone consisting of three interconnected basins: the Santo Domingo, Albuquerque, and Belen basins. They provide a relatively broad, interconnected valley for the river and its deposits. In the Santo Domingo Basin, the erodible Santa Fe Group again crops out near the river.[37] Erosion of the unit injects large quantities of sediment into the main valley, where some remains as channel and flood-plain deposits. An upper mid-Pleistocene pediment, known locally as the Oriz Geomorphic Surface, grades toward the Rio Grande at about 150 m above river level.[38] Numerous arroyos excavate the surface and conduct sediment to the modern channel and flood plain. Natural uranium occurs in sandstone of the Galisteo Formation in the Hagan Basin area of Sandoval County, east of the Rio Grande and south of Galisteo Creek.[39] The deposit was of ore quality, so natural erosion may have contributed some radionuclides to deposits downstream in the Rio Grande. Plutonium is not involved.

The Albuquerque–Belen Basin contains fluvial sediments 6,000 m in depth.[40] Sediment eroded from this material in tributary arroyos, mostly from dissected Quaternary terraces, contributes to the load of the main stream and produces aggradation in the channel and on the flood plains. Although the literature provides a variety of labels for them, the geologic formations that crop out in the basin are similar to the Santa Fe Group in age, particle size, and mobility.[41] The result is the infusion of large amounts of sand and silt into the Rio Grande system from nearby elevated terraces. The Jemez River—deriving most of its water from the Jemez Mountains and most of its sediment from poorly consolidated Tertiary materials south of the mountains, including the Santa Fe Group[42]—is a major component of the sediment system near the junction of the Santo Domingo and Albuquerque basins.

At the southern end of the Belen Basin, the Rio Puerco and Rio Salado join the Rio Grande from the west. The two tributaries drain several hundred square kilometers of Colorado Plateau terrain, consisting of erodible sandstones and shales, so that the two streams carry high sediment loads. The Rio Puerco, for example, commonly contains 400,000 ppm sediment in floods, with a maximum record of 680,000 ppm; 75 percent of the load is sand.[43] Most of the materials in the Rio Puerco are fine grained: fine sand, silt, and clay.[44] The tributaries derive additional sediment from erosion of the Santa Fe Group formations commonly found in the southwestern portion of the Albuquerque–Belen Basin.[45] In addition to the plutonium related to global fallout, the Rio Puerco transports in its sediments some natural uranium from mining activities in the Grants area.[46]

Between the southern end of the Santo Domingo–Albuquerque–Belen Basin and Elephant Butte Reservoir, the Rio Grande flows through two relatively narrow troughs: the Socorro and San Marcial basins. Volcanic rocks in the form of cinder cones, basalt flows, and tuffs are common along the margins of the Rio Grande Valley,[47] but their erosion contributes little to the river's sediment and contaminant loads.

Climate

The climate of the Northern Rio Grande varies greatly with elevation, ranging from warm dry deserts in the south to alpine climates in the San Juan Mountains. Climate is important to understanding the dynamics of plutonium in the Northern Rio Grande because precipitation delivers fallout to the surface and provides the mass and energy driving the kinetics of the river systems. Within the watershed of the Northern Rio Grande above Elephant Butte Reservoir, precipitation ranges from an annual minimum of less than 400 mm near Socorro to a maximum of over 2,000 mm on some of the peaks of the San Juan Mountains.[48] The precipitation has an unequal distribution, with the three areas of maximum values concentrated in the mountains: the San Juan Mountains with the highest values, then the Sangre de Cristo Mountains, and finally the Jemez Mountains (Figure 3.6). This distribution of precipitation shows why the water sources for stream flow are severely restricted. Along its course through New Mexico, the Rio Grande receives little additional water because precipitation decreases from north to south.

The variation of precipitation is also highly seasonal, with most of it in the mountains in the form of snow. The maximum annual recorded snowfall in the basin is at Wolf Creek Pass in the San Juan Mountains, which often exceeds 9 m.[49] In the northern portion of the basin, the meltwater from the winter snowpack provides the annual peak of runoff in late spring. In the central portion of the basin, an additional, smaller peak often results from convective precipitation during the summer. In the southern extremities of the area, the wettest months are in late summer.

Precipitation also varies on decadal and century-long time scales. Although the annual precipitation record for Santa Fe shows no long-term trend, the frequency of rainy days and the mean daily rainfall have varied systematically

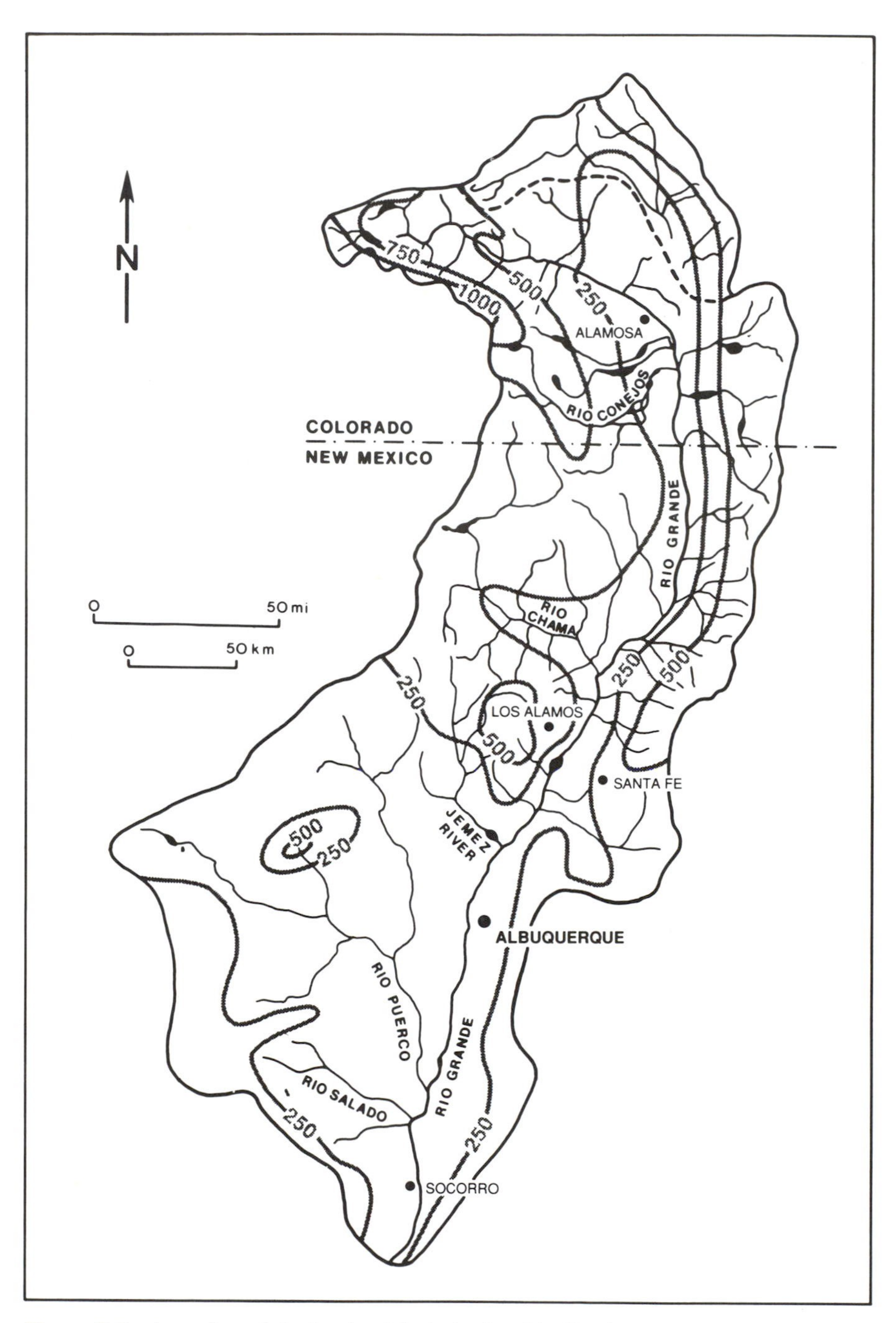

Figure 3.6. Annual precipitation (mm) in the basin of the Northern Rio Grande. (Modified from J. L. Williams, ed., *New Mexico in Maps*, 2nd ed. [Albuquerque: University of New Mexico Press, 1986]; and K. A. Erickson and A. W. Smith, *Atlas of Colorado* [Boulder: Colorado Associated University Press, 1985])

since records were first kept in the 1850s.[50] Rainfalls greater than 12.5 mm became progressively more frequent throughout the late nineteenth century, with the trend peaking during the 1920s. Since that time, despite large interannual variation, the frequency of intense rainfalls has generally declined. During the same century-long record, the mean daily rainfall dropped during the late nineteenth century, reaching a minimum during the 1920s before beginning an increase later. Thus during the late nineteenth century, rain fell more often, but in smaller amounts during each event. In recent decades, there have been fewer precipitation events, but those that do take place have issued larger amounts of rain.

The implications of this variation for the transport of sediment and associated contaminants are unclear. It is possible that the arroyo cutting during the late nineteenth and early twentieth centuries was partly a response to high-intensity, low-frequency rainfalls.[51] If this hypothesis is correct, more recent trends may signal a return to the highly erosive conditions in arroyos seen a century ago. The behavior of the main river, however, probably responds indirectly to this control, because the erosion of tributary arroyos inundated the main channel with sediment, causing aggradation and the development of braided conditions, a phenomenon observed during and immediately after the last arroyo-cutting episode in the late nineteenth and early twentieth centuries. The hydrologic behavior of the main stem of the Rio Grande is more closely tied to precipitation events in the mountain source areas, and their temporal variability is not clearly known.

Vegetation

Vegetation influences the dynamics of plutonium in the regional river systems because the nature of the plant cover influences the amount and timing of the runoff. Vegetation also influences the amount of surface erosion and the production of sediment available for transport in the streams. The vegetation of the Northern Rio Grande Basin occurs in two distinct geographic distributions, upland and riparian. The distribution of the basin's upland vegetation reflects geologic, pedologic, precipitation, and geomorphic controls. Montane conifer forests grow in the well-watered highland areas of the San Juan, Sangre de Cristo, Jemez, and Cebolleta mountains (Figure 3.7). At lower, drier elevations, the forests give way to Great Basin conifer woodland.[52] In the Northern Rio Grande Basin, these woodlands consist mostly of pinyon pine (*Pinus edulis* Englem.) and juniper (*Juniperus scopulorum* Sarg.), which often occur as widely spaced trees with seasonally barren ground between them. This barren ground is susceptible to rapid erosion during infrequent severe storms.[53]

In those parts of the basin too arid to support even the pinyon pine and juniper are the Great Basin grasslands. This vegetation community is particularly widespread in the southern half of the basin and often is found on the highly erodible Santa Fe Group. In the northern half of the area above Elephant Butte Reservoir,

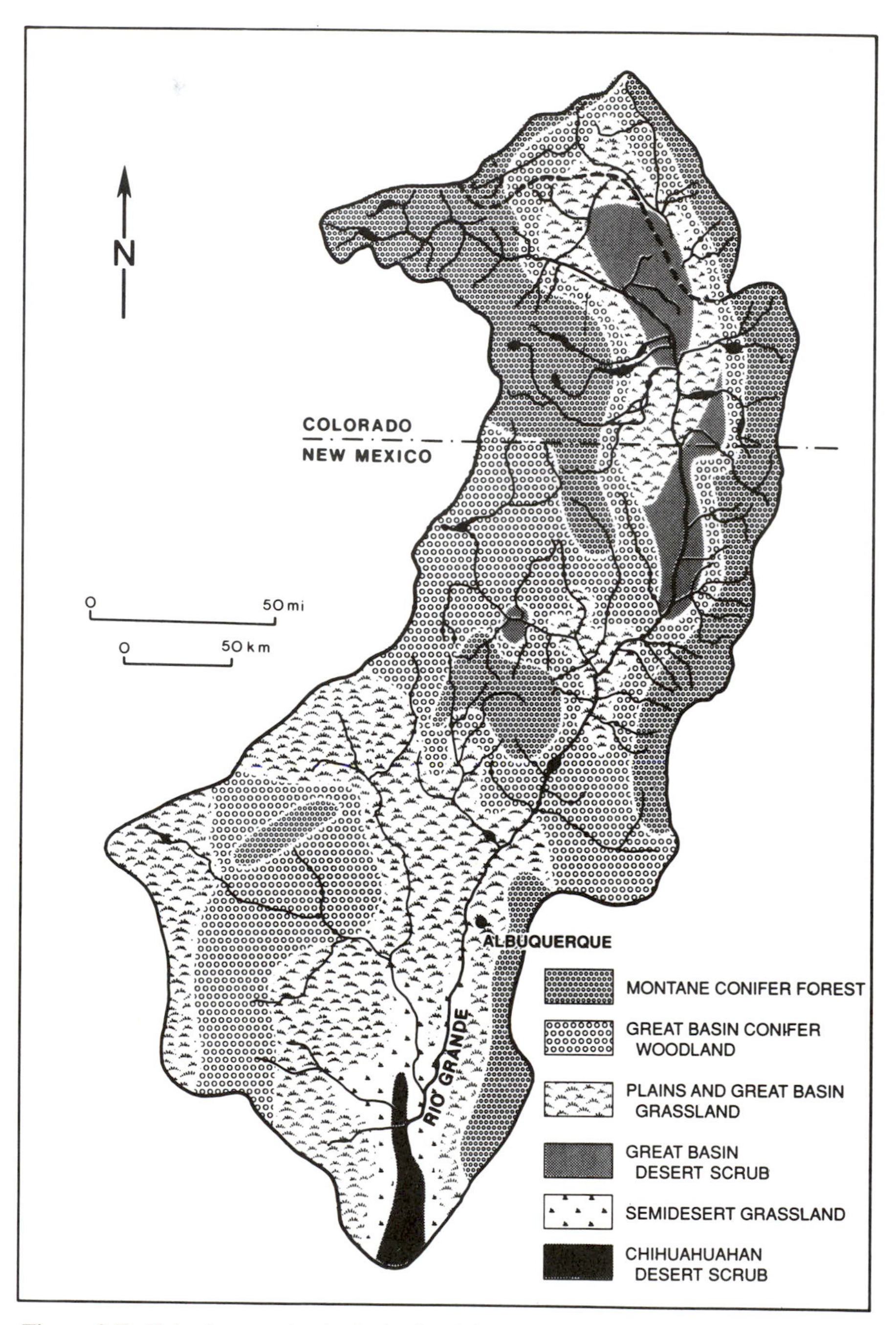

Figure 3.7. Upland vegetation in the basin of the Northern Rio Grande. (Modified from D. E. Brown, C. H. Lowe, and C. P. Pase, *A Digitized Systematic Classification for Ecosystems with an Illustrated Summary of the Natural Vegetation of North America*, U.S. Department of Agriculture, Forest Service General Technical Report RM-73 [Fort Collins, Colo.: U.S. Forest Service, 1980])

the Great Basin Desert scrub communities cover dry areas with a mixture of grass and low, woody-stemmed plants. At the southern extremities of the area, high annual temperatures and little rain produce semidesert grassland and Chihuahuan Desert scrub, communities that provide little stability for erodible soils on terraces and pediments near the river. Riparian communities owe their characteristics to the water-rich environments along the streams and are substantially different from the surrounding upland vegetation. As with upland vegetation, riparian communities also vary with elevation in the Northern Rio Grande Basin.[54] Because the riparian vegetation directly influences river processes, it is discussed in some detail in several subsequent chapters.

The general environment of the Rio Grande above Elephant Butte Reservoir therefore is highly variable in terms of its landscape, geology, water, and vegetation. The geography of these features influences the contributions of fallout plutonium to the Rio Grande and controls the flow of water through the streams that provides a transport mechanism for contaminants. The interactions of all these factors characterize the behavior of water in the Northern Rio Grande system.

Sources of Hydrologic Data

To measure water and sediment discharges, the U.S. Geological Survey has maintained an extensive network of stream gages for the Northern Rio Grande. The river is one of the most extensively instrumented in the world, and we have higher-quality data on it than on any other arid–semiarid drainage basin of similar size. Interest in the adjudication of water rights and distribution of the resource led to the establishment of the first long-term stream gage in the United States at Embudo, New Mexico, in 1885 (Figure 3.8).[55] This gage is the longest-running measurement site in the country. The construction and maintenance of reservoirs later led to an interest in sediment transport rates, and from the late 1940s to the present, some gages have produced sediment information in addition to water discharge data (Figure 3.9).

Scores of stream gages have operated within the basin at various places and times, but the construction of a regional budget for water and sediment depends on a few long-running, high-quality gages. Fourteen such sites span the basin from the San Juan Mountains to San Marcial (Figure 3.10 and Table 3.1). But even these main gages have variable lengths of record, and not all provide sediment data. Nonetheless, taken as a group, they represent the best source of information about the surface-water hydrology of the Northern Rio Grande.

Three closely related repositories store the data from the gages. First, the U.S. Geological Survey Water-Supply Paper series provides a permanent, paper-based record for the information. Most major research libraries stock the series, which

a

b

Figure 3.8. The U.S. Geological Survey stream-gaging station on the Rio Grande at Embudo, New Mexico, was the first such station established in the United States, with the first measurements made in 1885. The small structure at the left, constructed from material from the river, houses the measuring equipment, and the A-frame and platform anchor a cable that crosses the river to provide access for velocity measurements. (*a*) Looking north, upstream, about 1945, with the abandoned railroad grade at the lower left near the telephone pole. (U.S. Geological Survey Water-Supply Paper 247, U.S. Geological Survey Photography and Field Records Library, Denver) (*b*) The same view in 1989. The sandbar at the right has completely disappeared, and the cottonwood trees at the left have become considerably larger and denser. (W. L. Graf, photo 75-24A)

Figure 3.9. Stream gage on the Jemez River immediately below the Jemez Dam site. The view shows the stilling well (vertical cylinder) for stage or depth-of-flow measurements and an overhead cableway for moving a current meter and sediment sampler across the channel. (G. P. Williams, photo 83, U.S. Geological Survey Photography and Field Records Library, Denver)

Table 3.1. Major stream gages in the Northern Rio Grande system

Site		Gage ID	Drainage Area (sq mi)	Drainage Area (sq km)	Water Data	Sediment Data
A	Rio Grande at Thirtymile Bridge near Creede, Colo.	2135	163	422	1923–	–
B	Rio Grande at Wason, below Creede, Colo.	2170	705	1,826	1907–1954	–
C	Conejos River near Lasauses, Colo.	2490	887	2,297	1921–	–
D	Rio Grande near Lobatos, Colo.	2515	7,700	19,940	1899–	1969–
E	Rio Grande at Embudo, N.M.*	2795	10,400	26,930	1889–	1970–
F	Rio Chama near Chamita, N.M.	2900	3,144	8,140	1912–	1948–
G	Rio Grande at Otowi Bridge near San Idelfonso, N.M.	3130	14,300	37,037	1895–	1947–
H	Rio Grande at Cochiti, N.M.	3145	14,600	37,810	1924–1970	–
I	Rio Grande at San Felipe, N.M.	3190	16,100	41,700	1925–	1975–
J	Jemez River below Jemez Dam, N.M.	3290	1,038	2,690	1936–	1966–
K	Rio Grande at Albuquerque, N.M.	3300	17,740	45,950	1941–	1969–
L	Rio Puerco near Bernardo, N.M.*	3530	7,350	19,040	1909–	1947–
M	Rio Salado near San Acacia, N.M.	3540	1,380	3,570	1947–	1947–
N	Rio Grande at San Marcial, N.M.*	3584	27,700	71,740	1895–	1946–

Note: Letters are keyed to locations on Figure 3.10.

* Combined data for two closely associated gaging sites.

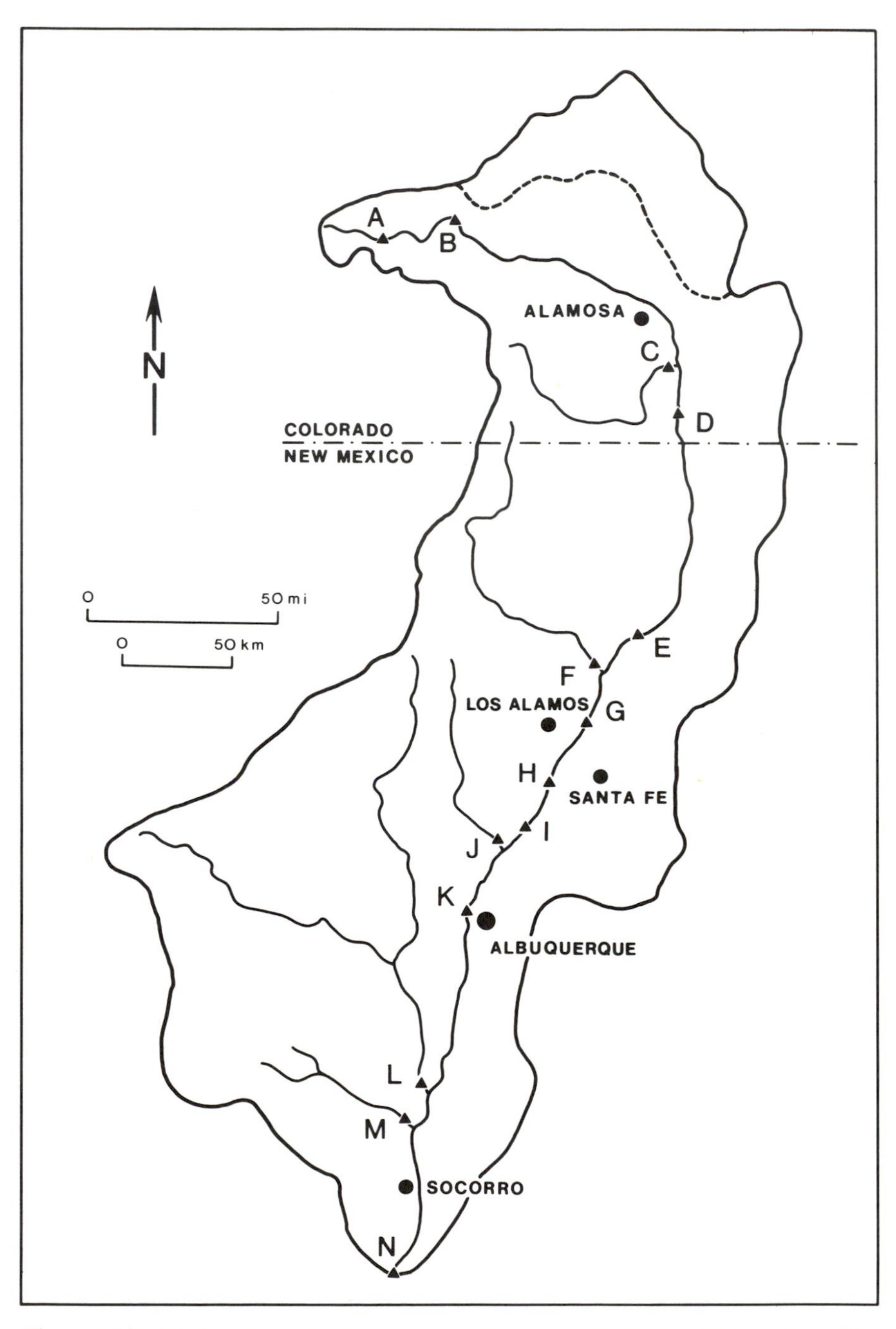

Figure 3.10. The locations of major stream-gaging sites in the basin of the Northern Rio Grande. For data, see Table 3.1.

includes some summaries in addition to the basic daily information. Second, the U.S. Geological Survey also maintains a computer-based system for hydrologic data that includes all the water and sediment information. This system, WATSTORE, is relatively complete but is difficult to use for those not in close and direct contact with the Survey. Finally, all the data in the first two sources are available in a personal computer-based system, HYDRODATA, marketed by EarthInfo, Inc., of Boulder, Colorado. HYDRODATA stores all the information about the gaging sites and their data on optical disks that can be read using personal-computer hardware and software. The stream flow and sediment data in this book are from either the Water-Supply Paper Series or HYDRODATA (Appendix B).

The gaging data provide three types of information useful for exploring the transport and storage of plutonium in the river system: annual water yield, the annual flood series, and annual sediment yield. For hydrologic purposes, the water year begins on October 1 and ends on September 30 of the following year. This arrangement is convenient because many mid-latitude streams, including the Rio Grande, experience an annual cycle of discharge that is at a minimum in early autumn. A date at the end of this season is therefore a useful starting and ending time for accounting purposes.

Annual water yield refers to the total amount of water that passes the gaging site in a given year. It is a measure of the total amount of water and therefore of the total energy available to perform work, including the transportation of sediment and contaminants. Likewise, annual sediment yield refers to the total amount of suspended sediment that passes the gaging site in a given year, and it is one measure of the work accomplished by the stream.

The annual flood series (also known as the annual maximum series) is the set of the peak discharges in each of the years of the entire gaging record.[56] There is one value for each year, representing the single largest discharge recorded during that year. Floods are important to assessing the geomorphologic and sedimentologic work in a river system, because most major channel changes and a large percentage of the sediment (and thus contaminant) transport take place during flood periods. The amount of work, change, or transport is often directly related to the magnitude of the annual flood (Figure 3.11),[57] which is also a reliable indicator of the timing of major system adjustments. The magnitude of the annual flood is often strongly correlated with the annual sediment yield (Appendix B3).

Water Yield and Annual Floods

The records for the river at Otowi Bridge and at San Marcial describe the time series for the water yield in the Northern Rio Grande. For Otowi Bridge, the data reflect the total amount of water available for work each year between 1895 and 1985 (with the exception of four years without data). Broad temporal trends in

Figure 3.11. A flood on the Rio Grande destroys a bridge during the 1930s. (Los Alamos Historical Society, photo R3458C)

water yield show that the first two decades of the twentieth century were a time of maximum yield, but thereafter a general decline is apparent (Figure 3.12). Two exceptionally high yield years occurred in the early 1940s. A minimum of yield was recorded in the late 1950s, and between the late 1950s and the present there has been a steady increase.

The trends at Otowi Bridge at the upstream end of the study area are similar to the trends at San Marcial at the downstream end (Figure 3.13). The differences between the two records are attributable to the hydrologic adjustments in stream flow between the two sites: irrigation withdrawals, flow loss through percolation into the bed, evaporation, additions from tributaries, and the closure of Cochiti Dam in 1973, which influenced the flow at San Marcial but not at Otowi Bridge.

Temporal trends in annual flood peaks for Otowi are similar to the water yield trends (Figure 3.14), in part because a substantial percentage of the annual yield is associated with the spring snowmelt flood, a months-long event that usually includes the highest discharge in any given year. The most recent major flood was in 1941, a year of widespread flooding throughout the southwestern United States. Since that time, the annual flood peaks gradually declined until the 1980s when a slight upward trend began. The annual flood series for San Marcial at the downstream end of the study area is similar to Otowi Bridge record, except that the attenuation of the 1941 flood relegated it to secondary status after events in 1930 and 1937 (Figure 3.15).

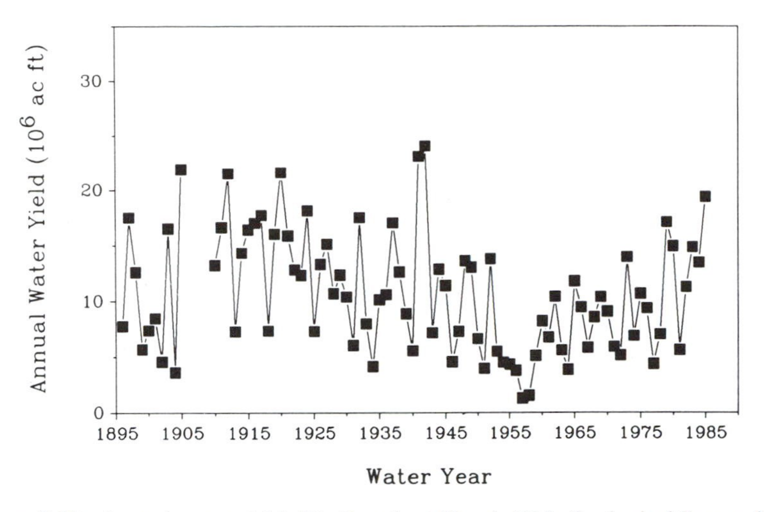

Figure 3.12. Annual water yield, Rio Grande at Otowi. (U.S. Geological Survey data)

Because flood flows generate high amounts of energy for sediment transport and river-channel change, the temporal trends of the annual flood series are important to explaining the transport and storage of plutonium associated with river sediments. Before plutonium appeared in the Rio Grande system, the river functioned at a higher energy level than it did after plutonium was available. At Otowi, the mean annual flood peak after 1944 was 66 percent lower than the mean values before that year; at San Marcial, the mean annual flood after 1944 was only

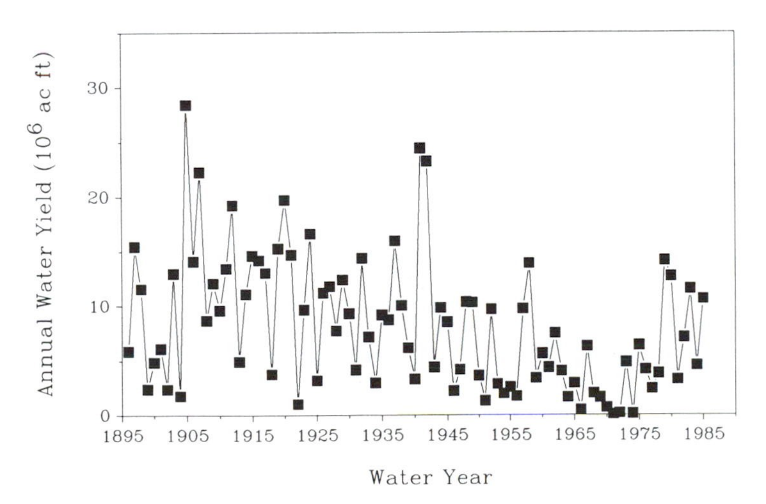

Figure 3.13. Annual water yield, Rio Grande at San Marcial. (U.S. Geological Survey data)

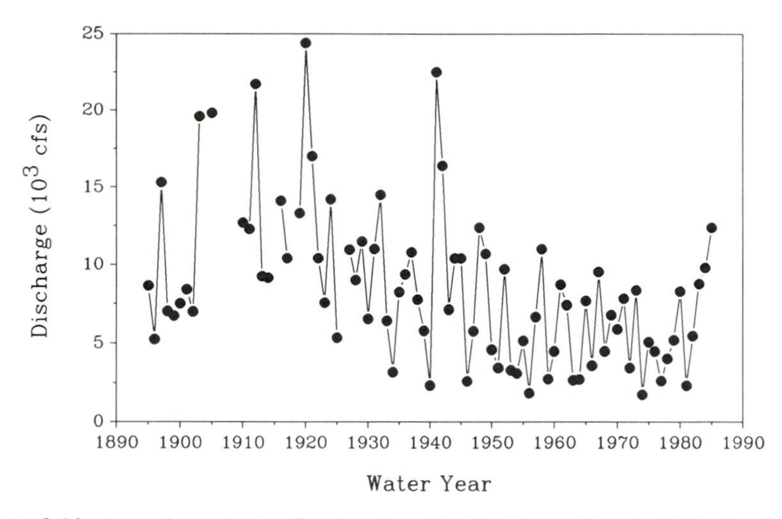

Figure 3.14. Annual maximum flood series, Rio Grande at Otowi. (U.S. Geological Survey data)

46 percent of the mean before 1944. These drastic declines suggest that from the mid-1940s to the 1980s the system experienced a diminishing energy regime and that the processes during the time when plutonium was in the system were different from the processes in the system before the element was present. After the mid-1940s, the channel was more stable and in general became progressively smaller.

The reasons for the temporal variation in water yield and flood peaks are obscure, but three explanations are commonly cited. First, during a period of

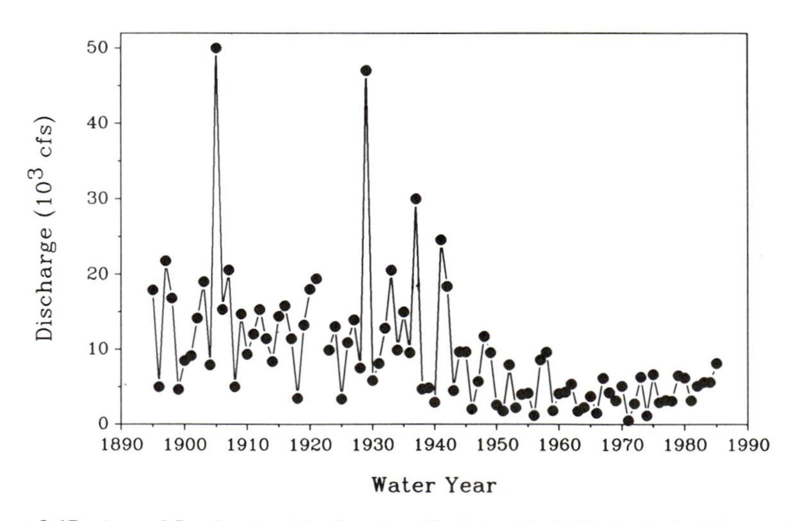

Figure 3.15. Annual flood series, Rio Grande at San Marcial. (U.S. Geological Survey data)

water yield decline, the federal and local governments erected numerous dams on upstream tributaries.[58] The reservoirs of these structures lose water through evaporation, decreasing the downstream yield. The magnitude of such a loss in water yield is not likely to be as large as the recorded decline, however, and the presence of the reservoirs did not prevent the increase in yield in the late record, so the construction of reservoirs was probably a minor influence. The influence of dams on flood peaks is more convincing. As more structures appeared, flood peaks clearly dropped downstream.

Second, land-management strategies may have influenced the water yield record. Overgrazing and extensive logging in the late nineteenth and early twentieth centuries may have accelerated runoff rates and produced high water yields in rivers, by removing vegetation from hillslopes. Later conservation measures may have reversed the trends.[59] Experiments on small western watersheds have demonstrated that increased grazing leads to higher water yields and larger floods.[60] There is no evidence, however, that these effects are cumulative in very large western river basins on the scale of the Rio Grande, where the effects of climate are more pronounced.[61] Some research has suggested similar significant limitations for the land-management explanation in eastern river basins.[62]

In addition to water yield, flood peaks on tributaries respond to land management in the Rio Grande Basin,[63] but the connection between local land management and the response in the main river is unproven. Historical records do reveal that some peaks in flow on the Rio Grande were related to logging activities in the nearby mountains in northern New Mexico. During the period 1909–1926, logging in the watersheds of the Rio Santa Barbara and Rio del Pueblo (tributaries of the Rio Grande north of Española) influenced the flow of the main river in the spring of each year.[64] During much of the year, timber cutters harvested logs from the mountain forests and stored them along the tributary streams. Temporary crib dams on the tributaries held the spring runoff until large volumes of water accumulated. The loggers then broke the dams to release the stored water, raising the discharges in the Rio Grande to levels that permitted their store of logs to float easily downstream. They floated the logs through the Rio Grande to an area near Cochiti Pueblo, where a tie boom across the river trapped them and made them available for nearby sawmills. The high flows generated by this mechanism contributed to the maintenance of the Northern Rio Grande as a wide, shallow, braided stream during the early twentieth century. After logging and with reforestation, the water contribution of tributary watersheds to the Rio Grande decreased and became less erratic.[65]

Finally, an important explanation for the variation in water yield and flood peak is that they respond to climatic changes, especially the magnitude and frequency of atmospheric circulation patterns that deliver moisture to the basin from the ocean.[66] Records of patterns over the western United States show considerable variation over the past century, with New Mexico precipitation

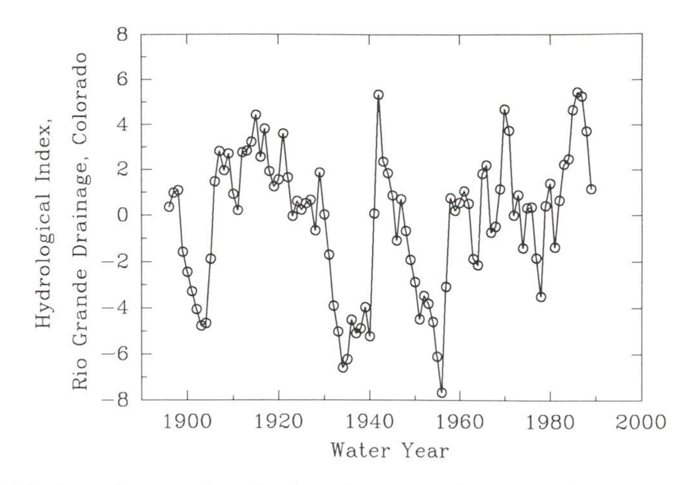

Figure 3.16. Annual mean values for the Palmer Hydrologic Drought Index for the water-source areas of the Northern Rio Grande. (National Climatic Data Center data)

showing marked decreases and increases coincidental with the changes in water yield and flood peaks.[67] Since 1900, small watersheds in northern New Mexico have experienced declines in runoff, primarily because of decreased winter precipitation and snowpacks. Summer precipitation (falling as rain during dry-soil periods that generate little runoff) has increased at the same time as the winter precipitation has decreased, but an overall decline results from the inability of the summer additions to compensate for the loss of snowmelt.[68]

The general influence of climate may be assessed through the standard measures of precipitation and temperature, but because the two mutually influence the availability of water in the surface hydrologic system, a combined measure is more useful. The Palmer Hydrologic Drought Index is a daily metric combining temperature, precipitation, and a lag factor that can have a strong connection to river hydrology.[69] Figure 3.16 shows that this measure of climatic variability, when calculated as mean annual values that are positive during moist periods and negative during dry periods, has had clearly defined trends throughout the twentieth century. Flucutations in river responses such as water yield and flood peaks appear to reflect closely the change in the drought index, showing the strong connection between moisture (and thus the climatic conditions that cause it) with the large regional river system. A combination of all three major influences on river hydrology (dams, land management, and climate) is probably required for a detailed explanation of changes in twentieth-century river flows in the Northern Rio Grande.

The implications of the water yield records for the mobility of plutonium in the Northern Rio Grande are that during the time period when plutonium was likely to be entering the system in greatest quantities from fallout and laboratory releases (the late 1940s to the early 1960s), the amount of energy in the system likely to be available to transport the plutonium was at or near a minimum. The reversal of these trends in the 1980s suggests that the sediments and associated plutonium may have remobilized.

Regional Stream Flows and Floods

The annual water yield and floods also vary geographically throughout the system. The spatial variation explains in part why transport or deposition occurs in particular places within the channel network. Although the drainage area consistently increases in the downstream direction through the study area (Figure 3.3), the annual water yield declines in the downstream direction (Figure 3.17). In the study area (and probably in the entire Rio Grande Basin), the maximum annual water yield is at Otowi Bridge, where inflows from the upper Rio Grande at Embudo combine with the inflows from the Rio Chama. Downstream from Otowi,

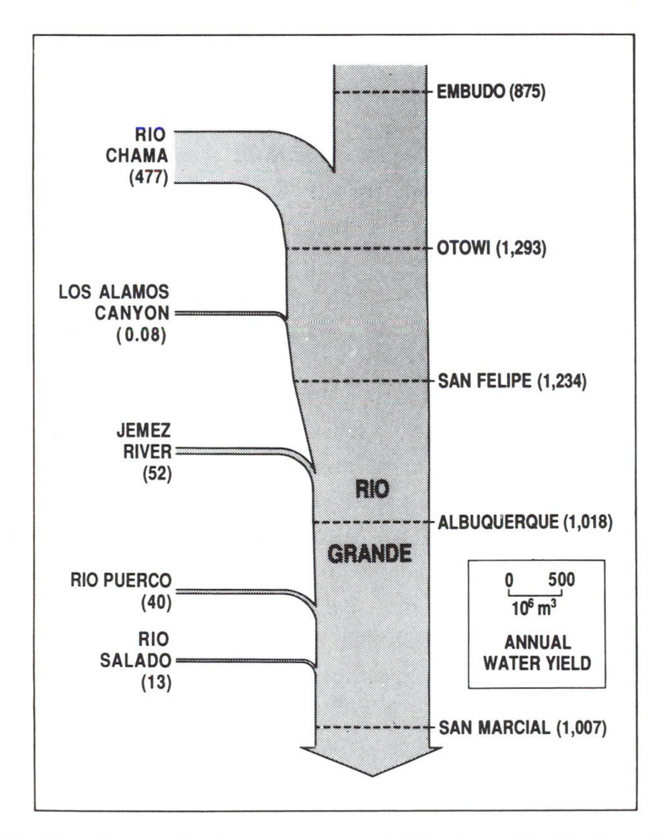

Figure 3.17. Flow diagram for the annual water yield budget of the Northern Rio Grande.

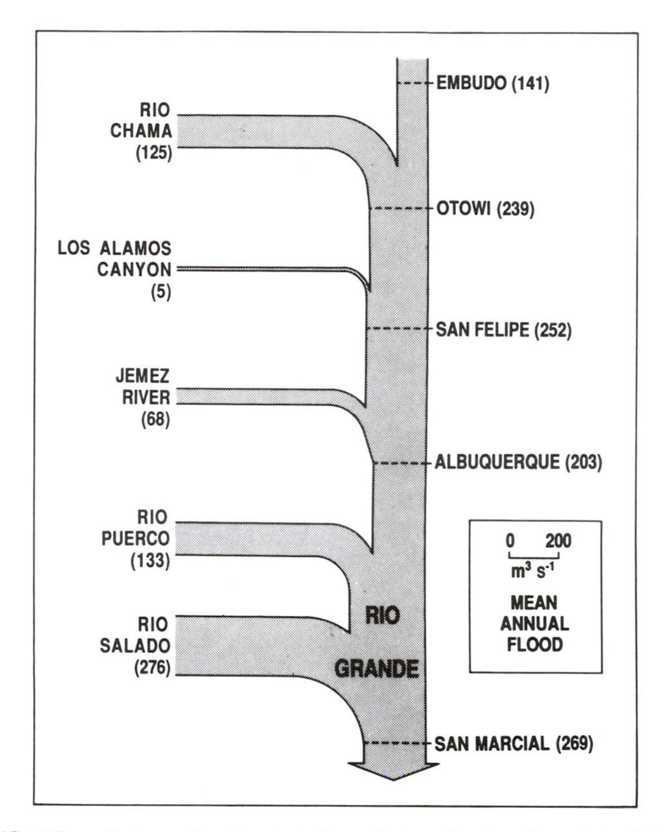

Figure 3.18. Flow diagram for the annual maximum flood series of the Northern Rio Grande.

the additions from tributaries are small, and the main channel loses water through evaporation, transpiration by riparian vegetation, irrigation withdrawals, and transmission losses through bed percolation. Between Otowi Bridge and San Marcial, the annual water yield has fallen by about 23 percent.

The geographic significance of this decline is that although sediment and plutonium contributions to the main stream increase with increasing drainage area, the total amount of stream flow available to transport the materials becomes progressively smaller downstream. It may be that the stream does not use all its available transport capacity in the northern part of the system, so that it can accommodate the increased load, but the downstream decline in transport capacity is one reason to expect the storage of sediments and plutonium in the system between Otowi and San Marcial.

The geography of annual flood peaks is not as simple as the water yield distribution. Annual floods on the Upper Rio Grande at Embudo and on the Rio Chama combine to create a minor maximum at Otowi Bridge (Figure 3.18). The flood maximum falls in the downstream direction through Albuquerque, but floods from the Rio Puerco and Rio Salado create the regional (and probably the basinwide) main-stream maximum at San Marcial.[70] The Rio Puerco has annual floods that are similar in magnitude to the annual floods of the Upper Rio Grande and Rio Chama, whereas those of the Rio Salado are similar to the Rio Grande floods at San Marcial, all despite the relatively small sizes of the tributaries. The Rio Puerco and Rio Salado drain semiarid basins that lack the dense vegetation cover of the larger basins, leading to flash floods that are often larger than those on the main stream.

Flood peaks are a gross measure of the streams' ability to move sediments, so these data indicate a broadly similar ability to entrain materials throughout the system. The result is that during the mean annual flood, seemingly small tributaries can move as much material as is moved in the main channel. Also, the water yield and the magnitude of the annual flood on the main stream are often not closely correlated with the annual sediment transport because of the large out-of-phase contributions of the tributaries (Appendix B3). This arrangement implies that as plutonium from mountain fallout zones and from Los Alamos moves downstream, it is likely to be diluted with tributary materials carried in floods. The arrangement also implies that internal storage of sediment and its plutonium is likely, because in the lower basin larger amounts of sediment are not matched by corresponding larger amounts of flood water to provide transport energy. Chapter 4 explores the nature of these sediments.

4

Fluvial Sediment, Forms, and Processes

Types and Sources of Sediment Data

Although there are numerous aspects of sediment that might be considered in conjunction with questions related to plutonium transport and storage in rivers, particle size is the critical characteristic. Information on the size distribution of particles in sedimentary deposits connects plutonium, sediment, and river processes. An explanation of the geography of plutonium in the regional river system requires knowledge of particle sizes, the distribution of those particle sizes, and their potential mobility in the regional canyons, rivers, and reservoirs. Some general data concerning the size characteristics of fluvial sediment in the Rio Grande system are available from published sources for a few locations, particularly Los Alamos Canyon.[1] Recent, previously unpublished field and laboratory investigations provide additional detailed information for the changing sedimentary environments associated with the river system (Figure 4.1). This chapter reviews the characteristics of river sediments in the Northern Rio Grande and presents a regional sediment budget from historical and geographical perspectives.

Almost 200 sediment samples from deposits of various ages near the channels of the Rio Grande and tributaries demonstrate the variability of the sediment particle sizes. The analysis had three parts: sample collection, sieving, and electronic particle analysis. In the sample-collection phase, collection sites represented identifiable sedimentary units or channel deposits. Each collection site yielded three samples to indicate local variability. Penetration of the surfaces to depths of 5 to 90 cm with a standard cylindrical soil probe provided masses of about 120 g each for laboratory analysis. Investigators kept only the split of the sample that included those particles with diameters of less than 2 mm (that is, sand size or smaller).

a

b

Figure 4.1. Channel changes on the Rio Grande near Corrales, New Mexico, are illustrated at the site of a wooden bridge. (*a*) The view southeast across the river toward the Sandia Mountains in about 1935, showing a single channel with little riparian tree growth along the opposite bank. (T. H. Parkhurst, photo 51466, Museum of New Mexico) (*b*) The same view in 1991, showing a split channel with an island in the center, with considerable amounts of sediment stored in the reach. A dense, mature riparian forest grows along the opposite bank. Although the bridge is no longer present, some footings and pilings remain to mark its location. (W. L. Graf, photo 104-8)

Laboratory procedures included sieving and electronic counting. Sieving divided each sample into masses consisting of those particles larger than 63 microns in diameter (the sand fraction) and those smaller than 63 microns (the silt and clay fraction). The weight of each fraction provided a standarized means of comparing the samples for this study and the results reported by other researchers. Analysis of the silt and clay fraction using a Coulter electronic particle analysis system permitted a detailed investigation of the frequency distribution of the particles in this restricted range. The system analyzes tens of thousands of particles in a brief period of time by passing a laser beam through a container filled with water and sediment. A photoelectric device detects the light passing through the mixture and computer processing of the resulting signal supplies information on the particles' sizes, volumes, and surface areas.[2]

On a larger scale, the stream-gaging network for the Northern Rio Grande includes several sites for collecting bulk sediment data (Figure 3.9 and Table 3.1). Although there are some measurements and estimates of sediment dynamics in the Rio Grande from the late nineteenth and early twentieth centuries, the most accurate and useful sediment data begin during the late 1940s at most locations. Even with some discontinuities, the regional data set is superior to information from most other systems in similar environments. The U.S. Geological Survey Water-Supply Paper series, the computer-based WATSTORE system, and the microcomputer optical disk system HYDRODATA contain all the relevant sediment records. Appendix B lists the data I used in my research for this book.

Two types of sediment are important to assessing contaminant transport and storage: suspended load and bedload. Rivers such as the Rio Grande transport fine sediments, those consisting of silt and clay (with a particle diameter of less than 63 microns), as suspended load.[3] Suspended materials move within the flowing water, giving it a turbid appearance. Because plutonium and other heavy metals preferentially adsorb onto the finest particles, fallout products are most likely to be associated with silt and clay in suspension. Bedload sediments are sand and gravel in the Rio Grande (sand with particle diameters of between 63 microns and 2 mm, gravel greater than 2 mm). They bounce, roll, and creep along the bed at the base of the flow. At flood discharges, the larger amounts of energy available permit the transport of sand in suspension. Sand-size particles are not as likely to contain as high concentrations of fallout products as the finer particles do, but in many systems the sand may be more abundant. When water spills out of the channel, it loses velocity and energy, and suspended materials settle out of the flow. Hence the flood plains along the Rio Grande contain mixtures of sand, silt, and clay, depending on the type of materials carried and the energy available to perform the hydraulic work of transport.

The U.S. Geological Survey's sediment records for the Rio Grande are almost exclusively for suspended sediments. Direct measurements for bedload are rare, so estimations for the amount of material moving at the bed must be based on a known or suspected ratio between suspended load and bedload. The total load—the sum of the suspended and the bedload portions—represents the total amount of sedimentary material moving through the system and available to transport contaminants.

Sediment Characteristics of the Rio Grande

In terms of particle-size distributions, the sediment in the Northern Rio Grande system is highly variable because of the sorting processes inherent in the deposition of the material (Figure 4.2). Although the mean silt and clay (here referred to as "fine" material) content of all the samples collected in this study was 25 percent, a significant variation appears when the samples are partitioned according to sedimentary environment (Table 4.1). Flood-plain (38 percent silt and clay) and reservoir sediments (44 percent) along the Rio Grande contain the greatest amount of fine materials. Tributary streams draining coarse-grained geologic outcrops have the least amount of fine material: channel sediments of the Rio Salado contain 7 percent fine materials, and those in Los Alamos Canyon contain 6 percent (Appendix C).

Figure 4.2. Boys from the Los Alamos Ranch School swimming in the Rio Grande near Otowi Bridge, in about 1938. The channel bar in the background is mostly cobbles and gravel in this view, but by the 1980s it was mostly sand, partly because of changes in river behavior and partly because of local gravel-mining operations. (Los Alamos Historical Society)

Table 4.1. Summary of particle sizes, sedimentary deposits in the Northern Rio Grande system

River	Deposit	N	Min.	Max.	Mean	St. Dev.	Sk.	Kur.
Rio Grande	Active channel	19	0.58	49.60	23.38	21.21	0.01	−1.82
Rio Grande	Flood plain	59	1.74	64.68	38.37	13.94	−0.22	−0.12
Rio Grande	Abandoned single channel	52	0.42	58.61	18.52	15.85	0.67	−0.37
Rio Grande	Abandoned braided channel	21	1.25	61.00	24.25	20.07	0.34	−1.41
Rio Grande	Bar	21	3.48	54.34	16.77	12.25	1.23	2.12
Los Alamos Canyon	Active channel	13	0.00	16.89	5.83	5.15	0.77	−0.14
Rio de Frijoles	Active channel	1	7.45	7.45	7.45	–	–	–
Cochiti Reservoir	Slack water deposit	4	41.00	49.85	44.89	3.77	0.29	−1.18
Rio Puerco	Active channel	3	26.59	29.59	28.56	1.76	−0.38	−1.50
Rio Salado	Active channel	3	6.71	7.15	6.94	0.22	−0.12	−1.50
All samples		196	0.00	64.68	25.00	18.34	0.25	−1.14

Note: Data are for percentage by weight of silt and clay. n, number of samples; min., minimum value; max., maximum value; mean, average value; st. dev., standard deviation; sk., skewness of the frequency distribution of values; kur., kurtosis of the frequency distribution of values.

The sediment characteristics of areas of the flood plain near the Rio Grande channel also exhibit substantial variability. Planners often consider all the relatively flat surfaces near the channel as "flood plain," but the various subcomponents of those surfaces have different geomorphic histories. Abandoned channels, for example, contain less than half the fine particles present in true geomorphic flood-plain areas (Table 4.1). Within the silt-size range alone (2.5–63 microns), different environments have different particle-size distributions (Figure 4.3). Not only do reservoir and suspended sediments contain more silt and clay than the other environments do, but their clay and silt also occur in the finer ranges of these size classes.

Fallout plutonium adsorbed onto silt and clay particles is therefore likely to have a specific geographic distribution in the near-channel environment and to occur in higher concentrations in flood-plain and reservoir sediments, with lower concentrations in abandoned channels. Because effluent released into Los Alamos Canyon entered the Rio Grande from a coarse-grained sedimentary system, it is likely to have a depositional geography different from that of fallout-related deposits. The mixing of sediments from Los Alamos Canyon with upstream

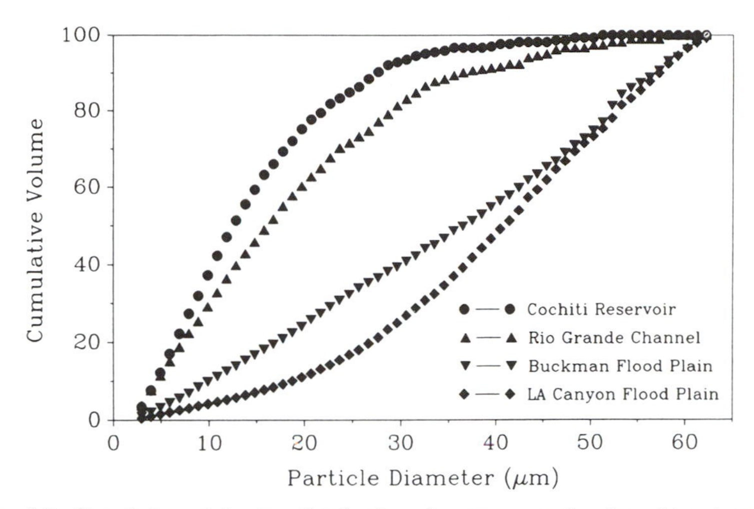

Figure 4.3. Detailed particle-size distributions for representative depositional environments in the Northern Rio Grande near Los Alamos.

materials in the Rio Grande presumably obscures the differences as distance increases downstream away from Los Alamos.

The suspended sediment in the Rio Grande consists of about 50 percent silt and clay, as indicated by the U.S. Geological Survey's gaging records. The samples that I collected for this study support this generalization, because the slack-water sediments from Cochiti Reservoir (near the Cañon de Rio de Frijoles, Bandelier National Monument), which essentially represent materials settled from suspension in the main river, contain 45 percent fines. The bedload for the Rio Grande is coarser, with about 25 percent fines. In the reach below Cochiti Dam, the bed materials have become coarser after the closure of the dam in 1973, a change observed in most channels below large dams.[4] Because the dam results in the release of water without its predam sediment load, erosion of the channel below the dam causes the evacuation of the finer materials, leaving the coarser, heavier particles as an armored layer.[5] The closure of Cochiti Dam caused the median diameter of coarse particles (those materials with particle diameters greater than 63 microns—all but the silt and clay) in the channel floor downstream from the dam to increase from 1 mm to as much as 5 mm.[6] Between 1973 and 1980, the "clear-water" flows from the dam had winnowed the smaller particles from the bed for a distance of about 32 km downstream from the structure (to a location a short distance upstream from the Jemez River). Processes after 1980 may have extended this armoring effect farther downstream, but studies of other rivers show that there is a definite limit to the distance of the impact,[7] and in other cases the impact has not extended much beyond a few tens of kilometers.[8]

The small tributaries of the Rio Grande in northern New Mexico transport sediment loads consisting of particle sizes larger than those in the main stream.

Table 4.2. Summary of particle sizes in sediment of the Los Alamos Canyon system

Location	N	Mean	St. Dev.
Acid and Pueblo Canyons	7	3.79	3.46
DP and Upper Los Alamos Canyons	6	2.59	2.62
Lower Los Alamos Canyon	3	4.01	0.48

Note: Data are for percentage by weight of silt and clay, recalculated for compatibility with the data in this present work by eliminating measurements of particles larger than sand.

Source: A. Stoker, A. J. Ahlquist, D. L. Mayfield, W. R. Hansen, A. D. Talley, and W. D. Purtymun, *Radiological Survey of the Site of a Former Radioacative Liquid Waste Treatment Plant (TA-45) and the Effluent Receiving Areas of Acid, Pueblo, and Los Alamos Canyons, Los Alamos, New Mexico,* U.S. Department of Energy and Los Alamos National Laboratory Report LA-8890-ENV, UC-70 (Los Alamos, N.M.: Los Alamos National Laboratory, 1981), pp. 126, 130, 133.

The samples that I collected from the Rio Puerco for this study show that the silt and clay content is only about 29 percent, a finding in agreement with previous investigations.[9] In some cases, the tributaries drain geologic formations that weather into coarse debris, such as that generated by the rocks of the Jemez Mountains. But the Los Alamos Canyon system contains very little silt and clay in either the channel or the flood plain (Table 4.1). The mean silt and clay content of the 13 samples in this study from Los Alamos Canyon near Otowi was about 6 percent, a value generally similar to those reported by other researchers (Table 4.2).[10]

Regional Sediment Budgets

The records of sediment transport at various gaging sites provide data for regional budgets for the movement of suspended and bed sediments in the Northern Rio Grande. Because the records vary in length throughout the region, the resulting budgets are approximations based on average values, but they provide a general framework in which the likely transport of plutonium may be considered. The suspended load data from direct measurement indicate that taken as a whole, the gaging period was one of some internal storage in the system (Figure 4.4). Summary diagrams illustrate this discussion of sediment budgets, and Appendix B contains the basic numerical data from the budget calculations. The amount of sediment contributed to the northern part of the system is that from the Rio Grande at Embudo, from the Rio Chama, and from minor tributaries. The total was more than 2.4 million Mg per year, yet only 1.7 million Mg per year passed Albuquerque. The remaining 0.7 million Mg per year was between Otowi and Albuquerque as flood-plain sediments and materials in abandoned channels and as lake sediments in the Cochiti Reservoir.

Despite its relatively small size (Figure 3.2) and small annual water yield (Figure 3.13), the Rio Puerco contributes more than twice the amount of sediment as the Rio Grande carries past Albuquerque. Between the late 1940s and the late 1980s, about half of the sediment from the Rio Puerco and the Rio Salado remained in near-channel deposits upstream from the gage at San Marcial (Figure 4.4). Any plutonium associated with the sediments is likely to have had a similar storage pattern.

The regional budget for bedload is less precise than the budget for suspended load, because except for experimental activities there are no direct measurements of the bedload sediment transport. Fluvial and hydraulic research suggests ratios between the two types of load for the various parts of the system, but the reliability of these ratios is unknown. The estimates of bedload are likely to be accurate to within an order of magnitude.[11] Given the amount of published research on the river, in the case of the Rio Grande the error is probably less than 50 percent. For the main stream, the bedload is 14 percent of the total load,[12] a value that is useful for the Rio Chama and the Rio Grande at Otowi, Albuquerque, and San Marcial. Although there are no sediment data for the Rio Grande at

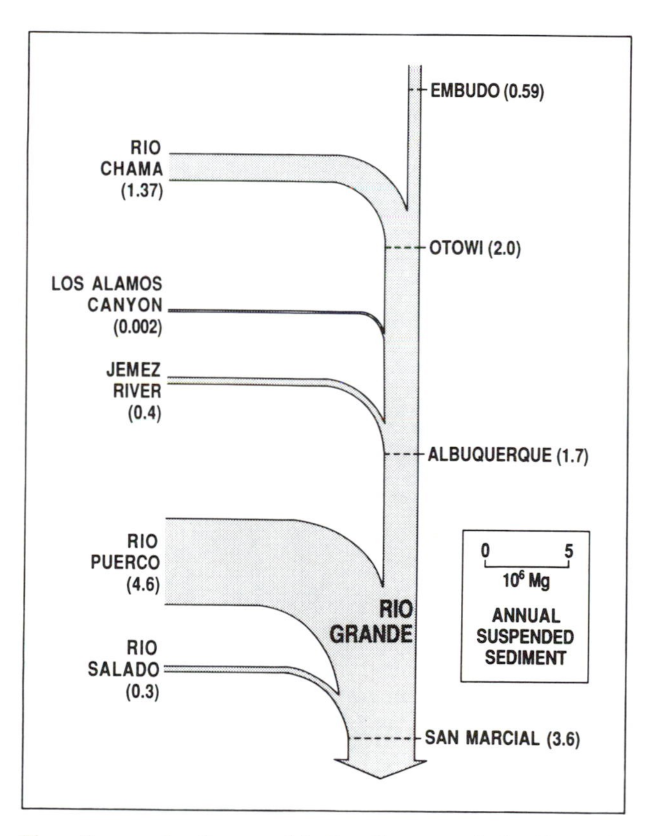

Figure 4.4. Flow diagram for the annual budget for suspended sediment in the Northern Rio Grande.

Embudo, the bedload can be estimated as the difference between the values recorded for the Rio Chama at Chamita and the Rio Grande at Otowi. The bedload for the Rio Puerco is 71 percent of the total load.[13] The assumption that the bedload is 50 percent of the total load of Los Alamos Canyon, Jemez River, and Rio Salado is reasonable.

The regional bedload budget is somewhat different from the suspended budget because of the overwhelming impact of the Rio Puerco (Figure 4.5). The northern part of the system carries relatively little bedload, but the arid and semiarid watershed of the Rio Puerco generates large amounts of bedload materials, two-thirds of which are stored in the near-channel environment once they enter the Rio Grande. The plutonium associated with releases of effluent into the alluvium of Los Alamos Canyon is connected to a tiny portion of the regional bedload budget, and if those materials were to move as far south as the Rio Puerco, their mixing with bedload sediments from the tributary would be likely to dilute the Los Alamos contribution to such a great degree that they would be unrecognizable.

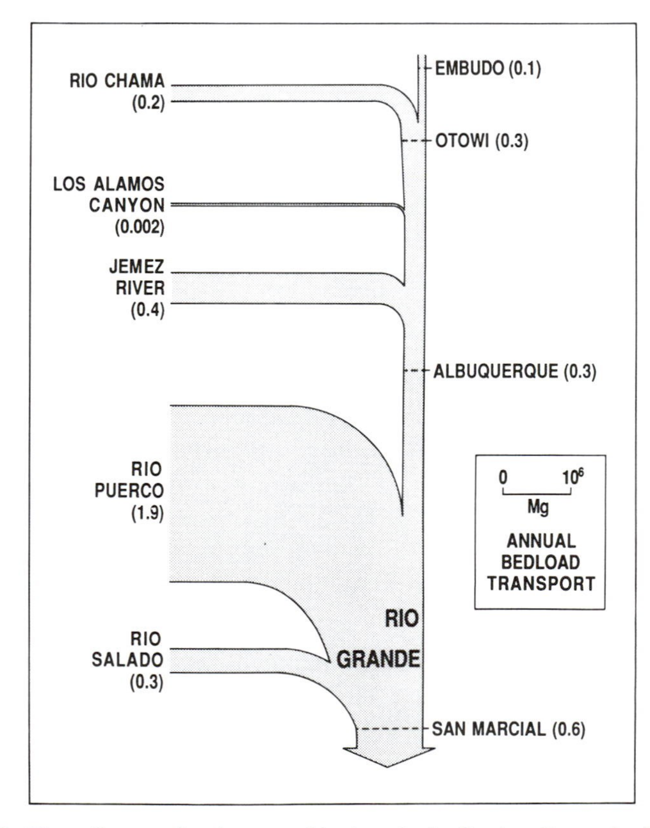

Figure 4.5. Flow diagram for the annual budget for bedload sediment in the Northern Rio Grande.

Time Series for Sediment Transport

The general regional budgets for suspended and bedload sediments provide a geographic perspective on fluvial transport processes, but there is considerable interannual variation in the rates of movement as recorded in the gaging records. The record of suspended sediment yield of the Rio Grande at Otowi began after the major flood of 1941, which probably would appear in the record as the year with the largest sediment yield in recent decades. Since 1947, when records were first kept at this location, the highest yield occurred in 1958 (Figure 4.6), a regional flood year in which the Rio Grande as well as local tributaries experienced significantly high discharges. Before 1958, there were two years, 1948 and 1952, when the sediment yield was greater than that in any year after 1958. After 1958, there was considerable variation from year to year, with a slight downward trend in the mean annual sediment yield.

At the lower end of the study area, on the Rio Grande at San Marcial, the record of sediment yield is different because of inputs from tributaries between Otowi and San Marcial, especially the Rio Puerco (Figure 4.7). The highest year recorded is again 1958, but later in the record, tributary floods either contributed large amounts of sediment to the system or remobilized sediments previously stored along the channel. Releases of sediment-free water from Cochiti Dam also helped increase the erosion of the channel and near-channel areas of the Rio Grande,[14] thereby raising the sediment yield of the river at San Marcial and Elephant Butte Reservoir.

Because of the plutonium releases by Los Alamos National Laboratory, the sediment yield of Los Alamos Canyon is of special interest in this study. Using

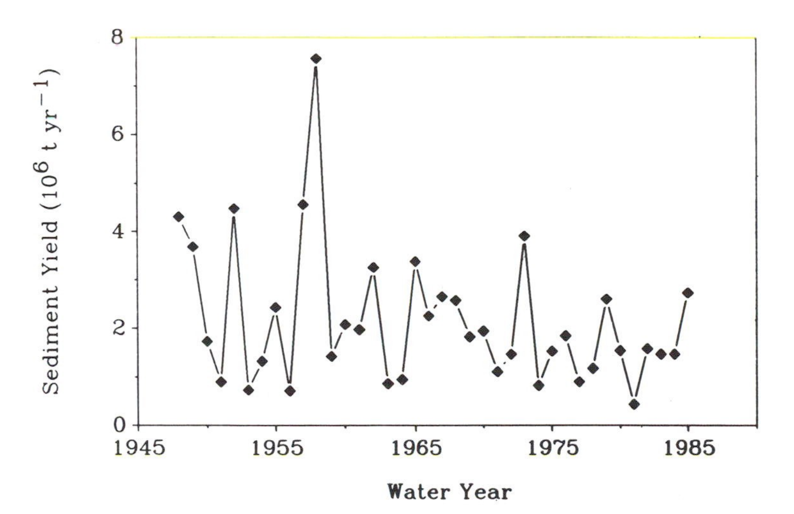

Figure 4.6. Annual suspended sediment discharge, Rio Grande at Otowi. (U.S. Geological Survey data)

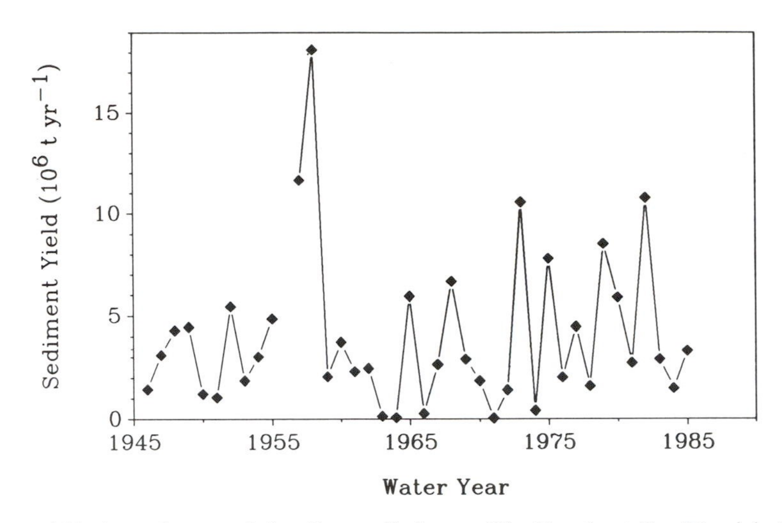

Figure 4.7. Annual suspended sediment discharge, Rio Grande at San Marcial. (U.S. Geological Survey data)

data from an intermittent gaging record on the stream and precipitation records at nearby locations, earlier researchers calculated the probable sediment yield from the canyon into the Rio Grande.[15] Figure 4.8 shows their unpublished results, the record from 1943 to 1980. The stream is less consistent in its sediment yield than the main river is, because the tributary is an intermittent stream, drains a smaller watershed, and lacks a large water source area at a high elevation. The years of high sediment yield from Los Alamos Canyon (1944, 1951, 1952, 1957, and 1968) are out of phase with the peaks on the main stream, because the Los Alamos stream responds to local events in the Jemez Mountains, whereas the Rio Grande responds to larger-scale climatological conditions in the San Juan Mountains (for example, compare Figures 4.8 and 4.6). Any plutonium associated with sediments in Los Alamos Canyon therefore entered the main system sporadically and was more or less diluted by main-stream sediment, depending on the annual yield of the Rio Grande for that particular year.

The storage component of the annual sediment budget has important implications for the dynamics of plutonium in the system. The stored sediments and the plutonium attached to them have a particular geographical distribution within the system, and the amounts of annual storage have varied considerably over time. The distribution of gaging sites on the Rio Grande defines four accounting areas within the larger general study reach from Otowi to San Marcial. The reaches of the Rio Grande and the gages that define them are as follows:

1. Otowi to Jemez River, gages at Otowi Bridge and on the Jemez River
2. Jemez River to Albuquerque, gages at Otowi Bridge and Albuquerque on the Rio Grande and on the Jemez River

3. Albuquerque to the Rio Puerco, gages at Albuquerque and on the Rio Puerco
4. Rio Puerco to San Marcial, gages at Albuquerque and San Marcial on the Rio Grande and on the Rio Puerco

The 1948–1985 record of total sediment (suspended load plus bedload) stored in the system as calculated from the gaging data for these reaches shows that the system has neither stored sediment continuously throughout the period of record nor stored it evenly throughout the length of the river. In the two reaches above Albuquerque, storage or losses have been relatively small, with storage dominating the system from the late 1950s to late 1960s and then again in the late 1970s (Figure 4.9). Losses were prominent in the late 1940s and early 1970s. Thus the years of major sediment yield from Los Alamos Canyon (probably including plutonium) coincided with a period of system storage. These variations may be related to flood events, especially in the late 1950s, when flows spilled out of defined channels and deposited much material on flood plains. The closure of Cochiti Dam in 1973 increased the erosion by clear-water releases from the dam, producing the observed sediment losses above Albuquerque. These losses were not recorded for the Otowi–Jemez River reach, which includes only a limited portion of the channel eroded below Cochiti. The dam affected all of the Jemez River–Albuquerque reach, as the record shows.

The two reaches below Albuquerque had more variable storage records than did those upstream (Figure 4.10). The 1958 flood resulted in significant erosion and a loss of stored sediment, and floods on tributaries affected the records for other years.

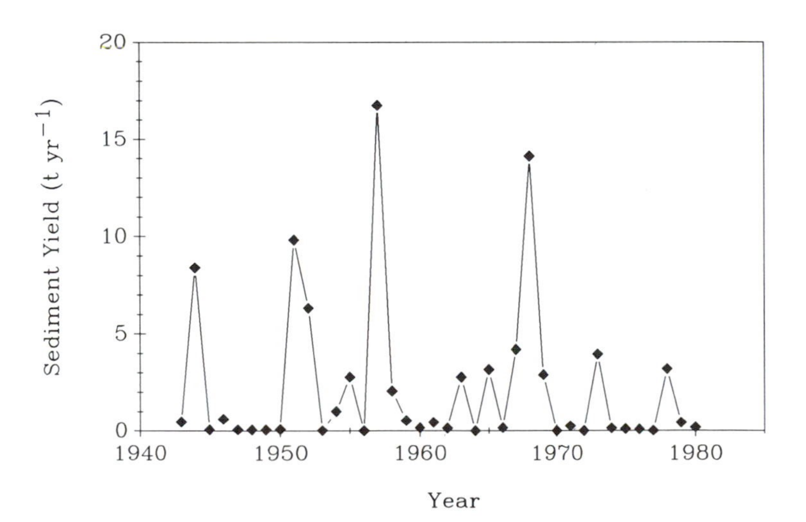

Figure 4.8. Annual sediment discharge from Los Alamos Canyon to the Rio Grande. (Data from L. J. Lane, Agricultural Research Service, Tucson)

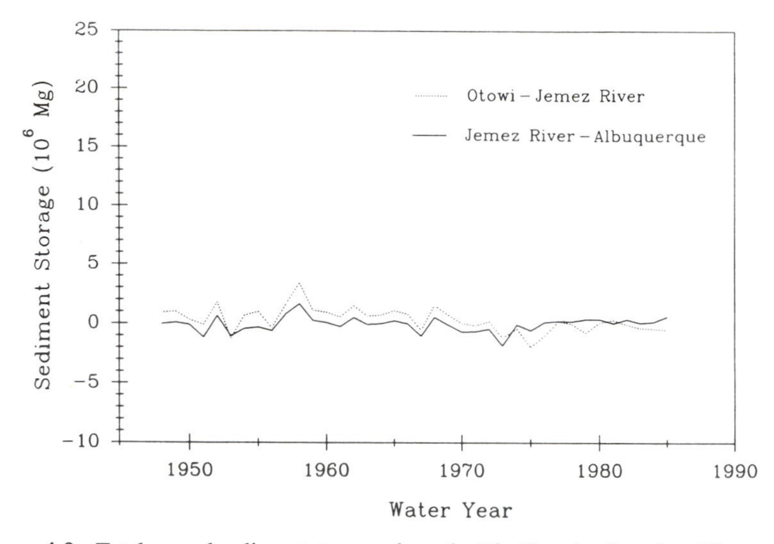

Figure 4.9. Total annual sediment storage along the Rio Grande, Otowi to Albuquerque.

Channel "improvements" (levees and channelization by artificial works) may have restricted the flow and caused some erosion after 1960 between Albuquerque and the Rio Puerco. The massive influxes of sediment during flood years on the Rio Puerco caused extreme variability in the record for the Rio Puerco–San Marcial reach. Unlike the reach immediately upstream, its record is one of nearly continuous storage.

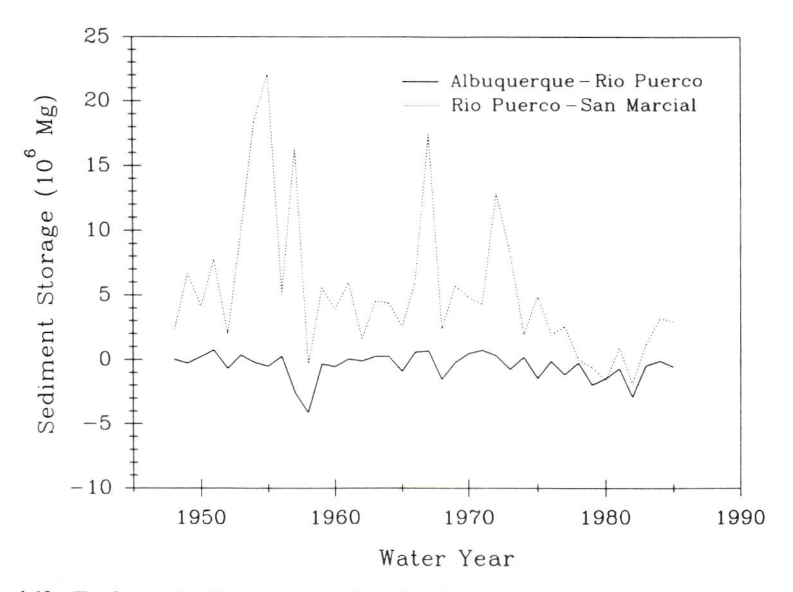

Figure 4.10. Total annual sediment storage along the Rio Grande, Albuquerque to San Marcial.

Annual Variation in the Regional Sediment Budget

A view of the total system of suspended sediment is a useful summary of the temporal changes that have affected the regional sediment budget. Because suspended load and bedload are related to each other, the trends for suspended sediment indicate change in the entire sediment system. The suspended sediment trends are also more reliable because they represent measurements rather than calculated values based on assumed ratios, as is the case for bedload. In a simplistic view, the total input for the study reach is the sum of inputs from the Rio Grande at Otowi and the Rio Puerco. This simplification ignores inputs from other tributaries, but they are relatively small (Figure 4.4). The total output from the system is the suspended sediment yield of the Rio Grande at San Marcial. This picture can be further simplified by smoothing the annual data by using a tenth-order polynomial to describe the time series of annual input and output (Figure 4.11).

The generalized approach shown in Figure 4.11 illustrates several important points about the temporal variation of the sediment transport system in the Northern Rio Grande from 1948 to 1985. Throughout this period, the sediment flux in the Rio Grande at Otowi gradually declined, largely because of smaller amounts of water passing through the river to provide a mechanism for transport. As Chapter 3 pointed out, this decline in water is the product of several factors, including human influences and climatological changes. Because it is in a different climatological and vegetation zone, the Rio Puerco first increased and then decreased its huge sediment contributions. Late in the record, the Rio Puerco

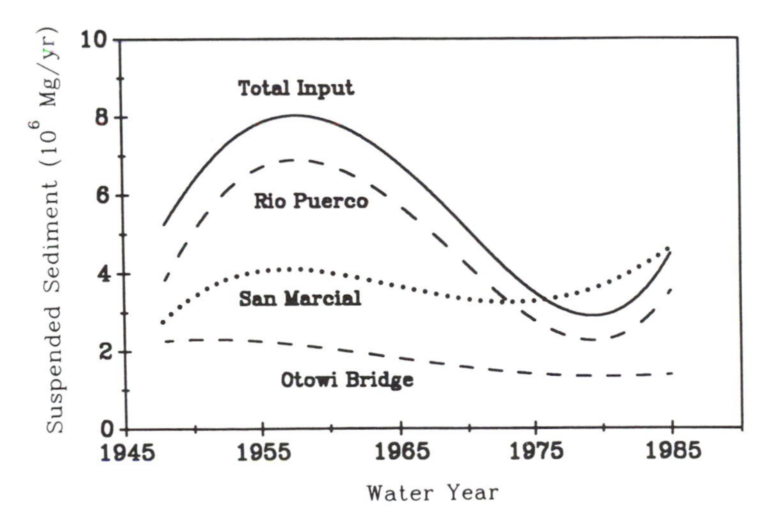

Figure 4.11. Systemwide annual sediment budget for the Northern Rio Grande, smoothed with tenth-order polynomials.

began to increase its inputs again. The sediment output from the Northern Rio Grande system at San Marcial fluctuated somewhat during these years, by reflecting trends in the Rio Puerco, but these changes were less dramatic than those in the tributary.

The generalized sediment budget also shows that between 1948 and 1975 much more sediment entered the system than left it. The internal storage of sediment by deposition on flood plains accounts for the difference. This well-defined storage period coincides with the time when maximum amounts of plutonium were likely to have entered the system from fallout and from Los Alamos Canyon. Much of the plutonium likely remains in internal storage in various deposits associated with the depositional period. For a relatively brief period after 1975, more sediment left the system than entered it, indicating an evacuation of sediments stored in internal flood plains. These remobilized sediments, containing varying amounts of plutonium depending on their ages, were probably transported by the relatively clear water discharges from Cochiti Dam, closed in late 1973. The sediments moved past San Marcial to their new deposition site, the pool of Elephant Butte Reservoir a short distance downstream from San Marcial. The general trends shown in Figure 4.11 suggest that by the 1980s this evacuation process had ended. The sediment budget calculations reach by reach (Figures 4.9 and 4.10) suggest that the 1975–1985 period of erosion removed only a small amount of the total mass of sediments stored after 1945.

The interactions of the landscape, climatic, and sediment systems that resulted in the storage of large quantities of sediment along the active channel of the Rio Grande created a suite of landforms associated with the river. The following sections focus on these landforms, particularly the channels and flood plains, as keys to understanding the arrangement of plutonium that their sediments might contain.

Sources of Geomorphic Data

The fluvial processes that move and store sediment in the Northern Rio Grande operate within a system of landforms consisting of the river channel and its associated flood plain. As the amounts of water and sediment change over time and from place to place and as the magnitude and frequency of floods change, the channel and near-channel landforms adjust to accommodate the new hydraulic conditions (Figure 4.12). The channel often changes its size, shape, pattern, or location in response to hydraulic controls, and in so doing it creates a suite of flood-plain features that include abandoned channels, terraces, active and inactive flood plains, and a variety of bars inside and outside the present active channel.[16] The fate of plutonium in this environment depends on the chronology and mechanism of these changes in the river channel, because sediment plus its associated contaminants make up the various landforms. In the case of plutonium in the Northern Rio Grande, identifying, mapping, and assigning dates to the

Figure 4.12. The Jemez River downstream from the Jemez Dam site shows significant channel adjustments, partly in response to the closure of the dam and the elimination of high flood flows. (*a*) Looking east, the 1936 channel is a broad, sandy ribbon in an unstable braided channel configuration. (G. P. Williams, photo 82, U.S. Geological Survey Photography and Field Records Library, Denver) (*b*) The same view in 1980, showing a narrow, single-thread, stable channel after the closure of the dam a short distance upstream. (G. P. Williams, photo 84, U.S. Geological Survey Photography and Field Records Library, Denver)

near-channel landforms and their sediments establish linkages with materials that entered the system during those years when contamination was most likely. The regional sediment budget shows that the river has stored large amounts of sediment between Otowi and San Marcial; an analysis of the fluvial geomorphology of the system can identify the locations and dates of the stored material.

Topographic maps and aerial photography provide basic location information for fluvial landforms. Topographic maps published by the U.S. Geological Survey depict the river and its environs at 1:24,000 scale and serve as base maps for plotting data and locating specific landforms or deposits. Appendix F provides a complete list of the topographic quadrangles that show the river from Española to San Marcial.

Aerial photographs offer basic mapping information on the location and extent of landforms and deposits at various times for the reach between Española and San Marcial. Through photogrammetric techniques, the photos yield quantitative data, particularly for area and distance measures. Most reaches of the river appear in aerial photographs at least once in each 10-year period from the late 1930s to the present. Since the mid-1940s, most reaches were photographed from the air at least every three years. Sources of aerial photography include the U.S. Army Corps of Engineers, U.S. Bureau of Reclamation, U.S. Geological Survey, U.S. Department of Agriculture, U.S. National Archives, U.S. Bureau of Indian Affairs, Los Alamos National Laboratory, New Mexico Highway Department, New Mexico State Land Department, county and city planning agencies, telephone and electrical utilities, and private aerial-photographic and engineering firms. This study used a representative selection of aerial photography from a limited number of sources of information on changes in river channels (Appendix G gives the addresses of these sources).

In some cases, historic ground photography supplemented the aerial-photographic record of channel changes. These historical photos are more sporadic in their temporal and geographic coverage than are their aerial counterparts, but they offer qualitative information and useful perspectives as well as reinforcing interpretations. Sources of historical photos of the Northern Rio Grande include the Colorado Historical Society, Denver Public Library, U.S. Geological Survey (Denver), Los Alamos Historical Society, Albuquerque Public Library, University of New Mexico Library, and New Mexico State Archives (Appendix H gives the addresses of these sources). For this book, I used only photos of areas of special interest rather than collecting all available images.

Channel Changes, 1940s–1980s

The temporal changes in water yield, flood series, and sediment yield in the Northern Rio Grande between the 1940s and the 1980s led to a predictable series of changes in the channel and its surroundings. Decreasing amounts of water produced a progressively smaller channel throughout most of the river from Española to San Marcial. The shrinkage of the channel resulted in larger flood-plain areas and the abandonment of some minor subchannels. The smaller annual floods contributed to this conversion from large to small channels. And the internal storage of large quantities of sediment helped expand the flood plains and filled some of the abandoned channels.

The overall character of the Rio Grande changed between the 1940s and the 1980s (Figure 4.13). Before the early 1940s, the channel was broad and shallow with numerous bars and subchannels, a classic case of a braided stream produced by high sediment loads, erodible banks, highly variable discharge, and high amounts of stream power.[17] As smaller water yields, sediment yields, and flood magnitudes forced the development of a smaller channel, the braided characteristic gradually disappeared, so that by the 1980s the river consisted of only a single thread.

The change from a braided to a single-thread pattern has occurred on most major streams in the western United States, including the Platte in Nebraska[18] and the San Juan in Colorado and Utah.[19] The shrinking trend is often directly related to the closure of dams,[20] and in some parts of the Rio Grande—specifically in the reach of the Jemez River downstream from Jemez Dam—this connection is obvious. The tendency toward single-thread channels must also derive from regionwide hydroclimatic influences, because the Rio Grande above Cochiti Dam also has contracted, and many other streams without major dams exhibit the same changes: the Fremont River in Utah,[21] the Paria River in Utah and Arizona,[22] the Little Colorado River in Arizona,[23] and the Upper Gila River in New Mexico and Arizona.[24] In the Rio Grande, engineering works have exaggerated the change and made it more lasting, by imposing on the system artificially designed pilot channels and levees.

As the Rio Grande channel decreased in width, it also became locationally unstable, moving from one position to another across the valley floor in those reaches where it was not confined by rock outcrops or levees. During the 1940s–1980s period, the main channel of the Rio Grande changed horizontal position by as much as 1 km. These changes came during floods when sediment plugged old courses and flows spilled over poorly consolidated banks to cut new courses. This migration is also common on dryland rivers, and the rate of change on the Rio Grande has been similar to that seen in other streams in the western United States.[25]

Near-Channel Landforms

Channel shrinkage, pattern change, and migration produce a characteristic series of landforms on the valley floor near the channel. Planners and engineers refer to this relatively flat area separated from the channel by banks in undisturbed environments as the flood plain.[26] But from a geomorphological and sedimentological perspective, this area consists of a variety of forms and deposits with different origins and characteristics. A true geomorphological and sedimentologi-

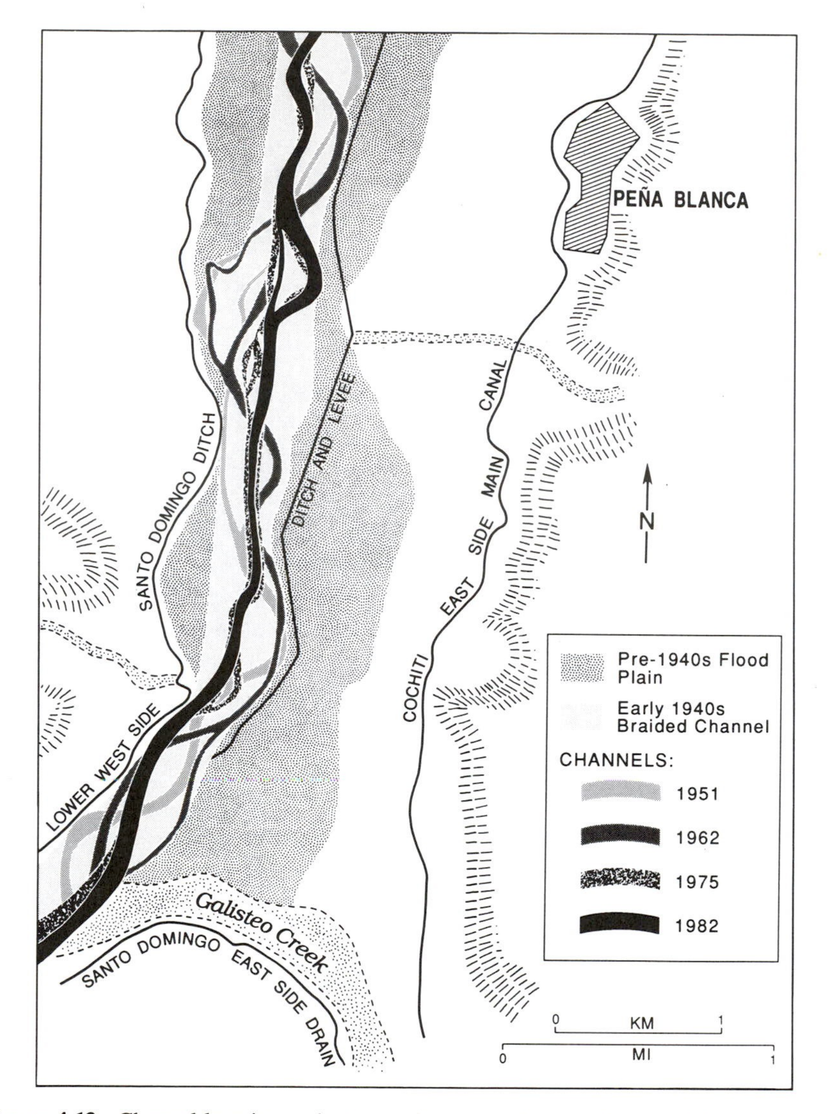

Figure 4.13. Channel location and pattern changes as deduced from aerial photography, Rio Grande near Peña Blanca, from the late 1930s to 1982.

cal flood plain is an alluvial surface next to a channel, separated from the channel by banks and composed of materials transported and deposited by the present hydrological regime of the river.[27] The sediments that make up the flood plain are usually finer than the bed sediments in the nearby channel, because when water spills out of the channel, its velocity slows and it deposits some of its fine suspended load. Small-scale features sometimes include extensive ripple patterns or occasional splays of gravels where flows exited the channel.

The near-channel areas of a changing river like the Rio Grande also include abandoned courses etched into the surfaces of the flood plain. Abandoned braided channels have sandy and gravelly beds, with linear depressions separated by ridges that reflect the original subchannels and bars. The bars often rise to levels similar to the level of the flood plain on either side of the abandoned channel area. Abandoned single-thread channels form linear depressions through flood plains and may contain coarser material than that on the flood plains, representing the bedload once carried in the channel. Many abandoned single-thread channels have had a more complex history, however, because after they were abandoned, flood flows temporarily filled them with slack water that deposited fine suspended sediment. If the abandoned channel is not completely filled with sediment, it will retain its depression characteristics and be a noticeable interruption of the planar flood-plain surface. But if the abandoned channel is completely filled with slack-water sediments, its surface may be coincidental with the surrounding flood plain, and indirect evidence such as vegetation may be the only obvious indication of its presence. Excavation reveals sedimentological variation that confirms the location of the channel and its filling.

Flood bars provide positive relief above some flood plains. As flood waters spill out of channels and over flood plains, turbulence in the lee of obstacles such as trees, buildings, bridge piers, or other structures produces localized accumulations of debris and sediment. Usually such features are less than a meter high and several meters long, but if they occur in great numbers as in a thicket or in artificial sediment traps, their total accumulation can be substantial.

Near-channel areas also include some deposits not directly related to processes in the main stream—deposits from tributary channels that interfinger with flood-plain and related materials near the valley margin. In some cases, tributary deposits cover the flood-plain materials in the form of alluvial fans extending from the mouths of tributaries in the valley side, across the flood plain, and terminating at the active channel of the main river.

In the Rio Grande between Española and San Marcial, the near-channel environment includes all these forms, but in many cases human activities have modified them. The construction of levees (reviewed in detail in Chapter 5) has eliminated flows from many geomorphic flood plains, and they no long receive infusions of sediment. This construction has disrupted other forms or obliterated them completely. In White Rock Canyon, the basalt and tuff walls constrict the valley floor so that flood-plain features are generally absent except for isolated deposits in small pockets in the canyon walls.

Near-Channel Deposits

The surface expressions of the near-channel forms have connections to subsurface variation, as seen in the characteristics of the sediments (Figure 4.14), with the

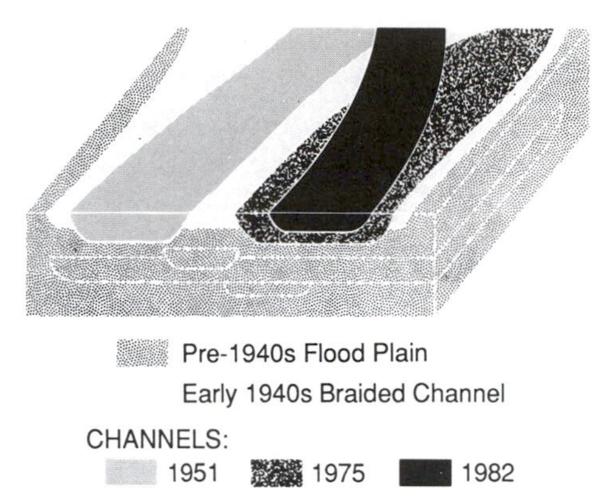

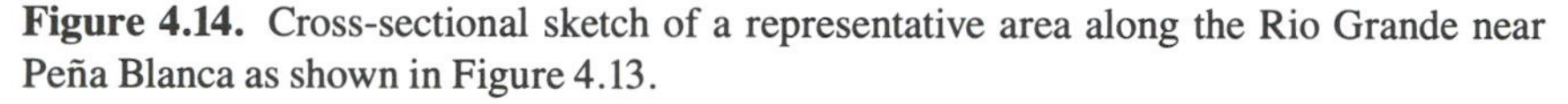

Figure 4.14. Cross-sectional sketch of a representative area along the Rio Grande near Peña Blanca as shown in Figure 4.13.

sediment in each form exhibiting a distinctive structure and particle size distribution.[28] Flood-plain deposits have finely laminated structures consisting of thin horizontal sheets. Along the Northern Rio Grande, the sheets are usually 1 cm or less in thickness and consist of very fine sand, silt, or, occasionally, clay. Abandoned braided channels contain materials that are relatively uniform in size, mostly sand and coarse sand in the Rio Grande, but prominent structures include cross-set beds formed by migrating dunes, bars, and sand waves when the channel was active.[29] Mid-channel bars sometimes appear in the deposits as gravel lenses. Abandoned single-thread channels may appear in deposits as linear accumulations of massively bedded sand, or they may contain large amounts of silt and clay if they were filled by slack-water deposits after their disconnection from the active channel system. Tributary deposits are almost always coarser than materials from the main channel, because the tributaries have steeper gradients and therefore generate more shear stress for the transport of larger particles. The deposits in alluvial fans tend to become coarser in successively higher layers,[30] but the deposits of the main river often become finer in the upward direction.

The implication of these arrangements for the movement of and storage of plutonium in the near-channel environment is that if sediment in the active channel carries plutonium adsorbed to its surfaces, if that sediment is deposited near the channel during a general storage period, and if the depositional forms and materials can be explained and mapped, then it is possible to deduce the ultimate distribution of the contaminant. Such knowledge can guide sampling and monitoring programs and can lead to a definition of those areas of likely concentration, especially given the affinity of plutonium for fine particles.

In the Rio Grande system, the differences among flood plains, abandoned channels, and bars are clear but subtle to the observer on the surface. Minor variations in elevation, surface irregularities, minor depressions or ridges, and small scarps, all less than 0.5 m high, provide clues to the origin and type of landform and the nature of sediments likely to be found beneath the surface. The surface forms are not usually visible directly on aerial photography, but because the various forms are connected to particular particle sizes and moisture conditions, vegetation communities visible in aerial photographs almost always reflect them.

The Distribution of Flood Plains

The distribution of active channels and flood plains (both active and those now isolated from the channel by engineering works) along the Northern Rio Grande varies from place to place. This distribution is connected to the distribution of potential storage sites for plutonium. A useful measure of the amount of flood plain area in a given reach of the river is the concept of unit area, the number of square kilometers of flood plain associated with each kilometer of river-channel length. If the unit area of flood plain in a particular reach is 1.0 sq km/km, then for each kilometer of channel length there is 1.0 sq km of flood-plain area. In this case, the flood plain might form strips 0.5 km wide on each side of the channel or a strip 1.0 km wide on only one side. A similar approach determines the unit area of active channel, the area of active channel per unit length of the river.

Data from aerial photography (1982–1985) enables us to find the distribution of flood-plain and channel areas along the Rio Grande, using the following method:

1. Define the scale of each frame of photography by measuring the horizontal distance between two prominent objects that appear in the photograph and on a topographic map of scale 1:24,000. Convert the map measurement M to the actual real-world distance it represents (A):

   $$A = 24{,}000\,M$$

   Measure the distance between the same two objects in the photograph, and determine the scale of the photograph by comparing the photograph measurement (P) with the actual distance:

   $$\text{Photo scale} = P/A$$

   The scales of photographs in this analysis ranged from about 1:20,000 to about 1:80,000.
2. Divide the length of the study river into convenient segments. The useful landmarks and geomorphic variation in the Rio Grande generate 42 segments ranging in length from 2 to 14 km, with a mean length of 6.8 km (Appendix D identifies the segments).
3. Using a computerized digitizing system (Jandel SIGMA-SCAN, Version 2.1), measure the length of each segment (L).

4. Using the same system, measure the area of active channel (*C*).
5. Using the same system, measure the area of flood plain (*F*).
6. Calculate the unit area of channel (*UAC*) for each segment by

$$UAC = C/L$$

and the unit area of flood plain (*UAFP*) by

$$UAFP = F/L$$

7. Plot the unit areas of channel and flood plain against the downstream distance of the center of each segment to produce a summary diagram.

The results of this procedure for the Rio Grande between Española and San Marcial show the exact distribution of channels and flood plains (Figure 4.15 and detailed data in Appendix D). Whereas tabular data describe the regional sediment budget in numerical terms, Figure 4.15 depicts the geography of the storage term in the total sediment budget for the main river system. Kilometer 0 is the Old Española Highway Bridge, and km 313 is the Southern Pacific Railroad Bridge near San Marcial. The unit area of flood plain changes radically from one segment of the river to another, mostly in response to the available space on the valley floor and the contributions of tributaries. Peaks in the distribution represent segments of the Rio Grande where the flood plain is exceptionally broad and where large amounts of sediment are stored. The first prominent peak is at about km 78, the confluence of the main stream with Galisteo Creek. Other areas of extensive flood plain are downstream from the confluence with the Jemez River (km 97), downstream from the confluence with the Rio Puerco, and wide valley areas near San Antonio (km 285) and near San Marcial (km 300).

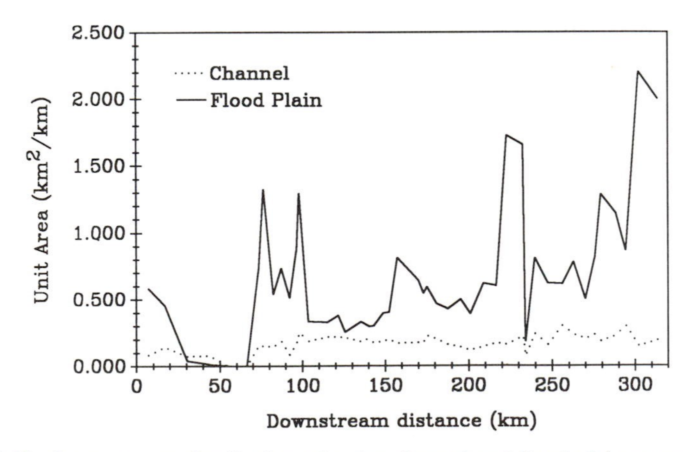

Figure 4.15. Downstream distribution of active channel and flood-plain areas along the Rio Grande from Española to San Marcial (313 km), as measured from aerial photography taken in the period 1982–1985.

The active-channel area is much less variable than the flood-plain area. Bedrock exposures constrict the channel severely in a few places, such as at km 91, upstream from the confluence with the Jemez River. The only segments where the channel is wider than the flood plain are those in White Rock Canyon, where the channel occupies most of the available space on the valley floor. The waters of Cochiti Reservoir drown the segments from km 50 to km 66.

The Depth of Deposition, 1948–1985

The regional sediment budget indicates the approximate amount of total load deposited in the Northern Rio Grande from Otowi to San Marcial in 1948–1985 and the area of that deposition, and so it is possible to estimate the depth of the deposition on flood-plain surfaces for the 1948–1985 period. The sediment transport data from gaging stations are by weight, necessitating the conversion to volume using an estimation of bulk density for the deposits. For flood-plain and related deposits, the following calculations used a bulk density of 1.6 Mg/cu m and the data reviewed in Appendices B (for sediment masses) and D (for areas of deposition):

1. Convert the total sediment stored as measured by its weight (W in Mg) to the total storage as measured by its volume (V in cu m):

$$V = W/1.6$$

2. Partition the area data for potential storage areas into reaches that match the reaches defined by the gaging data for sediment, and determine the total available storage area within each reach.
3. Determine the mean depth of storage (D in m) in each reach by dividing the volume (V in cu m) of storage by its area (A sq m):

$$D = V/A$$

These calculations assume an equal distribution of deposition over the available area, and so they produce only mean values. Field observations indicate that the depth of deposition varies from place to place, but not by more than a factor of about 2. The reach-by-reach summary (Table 4.3) shows that the Otowi–Jemez River and Albuquerque–Puerco River reaches had negative depositions—that is, a net loss of stored sediment through erosion. In the Otowi–Jemez River reach, White Rock Canyon had little storage and erosion after the closure of Cochiti Dam ensured the negative value. In the Albuquerque–Puerco River reach, channelization and levee construction restricted the channel and enhanced erosion. The Jemez River–Albuquerque and Puerco River–San Marcial reaches had net gains in storage, thereby raising the elevations of the surfaces of their flood plains and channels.

Table 4.3. Summary of calculations for mean depth of deposition of stored sediment

Calculated Item	Otwoi– Jemez River	Jemez River– Albuquerque	Albuquerque– Rio Puerco	Rio Puerco– San Marcial
Weight of stored sediment (tons)	−2,725,773	12,919,299	−20,653,569	220,331,627
Weight of stored sediment (mg)	−2,472,821	11,720,388	−18,736,917	199,884,852
Volume of deposition (cu m)	−1,545,513	7,325,243	−11,710,573	124,928,033
Area of deposition (sq m)	29,395,800	38,166,500	58,133,400	151,235,700
Depth of stored sediment (m)	−0.0526	0.1919	−0.2014	0.8260
Annual rate of sedimentation (cm/yr)	−0.1452	0.5186	−0.5443	2.2234

In those areas where there was net deposition between 1948 and 1985, the mean depth ranged from 0.19 to 0.82 m. The overall average from the entire river from Otowi to San Marcial was 0.53 m. Assuming the local variability to be within a factor of 2, the expected range of depths is therefore about 0.1 to 1.6 m. Detailed surveys of selected cross sections between Cochiti Dam and Isleta Diversion Dam south of Albuquerque confirm these estimates.[31] The importance of these estimates for the analysis of plutonium storage is that almost all the plutonium in the system must have entered the main stream during this period. It is therefore associated only with the sediment stored during the same period, and that sediment occurs in deposits that average less than 1.6 m deep. Given the tendency of plutonium not to migrate vertically within soil profiles, it is likely that the contaminant stored in flood plains along the river is within about 1.6 m of the surface. Local distribution of the materials may be strongly influenced by engineering works, the subject of Chapter 5.

5

Engineering Works

Historical Background

The hydrologic, sedimentologic, and geomorphic processes of the Northern Rio Grande as outlined in the previous chapters do not operate under natural, undisturbed conditions. Numerous engineering structures and activities have modified the processes and forms (Figure 5.1), and so an explanation of the movement and storage of contaminants in the system requires knowledge of the channelization and dam construction in the region. Channelization works are usually directed toward controlling the horizontal position of the channel, keeping it aligned in an economically advantageous arrangement, and maintaining a clear path for floodwaters to prevent them from spilling over the banks. The imposition of an artificial, stable channel on a naturally unstable system is rarely completely successful, but even with partial success, the newly defined system is a radical departure from the natural one.[1] Floodwaters usually flow through modified channels at higher velocities than they do through natural channels, and so they may transmit more sediment in the channel. Low flows, however, may deposit sediment in the engineered channel, thereby reducing its efficiency and raising its bed. The abandonment of previously active minor channels or braided sections provides new areas of colonization for riparian vegetation, which may enhance sedimentation when flows exceed the capacity of the designed channel. The construction of dams obviously disrupts river processes in the reservoir area but has indirect effects throughout the river system because of newly instituted controls on flood flows, normal low flows, and sediment discharges.[2]

The first engineering structures on the Rio Grande probably appeared about A.D. 1200. With the collapse of irrigation societies in the Salt and Gila River valleys in Arizona and in tributaries of the San Juan River in Colorado and New Mexico, migrants moved into the Rio Grande Valley.[3] By the time of the Spanish

Figure 5.1. Engineering works near Cochiti Pueblo have made great changes in the Rio Grande. (*a*) Looking east across the crest of the Cochiti Diversion Dam in about 1935. (Unknown photographer, photo 59142, Museum of New Mexico) (*b*) Looking west across the site of the dam in the upper photograph in 1992. The small diversion work is submerged below the reservoir impounded by Cochiti Dam, the large flood-control structure along the left side of the view. (W. L. Graf, photo 105-11)

incursions in the middle and late sixteenth century, the native population had developed extensive irrigation systems along the entire Northern Rio Grande to support numerous pueblos.[4] Diversion works on the main stream probably consisted of brush and boulder structures that directed the water into canal entrances through the low banks. These structures probably washed away with each spring flood.

Spanish and Mexican immigrants arrived in the sixteenth century and later built more extensive canal systems on flood plains and terraces near the river, with an extensive legal system for water management.[5] Still, their influence on the main river channel was probably limited. Hispanic settlers and later Anglo-Americans who arrived in greater numbers after 1848 did little to improve on the pueblo dwellers' efforts to control the main river until the late nineteenth century when diversion structures became more numerous and elaborate. The amount of irrigated acreage expanded until about 1880.

After about 1880, the amount of irrigated acreage fell in the Northern Rio Grande because of drainage problems in the fields and because the river was aggrading.[6] The aggradation resulted from at least two causes: changing hydroclimatic conditions and erosion of the tributaries. The late nineteenth century was a time of increasingly intense rainfall, as previously discussed, and the general erosion of the landscape surrounding the river increased. Land use in tributary watersheds also probably contributed to the development of gullies and arroyos, increased the erosion of the hillslopes, and generally added to the amount of sediment in the main stream, which was unable to transport all of the new load.[7]

As part of a regional assessment in the 1890s of the possibilities of water resource development throughout the American West, Congress and the executive branch of the federal government began to consider larger and more permanent diversion works for the Northern Rio Grande.[8] Problems associated with controlling the river, establishing useful diversions, and draining the fields led to the establishment of Middle Rio Grande Conservancy District in 1925. Within 10 years, the district, with considerable federal assistance, had completed diversion dams at Cochiti, Angostura, Isleta, and San Acacia as well as 290 km of riverside drains and 260 km of interior drains.[9] The drain projects included linear heaps of dredge spoil that separated the drains from the river as it was aligned during the 1930s. The area between the levees that included the active channel was called a floodway. Later engineering works followed the same alignment. Therefore, significant parts of the drainage system and main channel as they exist in the late twentieth century owe their geometry to a river-channel alignment that was naturally established more than 50 years ago.

As part of the general economic rehabilitation efforts of the New Deal in the administration of President Franklin D. Rooselvelt, several federal and local agencies conducted investigations of the Rio Grande during the late 1930s. The U.S. Department of Agriculture evaluated the consumptive uses of stream flow

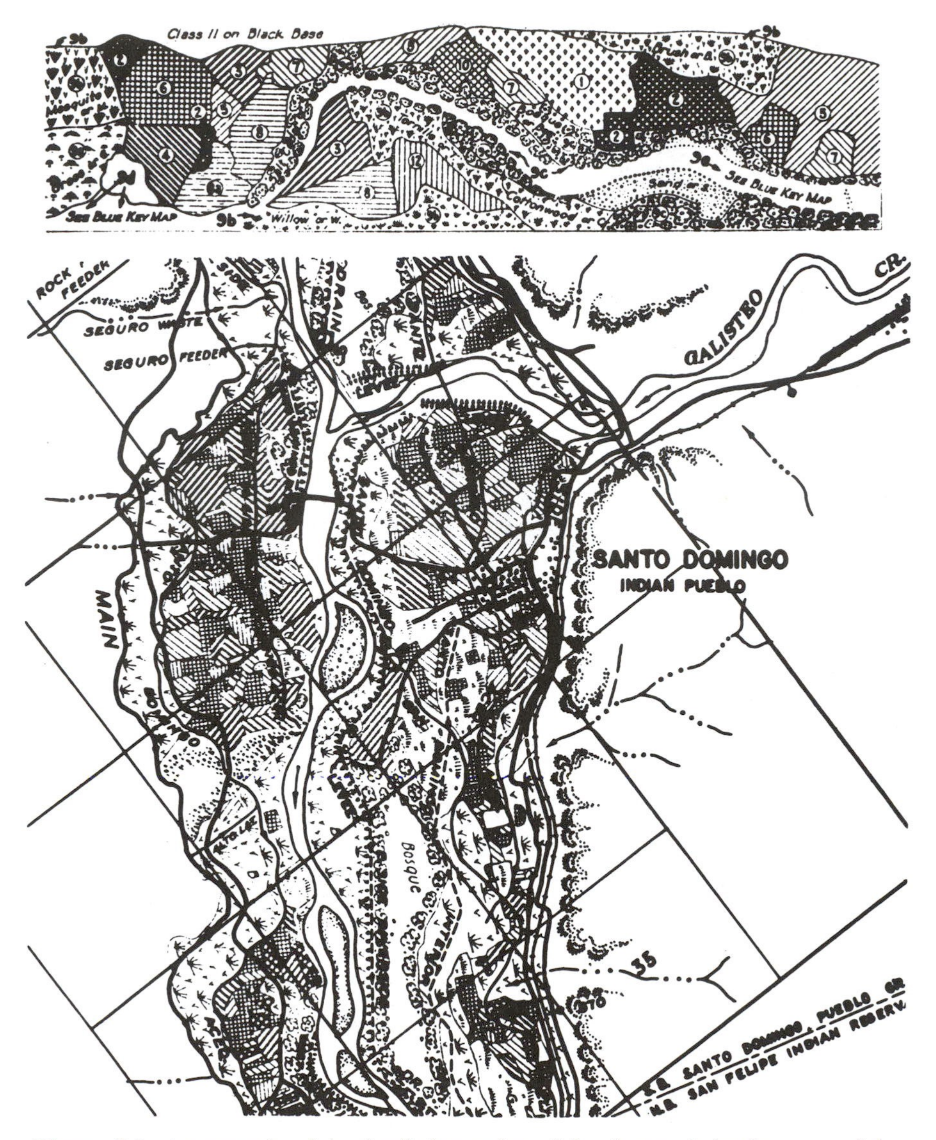

Figure 5.2. An example of the detailed mapping of the river and riparian zone of the Northern Rio Grande by planners during the New Deal era of the mid-1930s, showing the intricate detail of the effort. (National Resources Committee, 1938)

along the main river and collated the available flow, consumption, and crop data.[10] These data supported and complemented the activities of the National Resources Committee, an interagency group at the cabinet level that oversaw regional resource planning for the Rio Grande, including a division of its waters among competing states.[11] The committee report included maps of the river and near-channel environment that are so detailed that they extend the record of aerial-photographic-quality images back to the mid-1930s (Figure 5.2). Although the 1930s investigations provided data and recommendations for extensive engineering and control structures, the outbreak of World War II diverted national funds to other uses.

During the war years, the original Middle Rio Grande Project structures failed to control the problem of excess sediment, and the main stream continued to aggrade. The channel's instability, exacerbated by the aggradation, continued, especially during floods. In the area around Albuquerque, a flow of 25,400 cfs in 1941 and 24,200 cfs in 1942 caused 27 breaks in the poorly constructed levees.[12] In conjunction with the Flood Control Act of 1948, the U.S. Army Corps of Engineers and the U.S. Bureau of Reclamation surveyed the river's management problems and began attempts to stabilize the channel and reduce the inflow of sediment.[13] By 1962, the agencies had rehabilitated the drain system, strengthened the levees, built conveyance and pilot channels, cleared away the vegetation, and installed sediment traps.[14] By the 1970s, Cochiti Dam on the Rio Grande and several other dams on tributaries provided additional flood and sediment control. These structures now function as part of the river system, and they partly control the fate of plutonium in the system.

Channelization Works

The purposes of channelization works are to control the channel location and to prevent flooding outside the channel area, by employing levees, conveyance channels, and pilot channels and clearing away vegetation. Sediment traps were constructed to stabilize the channel margins. Levees in the Northern Rio Grande system are of two types: spoil heaps and designed levees. Spoil heaps are accumulations of sediment dredged from canals and drains. Placement of the heaps between the river and the canal or drain affords some protection to the ditch, but the heaps are not designed to any specifications. Vegetation stabilizes their slopes, and they often carry unpaved roads on their upper surfaces. They are easily eroded if channel migration forces flood flows against their outer banks. Engineered levees meet design criteria for size and resistance to erosion, and they are sometimes protected by riprap—waste rock less susceptible to erosion than soil material is.

Conveyance channels conduct low flows of the river in designed, confined, stable channels. Relatively steep gradients ensure the continued movement of sediment through the conveyance channel. During flood periods, the natural channel conducts water as well. Indeed, pilot channels are artificially aligned portions of the original, natural channel. The pilot version is usually straighter than the natural version, in an attempt to reduce the channel migration that results from meandering and bank erosion.

Vegetation and sediment management along the Northern Rio Grande may increase the stability of pilot channels and levees. If riparian vegetation can grow between the levee and the channel, the levee is less likely to be destroyed, because if water spills out of the channel, the vegetation will introduce considerable hydraulic roughness to the flow, thereby reducing its velocity and increasing the probability of sedimentation. If vegetation in other parts of the system is cleared away, this may enhance the channel's locational stability. But if vegetation does not grow in the pilot channel or the natural channel, floodwaters are more likely to flow by these routes than they are to carve new courses. Clearing projects therefore focus on maintaining a vegetation-free floodway in the most desired alignment.

Finally, structures introduced into the channel and flood-plain area between the levees may stimulate sedimentation, which protects the levees and stabilizes the channel.[15] The U.S. Army Corps of Engineers attempted to use pile jetties to protect levees in 1944, but without success. The jetties caused turbulence in flood flows that scoured the channel bed and washed out the jetties. Between 1953 and 1962, the corps installed jetty fields made of Kellner "jacks."[16] Developed for small Kansas streams and first used by the Santa Fe Railroad in the Rio Grande system in 1936, these jetties consist of jacks made of connected lengths of steel rail and a latticework of wire. Cables connect the individual jacks to one another and to anchoring posts outside the active-channel area. The jacks create enough turbulence in flows to cause sedimentation, but at high discharges, their porosity and flexibility diminish the scour. Since the early 1950s, river-management agencies have placed jack fields in those areas where the development of new channels is undesirable, leaving the preferred channel areas free of jacks. Field reconnaissance shows that by the late 1980s the jacks had accumulated 0.1 to 1.0 m of sediment and were the sites of dense riparian vegetation.

Throughout the Northern Rio Grande, the application of these measures varies from one segment of river to another. Between Española and Cochiti, engineering structures are rare, and the mining of sand and gravel has affected channel conditions in some reaches (Figure 5.3). Between Cochiti and the southern edge of Albuquerque, designed levees restrict the floodway on both sides. Inside each levee is a zone of dense vegetation, and jack fields are common but relatively small. South of Albuquerque, cleared channels and large jack fields supplement channelization efforts, and in the vicinity of Bernardo a conveyance channel removes all the flow of the river during low-discharge periods. A pilot channel appears in several reaches south of Albuquerque. All the major channelization works that currently affect the river were completed between 1950 and 1962 (Figure 5.4).

The success of these efforts is problematic. Although the jack fields have successfully stimulated moderate sedimentation, declining flood flows during the later period of record suggest that sedimentation might have formed in any case. Maintaining the vegetation in order to protect the levees has produced dense riparian forests in some areas, but clearing the pilot channels has been expensive and in some cases ineffective. In the Albuquerque area, for example, the original

Figure 5.3. Sand and gravel mining in the Rio Grande near Otowi Bridge introduces significant changes in channel geometry in about 1955. (Los Alamos Historical Society, photo EC-1133C)

design was for the floodway to conduct discharges of up to 42,000 cfs, but the present filled channels and vegetation growth probably restrict the capacity to less than 25,000 cfs. Elsewhere, the original design capacities were 20,000 cfs but are now only 10,000 cfs.[17] The sediment (with its adsorbed contaminants) that is stored in these reaches is part of the explanation of the regional plutonium budget.

Dams

North of Truth or Consequences, the Rio Grande Basin contains thousands of minor retention works for control erosion and develop stock water. Thirteen large structures directly affect the flow of the river and its sedimentary load (Figure 5.5 and Table 5.1). The large structures are products of three distinct eras of dam construction: an agricultural-development period in the 1910s and 1920s, the

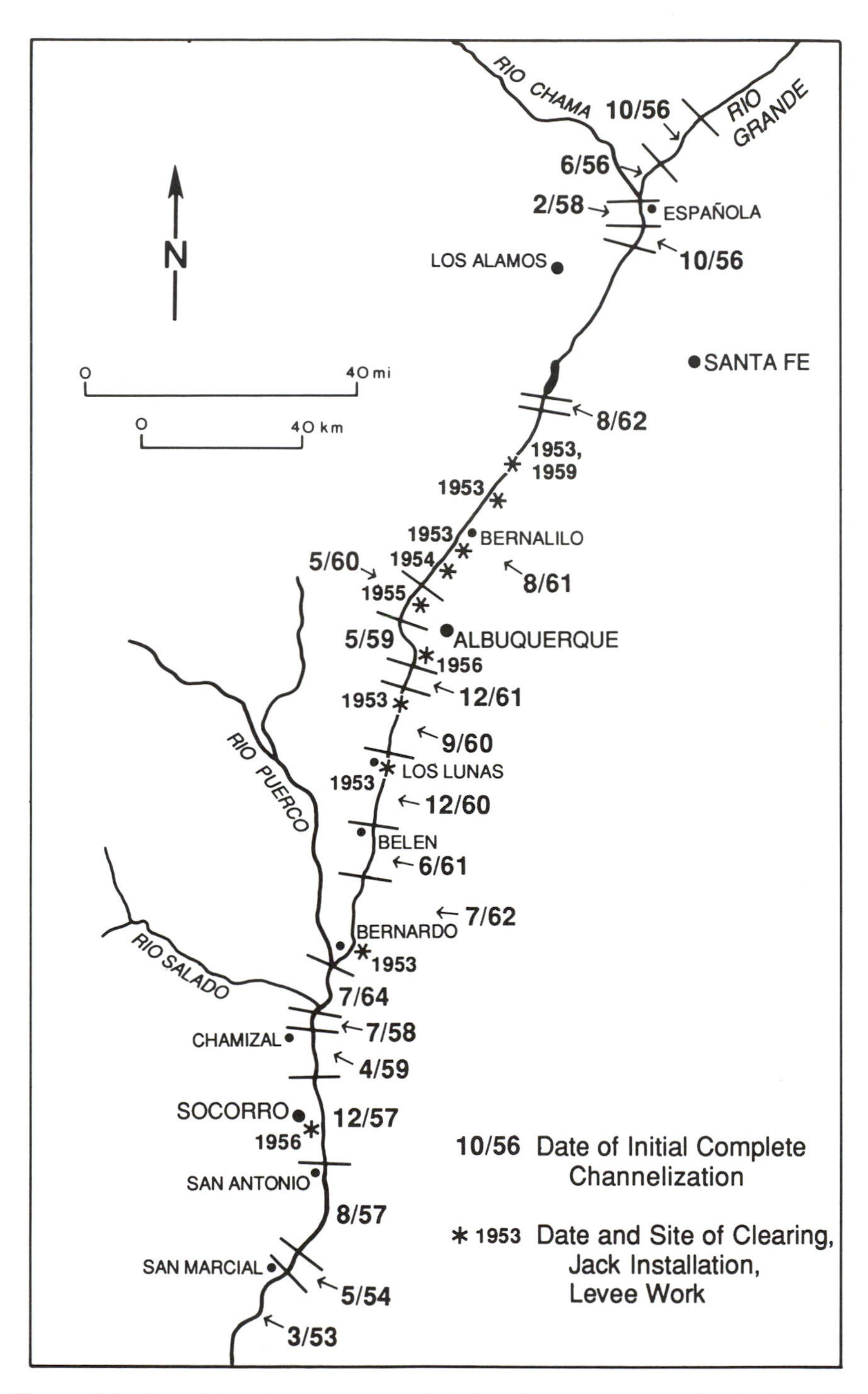

Figure 5.4. Completion dates for channel engineering works along the Northern Rio Grande after 1948.

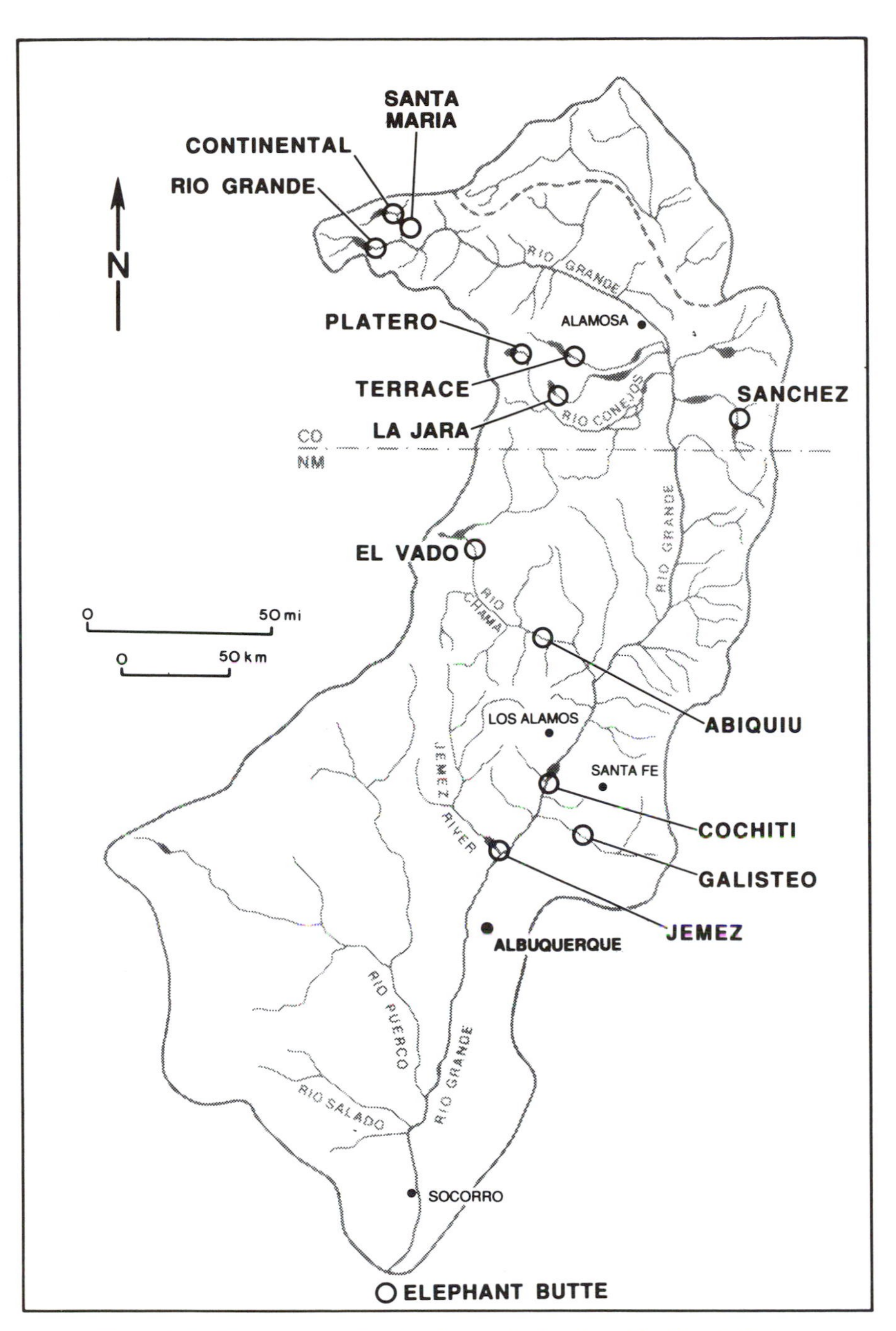

Figure 5.5. The locations of major dams in the Northern Rio Grande Basin.

Table 5.1. Major dams in the Northern Rio Grande Basin

Year	Dam	Stream	Capacity (ac ft)	Capacity (cu m)
1910	La Jara	La Jara Creek	14,100	17,189,000
1911	Sanchez	Culbebra Creek	104,000	127,214,000
1912	Terrace	Alamosa Creek	17,700	21,651,000
1913	Rio Grande	Rio Grande	51,500	63,000,000
1913	Santa Maria	Gooseberry Creek	42,000	51,375,000
1916	Elephant Butte	Rio Grande	2,155,000	2,636,000,000
1928	Continental	North Clear Creek	26,700	32,679,000
1935	El Vado	Rio Chama	198,700	243,000,000
1951	Platero	Conejos River	60,000	73,392,000
1952	Jemez	Jemez River	113,800	139,250,000
1963	Abiquiu	Rio Chama	1,221,000	1,493,000,000
1970	Galisteo	Galisteo Creek	90,600	110,770,000
1973	Cochiti	Rio Grande	607,100	742,567,000

Note: Dates are for initial closure; some structures were subsequently modified.

Sources: International Commission on Large Dams, *World Register of Dams* (Paris: International Commission on Large Dans, 1976); T. W. Mermel, *Register of Dams in the United States: Completed, Under Construction and Proposed* (New York: McGraw-Hill, 1958).

Middle Rio Grande Conservancy District period between 1925 and 1936, and the Middle Rio Grande Project years after 1948.

In the northern part of the basin, agricultural development in the San Luis Valley, Colorado, led to the construction of several dams in the 1910s. Although individuals and corporations established large numbers of artesian wells in the valley, water from the river also contributed to a land-investment boom early in the century.[18] Valley-based water districts constructed dams in the San Juan and Sangre de Cristo mountains to store excess spring runoff for release during the summer dry period. By 1928, six dams in mountain locations influenced the upper river's hydrology by reducing the spring flood peak and extending low flows at higher discharges into the summer months. Later construction added a seventh structure.

During this early period, attempts to control the flow of the river in central New Mexico resulted in the construction of Elephant Butte Dam, completed in 1916. Elephant Butte Reservoir, the downstream terminus of the study area, is a storage pool for the gradual release of water into irrigation lands in southern New Mexico and the El Paso, Texas, area. The U.S. Reclamation Service (which later changed its name to the U.S. Bureau of Reclamation) also built the dam to meet international treaty obligations for water delivery to Mexico.[19]

The Middle Rio Grande Conservancy District produced several diversion structures during its initial period of activity (1925–1936). The project built low structures designed to divert the flow of the main river efficiently into the headgates of extensive canal systems at Cochiti, Angostura, and Isleta. Local interests had constructed a diversion work at San Acacia in 1920, so that as a group, these four structures constituted a firm basis for withdrawing water. El Vado Dam on the Rio Chama provided further control of the spring melt flood in the main stream. In the late 1950s, the Bureau of Reclamation and the conservancy district agreed to an arrangement whereby the bureau assumed management control and maintenance of the structures.[20]

After World War II, the Middle Rio Grande Project, a joint venture of the U.S. Army Corps of Engineers and the U.S. Bureau of Reclamation, encouraged the construction of additional large structures, principally for flood and sediment control. Because of the high-value urban investments in the Albuquerque area, the benefits of these new structures emphasized flood protection for the city and retention of sediments to prevent channel aggradation through the metropolitan area. Jemez, Abiquiu, and Galisteo dams on large tributary streams control sediments control functions and prevent catastrophic summer flash floods.

Cochiti Dam, closed in December 1973 on the main stream, is the primary flood-control structure for the Albuquerque area. Built near the site of the old Cochiti Diversion Dam, the new structure has a permanent pool that fluctuates drastically in response to annual inflows. Normal storage in the early years of the reservoir's history was about 55,000 ac ft, but in the spring of 1979, high runoff from the upper basin left 189,000 ac ft in the pool.[21] During the mid-1980s, a series of floods brought the reservoir close to its capacity of slightly over 600,000 ac ft. The reservoir also stores huge quantities of sediment that otherwise would be deposited on flood plains and in channel areas south of Cochiti.[22]

Implications for Plutonium Mobility

The extensive engineering works in the Rio Grande system alter the natural behavior of the river and influence the mobility of plutonium by changing the sediment transport processes. Because most of the plutonium in the system is attached to sediment, the fate of the sediment is also the fate of the radionuclides. Engineering works partly control the location of deposition, and they therefore also partly control the sites for plutonium storage in the system.

Channelization works that direct the discharge into predetermined alignments generally have a higher velocity than do the original natural flows, thereby facilitating the transport of sediment in a limited part of the valley and channel cross section. The fine sediment that is likely to have the highest concentration of plutonium is rarely deposited under such conditions. The course of the river through much of Albuquerque has channels of this type. However, adjacent to the channel are sediment-trapping areas with dense riparian vegetation and jack fields where sediment of all sizes accumulates. Through the Albuquerque area these depositional zones are relatively narrow, but elsewhere the accumulation areas occupy substantial portions of the valley cross section between the levees. The width of this accumulation zone may be more than 1 km, providing primary storage space for the plutonium-bearing deposits that are almost always associated with riparian forest. A detailed map of these zones and precise dates of the implantation of the deposits can show the likely distribution of plutonium in local areas along the river.

Dams have a major impact on the mobility of sediment and attached plutonium in the Rio Grande. With the exception of Platero Dam on the Conejos River, all the mountain reservoirs were in place before plutonium entered the drainage system. Any fallout plutonium eroded from mountain slopes is likely to be stored in these mountain reservoirs, and the only fallout plutonium entering the lower reaches of the river (south of the Colorado–New Mexico boundary) comes from those slopes downstream from the mountain reservoirs. The possible effect of this arrangement on the regional plutonium budget is unclear, because the reservoirs contain water, sediment, and plutonium from the alpine zones where the accumulation of fallout is likely to be the greatest. The sediments of the Rio Grande Reservoir contain relatively high amounts of plutonium as a reflection of this process.[23] Alternatively, the mountain drainage areas downstream from the dam sites are larger than those upstream from the dams and their pools. Without further research, the specific plutonium contributions of these areas are unclear.

Those dams and reservoirs on tributaries of the Rio Grande that are not in the high mountains trap sediments from landscapes that probably do not receive as much fallout as the alpine zones do. Occasional sampling of the materials stored in Abiquiu, El Vado, and Platero reservoirs show that the concentrations of plutonium are lower than those in Rio Grande Reservoir.[24] The total amount of plutonium stored in the low-elevation tributary reservoirs may eventually exceed the amount stored in the alpine reservoirs, however, because the tributary pools drain those landscapes that produce more sediment.

Cochiti Reservoir is especially significant to the analysis of plutonium mobility because of its size and location. Its capability to store almost all the sediment that enters the lake make Cochiti Reservoir the ultimate storage site for almost all the river sediment generated in the basin north of Cochiti. The location of the dam on the main stream of the Rio Grande downstream from Los Alamos National Laboratory ensures that the sediments of the reservoir store plutonium from both fallout and laboratory sources. Before December 1973, the plutonium in river sediment could be stored across thousands of square kilometers of flood plains and could be diluted by tributary sediments from streams south of Cochiti, including the Galisteo Creek, Jemez River, Rio Puerco, and Rio Salado. After December 1973, the plutonium from the laboratory in fluvial sediment has been deposited in the restricted reservoir basin behind the dam, and concentrations of the metal, although variable, have recently appeared to be in the same range as those in Rio Grande Reservoir.

6

Riparian Vegetation

Riparian Vegetation Communities

The interaction among water, sediment, landforms, and human environmental manipulation on the Northern Rio Grande has produced a distinctive assemblage of plants in the riparian (or near-channel) community. The fluvial landforms and the sediment of which they are composed are often not immediately visible in field investigations because of the dense cover of riparian vegetation. In aerial photography—the primary source of data for historical river-channel change and sedimentation—riparian vegetation is often the only aspect of the near-channel environment that is amenable to interpretation and mapping. Vegetation also provides information about the date of emplacement of the sediments on which it grows, information useful in tracking contaminants introduced into the system during known time periods. Vegetation communities therefore provide useful keys to identifying the distribution of near-channel sediments and the contaminants they contain. This chapter briefly reviews the origin and changes in riparian vegetation in the study area, including its connections with geomorphic systems.

Almost all major rivers in the American Southwest have undergone considerable geomorphic and vegetation change since the early nineteenth century when channel margins were the sites of bogs, lakes, abandoned meanders (sloughs), and marshes (Figure 6.1).[1] Most major rivers had broad, sandy channels with braided configurations and meandering low-flow channels.[2] Even small tributaries had marshy areas created by beavers.[3] The riparian vegetation originally evolved in association with frequent extensive flooding.[4] Removal of the beavers, the development of gullies and arroyos, land-management schemes, changes in climate, and the construction of dams changed the streams into single-thread or compound channels that flooded less often.

Figure 6.1. Variations in vegetation cover in sand dunes immediately upstream from the San Acacia Diversion Dam illustrate changes in plant communities of the Rio Grande Basin. (*a*) Looking east in 1905, with the small pool behind San Acacia Dam in the middle distance, showing the sparse cover and an unstable sand surface. (R. H. Chapman, photo 178, U.S. Geological Survey Photography and Field Records Library, Denver) (*b*) The same view in 1988, showing dense grass cover with shrubs and a stable sand surface. (W. L. Graf, photo 77-1)

The Rio Grande's recent history is typical of the larger region except for the extensive recent engineering works that restrict the active channel and flood plains. There are few detailed descriptions of the channel and riparian vegetation before major human intervention, but generally, most firsthand observers indicate that the Northern Rio Grande was broad and shallow, with meandering subchannels frequently altered by flooding.[5] After channel migration, cottonwood, willow, and cattail colonized the newly exposed alluvial surfaces.[6] Early in the twentieth century, the cottonwood groves near the river rarely developed trees more than about 10 m high before more changes in the channel destroyed them.[7] As elsewhere in the Southwest, a high water table existed close to the river, creating marshy conditions and lakes; irrigation contributed to this natural condition until the Middle Rio Grande Conservancy District began its engineering work in the late 1920s and 1930s. Before this, lakes, marshes, wet meadows, and a mixed woodland (including cottonwood and willow trees) fringed the river, but after the construction, the lakes, marshes, and wet meadows disappeared. Willows, once common everywhere in the riparian environment,[8] also disappeared except for small scattered shrubs.

The absence of large floods since 1941, the beginning of the Middle Rio Grande Project in 1948, and the closure of flood-control dams in the 1960s and 1970s completed the changes in the vegetation system.[9] Spring floods were smaller than before, moderate flows extended longer in the summer season, and the channel location became much more stable. As a result, cottonwood trees now grow 20 to 30 m high, and they no longer regenerate through flood damage. Instead, human disturbance by construction and fire lead to new growth in the riparian forest. The Kellner jack fields, erected on abandoned braided channels and flood plains, became the sites for rapid colonization by cottonwood, willow, tamarisk, and russian olive.

Tamarisk (*Tamarix chinensis* Lour.) and russian olive (*Eleagnus angustifolia* Lour.), especially useful in mapping river forms and deposits, are exotic species that may be gradually replacing the native species. The tamarisk, a tree or shrub native to the Mediterranean region, was part of an international seed-exchange program and appeared in 1852 in southern California nurseries as an ornamental plant.[10] By 1900, the plant had escaped cultivation and grew along sand bars and channel margins of the major rivers in the Sonoran Desert.[11] It spread throughout the Southwest between 1900 and 1940. It was a domestic ornamental plant in Albuquerque early in the twentieth century but became widely naturalized throughout the valley by about 1935.[12] The plant's prodigious production of airborne seeds, long germination period, and rapid growth allowed it to compete favorably with native trees in colonizing newly stabilized sandy or silty surfaces.[13]

Farmers probably introduced the russian olive to Albuquerque between 1900 and 1915, and by 1935 it too had become common in the Rio Grande's riparian communities.[14] Like the tamarisk, the russian olive easily colonizes newly exposed river deposits, where it grows well in shade and invades existing woodlands.[15] By the late 1980s, russian olives were frequently found in linear groves along the margin of the channel, and individual trees appeared in cottonwood areas. It also appeared as new growth in many channel areas that were artificially cleared of vegetation in an effort to control floods.

Along the Northern Rio Grande in the late 1980s, there were eight primary community types,[16] with cottonwood (*Populus fremontii* S. Wats., *Populus fremontii* var. *wislizenii* [Torr.] S. Wats., and *Populus angustifolia* James) dominating three of the most common riparian communities. In the cottonwood and coyote willow community, coyote willow (*Salix exigua* var. *nevadensis* [Wats.] Schneid.) is the most abundant understory plant, but tamarisk (also known as salt cedar or saltcedar), russian olive, and seepwillow (*Baccharis salicina*) also are common. Grasses and forbs complete the ground cover.

The cottonwood and russian olive community has an understory consisting almost exclusively of russian olive or, in the area between Española and Albuquerque, of New Mexico olive (*Forestiera noemexicana*) without herbaceous growth. The cottonwood and juniper community also lacks significant herbaceous growth, and instead of olive, the dominant understory plant is the one-seed juniper (*Juniperus monosperma*). Russian olive communities, which dominate in narrow strips along the main river channel, include a few seedlings of other trees. Tamarisk communities can be found most often in the southern portions of the area, where the tamarisk plants grow to the exclusion of other tree species.

Three nonarborial communities also make up the riparian environment of the Northern Rio Grande in New Mexico. Cattail marsh communities with cattails (*Typha latifolia*) and bulrush (*Scirpus acutus*) grow in some recently abandoned channel areas or in low, waterlogged portions of the flood plain. Two additional community types are related directly to the river. Sandbar communities have mostly barren, sandy surfaces but occasionally support sparse grass, annuals, and various seedlings that do not survive flooding. The river channel is sometimes dry and may briefly support some grasses.

Compared with the upland vegetation, riparian communities are relatively unstable,[17] responding annually to changes in the river landscape. The importance of these various vegetation communities to our investigation is that each community is associated with a particular type of river landform or sediment. The tamarisk communities, for example, colonize abandoned river channels, and so a map of that particular community is essentially a map of abandoned channels and sedimentary fillings. Mapping the vegetation is often more efficient than mapping the landforms and sediments directly, because the vegetation is clearly visible on aerial photography. Determining the age of the vegetation, either through photographic evidence or by direct physical evidence, provides a minimum age for the deposition of sediments and their associated contaminants.

The Connection Between Vegetation and Geomorphology

The distribution of riparian vegetation communities is closely tied to the distribution of near-channel landforms through historical associations, influence of fluvial processes, the variation in sediment characteristics, and availability of groundwater. The river channel creates and abandons portions of the landscape that then become seedbeds where colonization by vegetation reflects the temporal changes in the river's course (Figure 6.2). Those forms created by the river before the advent of tamarisk and russian olive, for example, are less likely to be dominated by them than are those forms created later when their seeds were available for colonization. In the Northern Rio Grande system, the areas most densely populated by tamarisk are those made available for seedling development during and after the 1950s when tamarisk had become established in the region.

In addition to the temporal factor, river landforms and processes directly influence vegetation communities. Cottonwood often grew along the margins of single-thread channels before the invasion of tamarisk and russian olive. When the river course changed, abandoning some single-thread channels, the cottonwoods remained along the alignments of the previously active water courses, resulting in

Figure 6.2. Forms and riparian vegetation on the Rio Grande flood plain near Santo Domingo Pueblo, showing a small abandoned channel across the foreground covered with grass and sedges, with tamarisk and cottonwood trees in the background along what was once the bank. (W. L. Graf, photo 76-24)

Figure 6.3. An abandoned channel lined by cottonwood trees in the Rio Grande flood plain near Santo Domingo Pueblo. (W. L. Graf, photo 76-21)

lines of trees across flood plains covered with other types of vegetation (Figure 6.3). Vegetation communities respond to flood processes as well, and so flood plains frequently inundated by fast-flowing water are likely to contain flood-resistant plant types such as tamarisk and willow rather than relatively brittle hardwoods such as elm and cottonwood.

Fluvial form and process changes are directly connected to the particle sizes of the resulting deposits, providing another link with the vegetation communities. Willow, for example, favors fine-grained soils and so is most likely to grow on flood plains rather than on coarse bar deposits. There are limits to this particle size and vegetation connection, however, because some species, including tamarisk, grow aggressively in soils dominated by particles ranging from silt to course gravel.

The availability of groundwater is a major determinant of the distribution of many riparian species that are phreatophytes, plants that have extensive taproot systems that allow them to obtain water directly from the zone below the water table.[18] In most areas near the channel of the Northern Rio Grande, the water table is within a few meters of the surface, and the minor variations in the depth to the groundwater may influence the nature of the vegetation communities on the surface. In those places where obstructions force the groundwater close to the surface, phreatophytes may gain a competitive advantage over other species. Near San Marcial, for example, a basalt flow constricts the river channel and the groundwater flow, forcing the groundwater close to the surface. Accordingly, tamarisk, an aggressive phreatophyte, grows so densely there that it keeps out almost all other species.

An ecological-sampling scheme reveals the general connection between geomorphology and vegetation. The following procedure provided a census of tree forms 2 m or greater in height from 32 sample plots scattered throughout the Northern Rio Grande:

1. Select a starting point for the plot in a representative landform–sediment area, such as a flood plain, a bar, or an abandoned channel.
2. From the starting point, define a straight sample line in a random direction for a distance of 100 m.
3. Define as the sample area the rectangle outlined by the area 1 m on both sides of the sample line, resulting in an area 100 × 2 m.
4. Identify and tally all woody-stemmed plants 2 m or greater in height in the sample area.

Of the 1,985 trees identified on the 32 plots, 66 percent were tamarisk; 10 percent, willow; 9 percent, cottonwood; and 8 percent, russian olive. The remaining 7 percent were saltbush (*Atriplex canescens*), Rocky Mountain maple (*Acer glabrum* Torr.), American elm (all juvenile, *Ulmus americana* L.), red mulberry (*Morus rubra* L.), and sagebrush (*Artemisia tridentata*).

Data from each plot, including soil samples, and identification of the fluvial landform permitted a test of the associations among the four most abundant species, sediment particle size, and landform (Table 6.1). The tree densities on the sample plots ranged from about 3,300 per sq km (an open stand of cottonwoods) to about 560,000 per sq km (a dense thicket of tamarisk). A review of the data shows that tamarisk is so common that it occupies all forms and soil types when considering the entire reach of the river in this study. On the scale of individual reaches, however, tamarisk is a significant discriminator of forms and sediment, with its distribution closely associated with abandoned channels, for example. Russian olive occurs almost exclusively in association with active channels and rarely elsewhere, except on abandoned bars that recently were next to active channels. Cottonwood is a flood-plain species, with the highest densities along abandoned single-thread channels. Willow grows on flood plains and in abandoned single-thread channels where fine-grained materials are common.

When viewed as communities, the four dominant species provide unique signatures of the various depositional environments along the Northern Rio Grande (Figure 6.4). Although tamarisk is the most common species in each environment, large stands of russian olive identify active-channel areas and bars.

Table 6.1. Vegetation data from sample plots, Northern Rio Grande

Reach	Map Unit	Landform	Tamarisk (%)	Cottonwood (%)	Russian Olive (%)	Willow (%)	Total Count (sq km)
Otowi	1B	Active channel edge	3	75	16	0	53,333
Otowi	2B	Abandoned flood plain	0	18	82	0	110,000
Buckman	2B	Abandoned slough	84	0	11	0	150,000
Buckman	2B	Abandoned slough	94	0	4	0	140,000
Buckman	3B	Abandoned flood plain	98	0	2	0	363,334
Peña Blanca	4A	Abandoned channel	0	100	0	0	3,333
Peña Blanca	4B	Abandoned flood plain	74	26	0	0	45,000
Peña Blanca	5A	Older abandoned channel	0	71	0	0	11,666
Coronado	2A	Abandoned channel	0	43	7	45	70,000
Coronado	3B	Abandoned braided channel	77	4	15	4	43,333
Coronado	3A	Older abandoned channel	4	17	2	77	88,333
Los Griegos	2A	Abandoned channel	66	3	3	24	145,000
Los Griegos	2B	Abandoned flood plain	0	60	0	0	25,000
Los Griegos	3A	Older abandoned channel	51	11	11	26	88,333
Los Lunas	2B	Abandoned channel	11	5	1	74	218,333
Los Lunas	3A	Abandoned braided channel	50	42	8	0	40,000
Los Lunas	3A	Abandoned braided channel	60	33	7	0	25,000
San Geronimo	1A	Active channel edge	7	0	93	0	90,000
San Geronimo	2A	Abandoned channel	85	15	0	0	130,000
San Geronimo	2A	Abandoned channel	100	0	0	0	6,666
San Geronimo	2B	Abandoned flood plain	100	0	0	0	13,333
San Geronimo	2B	Abandoned flood plain	98	0	2	0	71,666
Chamizal	1B	Active flood plain	93	0	0	2	70,000
Chamizal	2A	Abandoned braided channel	20	27	0	0	25,000
Chamizal	2B	Abandoned flood plain	89	5	0	5	123,333
Chamizal	2B	Abandoned flood plain	76	19	0	5	70,000
Chamizal	2B	Abandoned flood plain	76	24	0	0	35,000
San Marcial	1A	Active channel edge	96	4	0	0	198,333
San Marcial	2A	Abandoned channel	81	0	0	19	26,666
San Marcial	2B	Abandonedflood plain	5	0	0	0	175,000
San Marcial	3A	Older abandoned channel	63	32	2	4	95,000
San Marcial	3C	Older abandoned bar	100	0	0	0	560,001

Note: Reaches and observation units are keyed to geomorphic maps in Chapters 9 and 10.

Flood plains contain a mixture of species similar to the mixture growing on abandoned single-thread channels, although the latter have more cottonwoods. Abandoned braided channels lack the willow found on flood plains and abandoned single-thread channels. When supported by field checks, these associations establish connections between the vegetation visible in aerial photography and the less visible underlying landforms with their sediment. The result is a rapid and efficient method of mapping large areas in the search for likely sites for the deposition of plutonium.

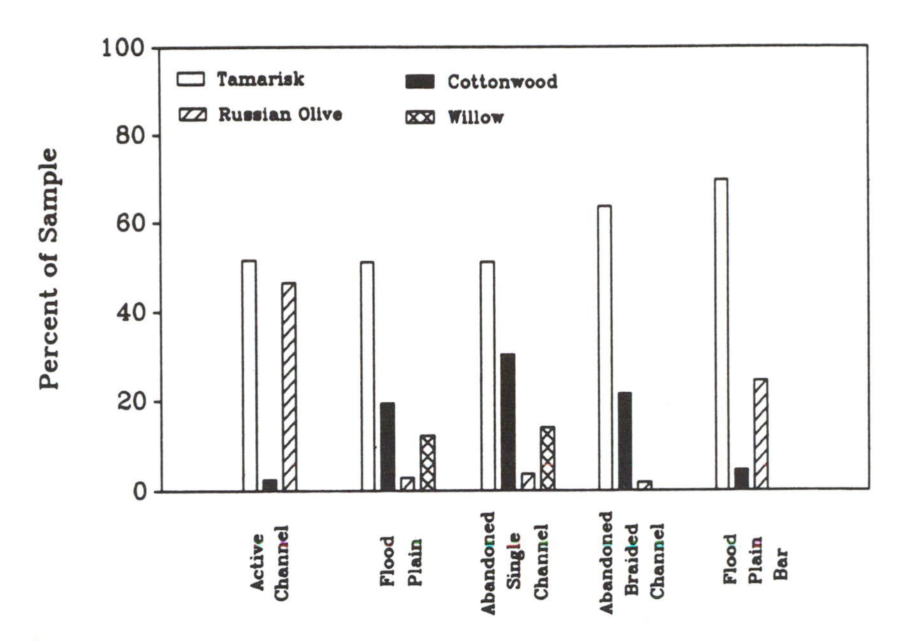

Figure 6.4. The statistical association between the composition of riparian vegetation and the underlying sediment–landform complex.

Dating Deposits with Vegetation

Once those landforms and bodies of sediment that might contain plutonium have been identified, the dates of their origination must be determined, in order to assess their potential as plutonium storage sites. Because plutonium entered the river system only after 1944, identifying those mapped deposits dated after 1944 is the method for connecting the deposits to potential plutonium loading. In this study, information from aerial photography and tree-ring ages provided the dates for the deposits.

Since 1935, aerial-photographic coverage of the Northern Rio Grande has been frequent enough to assign a specific date of origination to most of the sedimentary bodies and landforms created during that time. Because the major channel changes came during large floods, the year of most likely change is usually obvious even if the interval between photographic coverage for a particular area is as much as three or four years. In a few cases, ground photographs or documentary evidence further narrows the range of possible dates for the landform and sediment body in question.

For those few features not dated and for small areas examined in the field, the vegetation itself provides a minimum age for the landform and sedimentary body on which it grows. After being created, through either deposition or abandonment by flowing water, the newly exposed surfaces provide seedbeds for the growth of

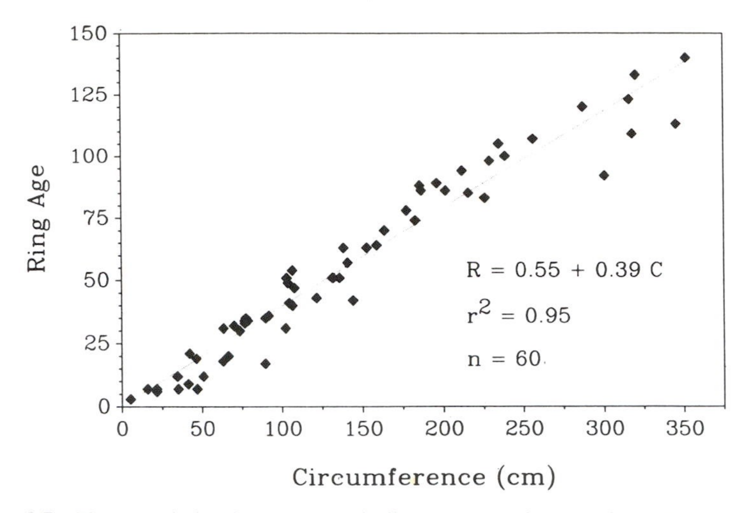

Figure 6.5. The association between trunk diameter and ring age for cottonwood trees along the Northern Rio Grande.

trees. Direct observation of the Northern Rio Grande shows that tamarisk, willow, and russian olive begin growing in new areas within one year, and cottonwood seedlings develop within two or three. Except for willow, each species records the age of individual plants by developing annual growth rings, usually one ring per year.[19] When exposed with a Swedish increment borer—a drill that extracts a cylindrical sample of the tree in cross section 1.5 m above the ground—the ring count indicates the age of the tree. Samples from several trees growing on the same form provide a minimum age for the surface and sediment.

Tamarisk, russian olive, and cottonwood yielded useful ring sequences for dating purposes. Cottonwood was the most consistent and provided the most easily interpreted rings, and most of the tree-based dates in this study were from this species. But the Swedish increment borer takes a lot of time to use, especially when several samples are needed to provide the most reliable date. Large trees also require more time for drilling. A more efficient approach is to measure the trunk circumference 1.5 m above the ground and to relate the circumference to age. Figure 6.5 shows the connection between circumference and ring age for 60 test cottonwood trees. The association was especially close for those trees less than 50 years old, the age of primary interest. After this initial calibration, the circumference measurement and the statistical relationship shown in Figure 6.5 can be used to estimate the ages of cottonwood trees on the Northern Rio Grande flood plains.

Vegetation as a Radionuclide Reservoir

The importance of riparian vegetation in analyzing the plutonium budget is as an indicator of the location and age of those bodies of sediment containing the contaminant. The vegetation itself is probably not a significant reservoir of plutonium in comparison with the amount stored in sediment. Although plutonium is found in plants, with the greatest quantities in their roots,[20] the total amount associated with biosystems is relatively limited and does not approach toxic proportions.[21] As with other heavy metals, the amount of plutonium found in soil and sediment is 10,000 to 100 million times greater than the amount found in associated plants.[22] It is therefore reasonable to disregard riparian vegetation as a significant component of the regional plutonium budget.

The rings in riparian trees may, however, contain useful information about plutonium in shallow groundwater. Near the active channel of the Rio Grande in the Albuquerque area, the rings of many cottonwood trees for the years 1965 and 1973 have a distinctive red coloration. These rings may contain an unknown material drawn from the water during their development. In 1965 and 1973, there were floods significantly larger than those in other years during the 1959–1978 period, but the reason for the rings' red color is not clear. An evaluation of individual rings for their chemical or radionuclide content might reveal a time series of contaminant concentrations in the shallow groundwater and stream water in those trees close to the channel, but this is beyond the scope of this book.

7

Plutonium in the Rio Grande System

Inputs from Fallout

The water, sediment, landform, and vegetation systems of the Northern Rio Grande provide the environmental framework within which plutonium moves and is stored. Plutonium enters the Northern Rio Grande from two sources: atmospheric fallout and releases from operations of Los Alamos National Laboratory that enter the main stream by transport through Los Alamos Canyon (Figure 7.1). This chapter describes the nature and timing of plutonium loading in the river's sediment system as a means of identifying those years when sedimentation is likely to have accumulated those deposits with the highest concentrations of plutonium. This chapter also discusses plutonium in river water, sediments in transit, and sediments deposited along and stored along the channel, as well as the various mean values of plutonium concentrations found in the region of Los Alamos. The review includes plutonium in the regional environments around Los Alamos, including the compartments of river water, active sediments, flood-plain deposits, and reservoir deposits, as well as the plutonium concentrations in the sediments of Los Alamos Canyon.

Most of the plutonium in atmospheric fallout is from the testing of nuclear weapons. Five nations have detonated a total of 484 nuclear devices in the atmosphere, 466 with known dates.[1] These explosions have injected plutonium into the general atmospheric circulation, resulting in a global distribution of fallout as the material returns to the surface. There are three types of fallout: local, tropospheric, and stratospheric.[2] Local or early fallout arrives within a day of the detonation and consists of particles 100 to 200 microns in diameter (fine sand) transported in the lower atmosphere and deposited within several hundred kilometers of the site of the explosion.

Figure 7.1. Lower Los Alamos Canyon at its confluence with the Rio Grande at Otowi, New Mexico, with the Rio Grande flowing from right to left in the foreground. The large bridge in the center is the highway bridge built during the early 1940s to serve Los Alamos; the smaller structure on the left is a remnant of bridge of the Chili Line railroad that serves as a support for a U.S. Geological Survey stream gage. (*a*) Looking west in the early 1940s. (Unknown photographer, photo M148396, Los Alamos Historical Society) (*b*) The same view in 1991, with the highway bridge abandoned and a new bridge and road grade constructed on the right. (W. L. Graf, photo 104-20)

Finer particles travel greater distances and disperse over greater areas. Tropospheric fallout arrives within a month of the detonation and consists of particles less than 100 microns in diameter (mostly silt size), transported in the lower atmosphere. The global atmospheric circulation transports tropospheric fallout around the world in a band about 30 degrees latitude wide, centered on the site of the explosion. Most of the tropospheric fallout delivers plutonium to the earth's surface in precipitation, with only about 10 percent occurring as dry fall.[3]

Stratospheric fallout requires several years to return to the surface and consists of particles less than 10 microns in diameter (mostly clay size) that the explosion has injected above the tropopause into the stratosphere. At these high altitudes (greater than 10,000 m), global atmospheric circulation distributes the material throughout the hemisphere of the originating detonation. Only in unusual circumstances when the direct injection exceeds 21,000 m in altitude does the circulation system transport the materials across the equator.

The Northern Rio Grande watershed has probably received all three types of fallout. Early fallout from detonations at the Nevada Test Site may have reached the San Juan Mountains, and tropospheric fallout certainly did, because the Rio Grande lies in the center of the latitudinal belt containing the test site. Stratospheric fallout has been greatest in the Northern Hemisphere, where the majority of atmospheric detonations took place, and so the Rio Grande has also received fallout from this source. Most of the detonations at the Nevada Test Site were at times when the prevailing wind was from a westerly direction,[4] and the San Juan Mountains are the first major mountain range exceeding 3,000 m in elevation downwind from the test site. The paths of debris clouds often crossed the axis of the Northern Rio Grande,[5] so it is likely that the river system now contains plutonium from the test site.

In addition to weapons testing, sources of atmospheric fallout of plutonium include the atmospheric disintegration of satellites fueled by plutonium and weapons accidents. Several satellites have fallen from decayed earth orbits and released their plutonium in the upper atmosphere, including the SNAP-9A navigation satellite lost in 1964.[6] Most of the plutonium from satellite debris is likely to be in the form of tropospheric and stratospheric fallout, with hemispherewide distribution and latitudinal bands of higher concentrations. Weapons accidents such as crashes of aircraft carrying nuclear weapons have released plutonium in restricted local environments, but because they were not in the western United States, they are not likely to have affected the Northern Rio Grande.

The total amount of fallout over the entire earth surface is about 24 kCi of plutonium 238 (16 kCi of which is from SNAP-9), 154 kCi of plutonium 239, and 209 kCi of plutonium 240.[7] On a global scale, the distribution of this fallout is uneven. Plutonium fallout from all sources is at a minimum (less than 0.005 mCi/sq km for plutonium 238 and 0.01 mCi/sq km for plutonium 239 and 240) in the 10 degrees of latitude surrounding the South Pole.[8] The maximum is in the band between 40 and 50 degrees north: 0.079 mCi/sq km for plutonium 238 and 2.2 mCi/sq km for plutonium 239 and 240. In the latitudinal band containing the Northern Rio Grande, fallout has been 0.042 mCi/sq km for plutonium 238 and 1.8 mCi/sq km for plutonium 239 and 240.

Atmospheric detonations that have contributed fallout plutonium to the surface have occurred irregularly since 1945. Figure 7.2 shows the time series of all known detonations, except 18 tests in the former Soviet Union between 1949 and 1958 for which there are no exactly known dates.[9] The amount of material injected into the atmosphere by each test depended on the altitude of the burst, the ground conditions beneath the burst, local weather conditions, and the nature of the exploded device, but the overall trends indicate the probable general plutonium loading of the atmosphere and the resulting fallout.

The importance of the fallout time series is that the amount falling on the Northern Rio Grande's surfaces were greatest during the late 1950s and 1960s. The years with the most detonations were 1958, with 96, and 1962, with 81. A moratorium on atmospheric testing resulted in no detonations during 1959 and most of 1960. The Partial Test Ban Treaty became effective on August 5, 1963, and thereafter the United States, the Soviet Union, and the United Kingdom confined their testing to underground sites. A composite time series for plutonium fallout from these tests does not appear in the general literature. However, the plutonium fallout probably followed a trend similar to the time series for strontium 90, which was also produced by the detonations.[10] Strontium 90 loadings in surface materials increased significantly from the early 1950s to 1961 and then rapidly from 1962 to 1965 in response to the numerous tests in the early 1960s (Figure 7.3). Assuming that plutonium followed a similar trend, the peak rates of fallout on the Rio Grande landscape probably were between 1954 and 1967.

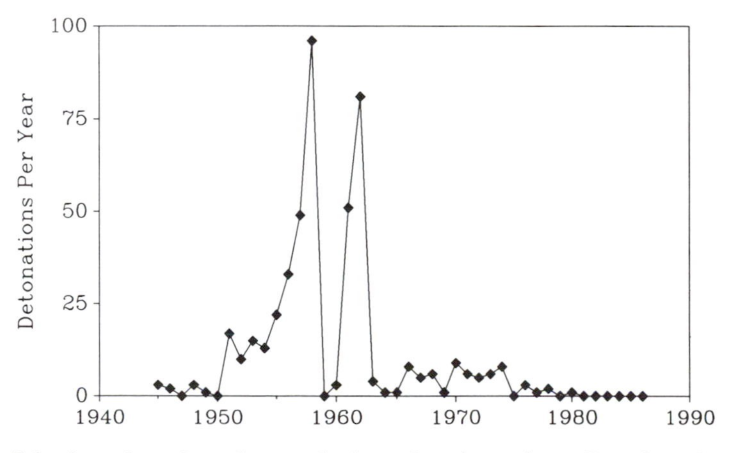

Figure 7.2. Annual number of atmospheric nuclear detonations. (Data from Stockholm International Peace Research Institute, 1987)

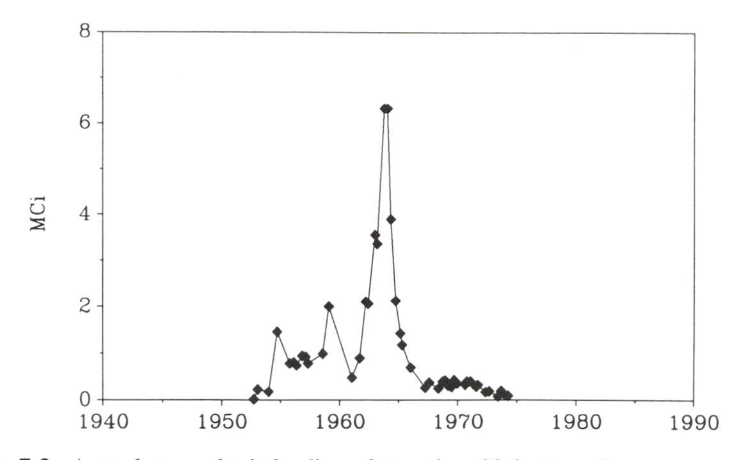

Figure 7.3. Annual atmospheric loading of strontium 90 from nuclear-weapons testing. (Data from S. Glasstone and P. J. Dolan, eds., *The Effects of Nuclear Weapons*, 3rd ed. [Washington, D.C.: U.S. Department of Defense and the Energy Research and Development Administration, 1977])

Once fallout plutonium reaches the surface, it quickly binds with soil materials and becomes part of the soil erosion and sediment transport system. "The movement of these . . . materials determines the fate of the radioactivity."[11] In a broadly based study using agricultural models, George Foster and Thomas Hakonson estimated the amount of the fallout plutonium that was likely to have eroded from earth surfaces and entered river systems.[12] Taking into account climatic variations, differences in land management, and vegetation cover, Foster and Hakonson attempted to account for the variability in plutonium mobility from one region to another. They estimated that about 0.02 percent of the total plutonium inventory in soils was delivered to streams each year in the eastern United States. The delivery rate was about 0.04 percent per year for the Northwest, 0.05 percent for the agricultural Midwest, and 0.08 for the sparsely vegetated Southwest. If the average fallout deposition on landscapes had been 1.0 mCi/sq km, the resulting concentration of plutonium in river sediments would range from about 0.01 pCi/g in the Midwest to about 0.04 pCi/g in the Southwest.

Foster and Hakonson's calculations provide a useful general picture of fallout plutonium in the Rio Grande system. Their calculated "enrichment ratio" for the system—defined as the ratio between the amount of plutonium in soils and the amount in river sediment—showed that the concentrations in river sediments were about 2.76 times the concentration in soils, or close to the national average (Table 7.1). The value of their "concentration index" for the Rio Grande—a measure of the concentration of plutonium on sediments in fluvial transport—was much above the national average. Given the estimation of a plutonium 238, 239, and 240 combined fallout burden for the latitudinal belt including the Northern Rio

Grande of 1.842 mCi/sq km,[13] Foster and Hakonson's method predicts about 0.0737 pCi/g total plutonium in river sediments. This value is the same order of magnitude as those values reported for reservoir sediments in the system.

Foster and Hakonson's analysis contains a theme central to my research. They estimated that in the Rio Grande system more than 99 percent of the sediment eroded from the landscape will remain in the river system for more than 20 years and that because almost all the fallout plutonium is attached to that sediment, more than 99 percent of the fallout also will remain in the system. The empirical data in my work indicate that the figure is probably 98 percent. Nonetheless, these high rates of storage lead to two important conclusions regarding plutonium in the Northern Rio Grande:

1. The distribution of the stored sediment and plutonium is highly variable geographically, with concentrations likely in some places but not in others. The purpose of my work is to define this geography.
2. The fate of the plutonium released by Los Alamos National Laboratory into the system must be the same as the fate of the fallout plutonium. Therefore, almost all the plutonium from the laboratory is, like the fallout, still in the system, mostly in river deposits.

The estimates by Foster and Hakonson necessarily contain significant simplifying assumptions.[14] Although they recognized that plutonium concentrations

Table 7.1. Plutonium and sediment processes in large basins

River Basin	Average Erosion Rate (kg/sq m/yr)	Sediment Stored (%)	Enrichment Ratio	Concentration Index (sq m/kg)	Plutonium Eroded in 20 Years (%)
Hudson	0.36	97.5	2.76	0.0125	0.271
Savannah	0.36	99.5	2.83	0.0070	0.151
Miami	0.83	97.0	2.74	0.0090	0.459
Ohio	0.88	94.5	2.49	0.0116	0.605
Tennessee	0.71	94.0	2.50	0.0111	0.574
Northern Mississippi	1.07	97.9	2.57	0.0064	0.438
Missouri	0.71	85.9	2.43	0.0044	0.461
Upper Mississippi	0.82	90.4	2.48	0.0040	0.488
Arkansas	0.67	98.0	2.71	0.0114	0.486
Lower Mississippi	0.85	92.7	2.33	0.0049	0.357
Total Mississippi	0.79	92.2	2.51	0.0094	0.476
Rio Grande	0.72	99.3	2.76	0.0135	0.570
Columbia	0.32	98.2	2.70	0.0141	0.263
Mean	0.70	95.2	2.60	0.0092	0.431

Source: Data and calculations based on data from G. R. Foster and T. E. Hakonson, "Predicted Erosion and Sediment Delivery of Fallout Plutonium," *Journal of Environmental Quality* 13 (1984): 595–602.

in nonagricultural soils decline exponentially with depth, for purposes of calculation they assumed uniform concentrations with depth. Thus in their calculations, the estimations of annual releases of fallout plutonium to streams were the same for each year. In reality, the plutonium losses to streams are probably relatively high immediately after the fallout and relatively low in later years. Fallout cesium (and almost certainly the associated fallout plutonium) losses from soil to river sediments was 50 percent greater in the year after fallout than in subsequent years.[15]

The time series for atmospheric nuclear detonations and the estimated delivery of fallout plutonium to streams from the landscape provide insight into the variability of plutonium in deposits. Because the fallout loading of plutonium in the Northern Rio Grande was at a maximum rate during the 1960s, the sediments deposited along the river during that period are likely to contain the greatest amounts of plutonium. Because the loss from soils to river sediments was greatest in the years immediately after the fallout, the introduction of plutonium from fallout into the river system peaked before the 1970s. Those sediments deposited during the 1960s may therefore reasonably be expected to contain higher amounts of plutonium from fallout than do those deposited in other periods. Because almost all the sediments in the Northern Rio Grande during this time were stored either on flood plains or in reservoirs, most of the fallout plutonium is still in the system.

Inputs from Los Alamos

Most of the plutonium reaching the Northern Rio Grande that was lost from research and development activities at Los Alamos National Laboratory entered the environment between 1945 and 1952. The first major technical facility constructed at Los Alamos for the Manhattan Project was the Main Technical Area, known to users as TA-1. Constructed near Ashley Pond at the old ranch buildings, TA-1 housed a general laboratory for process chemistry that generated relatively large quantities of liquid wastes containing strontium, cesium, uranium, plutonium, americium, and tritium (Figures 2.3 and 2.4). Acid sewers, consisting mostly of buried pipes with some lengths above ground, carried the untreated liquid across the mesa top at Los Alamos through a main line that ended on the north edge of Acid Canyon (Figures 7.4 and 7.5). Acid Canyon is north of Los Alamos; a short distance downstream, it joins Pueblo Canyon. Photographs of the pipe taken in 1947 show that it ended with an open outlet, with liquid spilling over the rock face of the canyon and eventually flowing into alluvial sediment on the canyon floor (Figure 7.6).[16] Water and materials have moved through Acid and Pueblo canyons to Los Alamos Canyon, which leads to the Rio Grande (Figure 7.7).

The laboratory, in conjunction with the U.S. Public Health Service, constructed a waste-treatment plant in 1951 (labeled TA-45), and thereafter the liquid that emptied into Acid Canyon contained relatively few radionuclides (Figure 7.8). The radionuclide content of the effluent varied somewhat because new facilities began

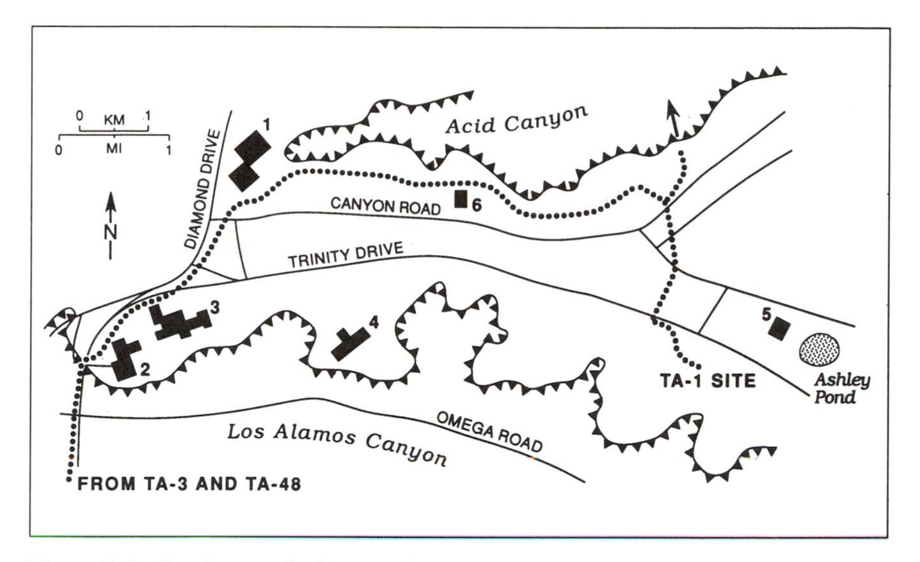

Figure 7.4. Sketch map of acid sewer lines on Los Alamos Mesa during the mid- and late 1940s. Numbers indicate building locations: (1) high school, (2) technical area-43, (3) medical center, (4) administration, (5) county building. (Redrawn from A. Stoker, A. J. Ahlquist, D. L. Mayfield, W. R. Hansen, A. D. Talley, and W. D. Purtymun, *Radiological Survey of the Site of a Former Radioactive Liquid Waste Treatment Plant (TA-45) and the Effluent Receiving Areas of Acid, Pueblo, and Los Alamos Canyons, Los Alamos, New Mexico*, U.S. Department of Energy and Los Alamos National Laboratory Report LA-8890-ENV, UC-70 [Los Alamos, N.M.: Los Alamos National Laboratory, 1981])

contributing to the materials being processed. In 1953 a new facility south of Los Alamos Canyon (TA-3), which included the Chemistry and Metallurgical Research Building, sent liquid wastes to the treatment plant. Also in 1953, the Health Research Laboratory (TA-43) began providing wastes for processing. In 1958 a new radiochemistry facility added more materials for treatment. Wastes from these sources contained highly variable amounts of radionuclides, and not all the material required processing to remove contaminants before being released.

In 1963 a new treatment plant (TA-50) south of Los Alamos Canyon began accepting wastes from the research facilities south of the canyon (TA-3 and TA-48). Its releases contain so few radionuclides that they are not an issue in the regional river system. In 1964 the original laboratory facilities at TA-1 and the original waste-treatment facility (TA-45) ceased operation, and so there were no more releases of effluent into Acid Canyon north of Los Alamos. One additional facility on the main Los Alamos mesa, a plutonium-processing plant (TA-21), released treated wastes into DP Canyon, a small tributary to Los Alamos Canyon, between 1952 and the early 1980s. These releases contained less plutonium than did those from the waste-treatment plant that emptied into Acid and Pueblo canyons, but the materials combined with the alluvium in the canyon floors.

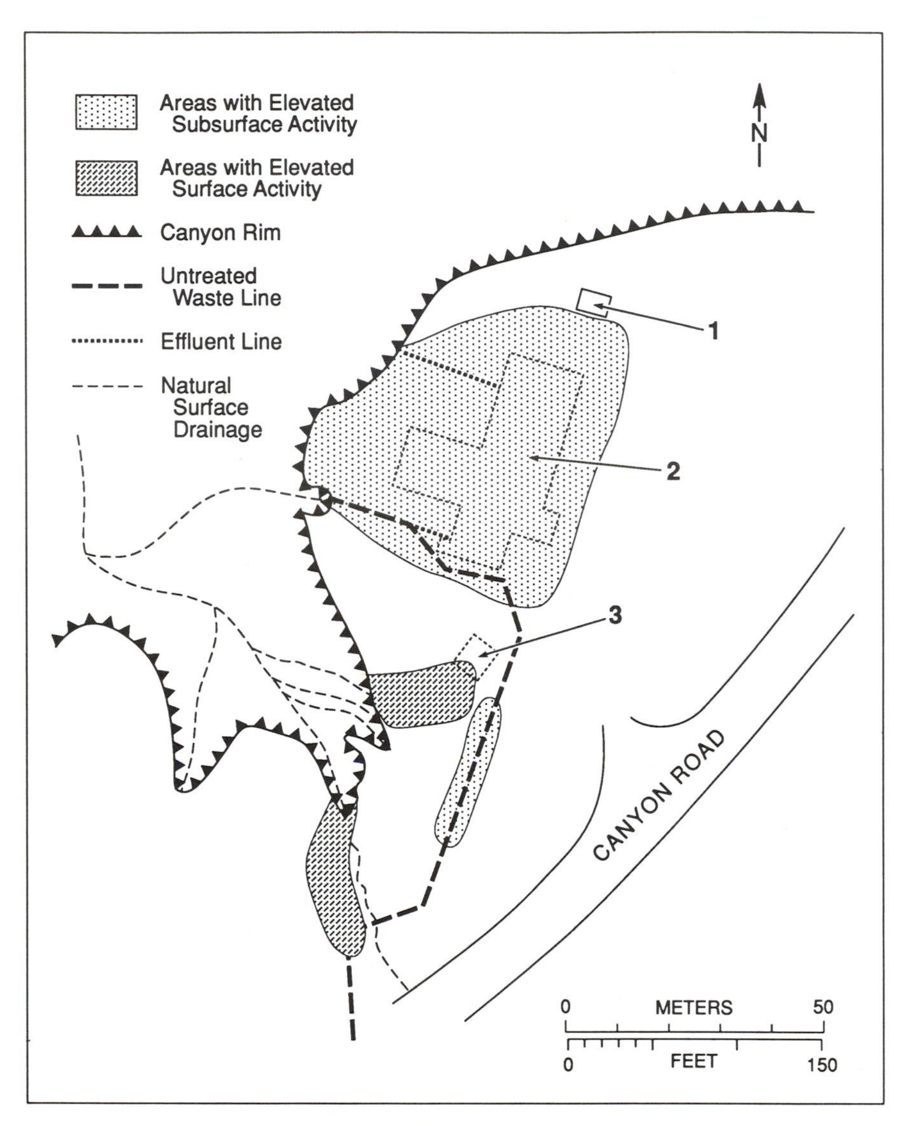

Figure 7.5. Sketch map of the Acid Canyon area, an enlargement of the area at the plutonium release site indicated by an arrow in the upper-right corner on Figure 7.4. Numbers indicate building locations: (1) sewage lift station, (2) treatment plant, (3) vehicle-decontamination facility. (Redrawn from A. Stoker, A. J. Ahlquist, D. L. Mayfield, W. R. Hansen, A. D. Talley, and W. D. Purtymun, *Radiological Survey of the Site of a Former Radioactive Liquid Waste Treatment Plant (TA-45) and the Effluent Receiving Areas of Acid, Pueblo, and Los Alamos Canyons, Los Alamos, New Mexico,* U.S. Department of Energy and Los Alamos National Laboratory Report LA-8890-ENV, UC-70 [Los Alamos, N.M.: Los Alamos National Laboratory, 1981]). The general location of this area is shown by a circle on the topographic map in Figure 7.7a.

Figure 7.6. Untreated liquid waste exiting Acid Sewer 1 into Acid Canyon in 1947. The location shown is the termination point of the heavy dashed line at the cliff's edge in Figure 7.5. (Los Alamos National Laboratory photo, document LAMS-516)

The original waste-treatment plant (TA-45) and its acid sewers that emptied over the cliff face were decontaminanted and decommissioned in 1966.[17] Workers demolished the buildings, collected the debris, dismantled the pipelines, excavated soils in the vicinity of the structures and under the pipeline route, removed the rock from the cliff face, and buried the materials in the solid-waste-disposal area at Los Alamos. They also removed some alluvium from the floor of Acid Canyon and buried it at the disposal site.

The amounts of plutonium introduced into the alluvium on the canyon floors in the Los Alamos Canyon system (which includes Acid, Pueblo, and DP canyons) by laboratory operations are not precisely known because the effluent was not continuously tested for radionuclide contents. For Acid and Pueblo canyons, there are two published estimates of the total plutonium releases between 1943 and 1950 when

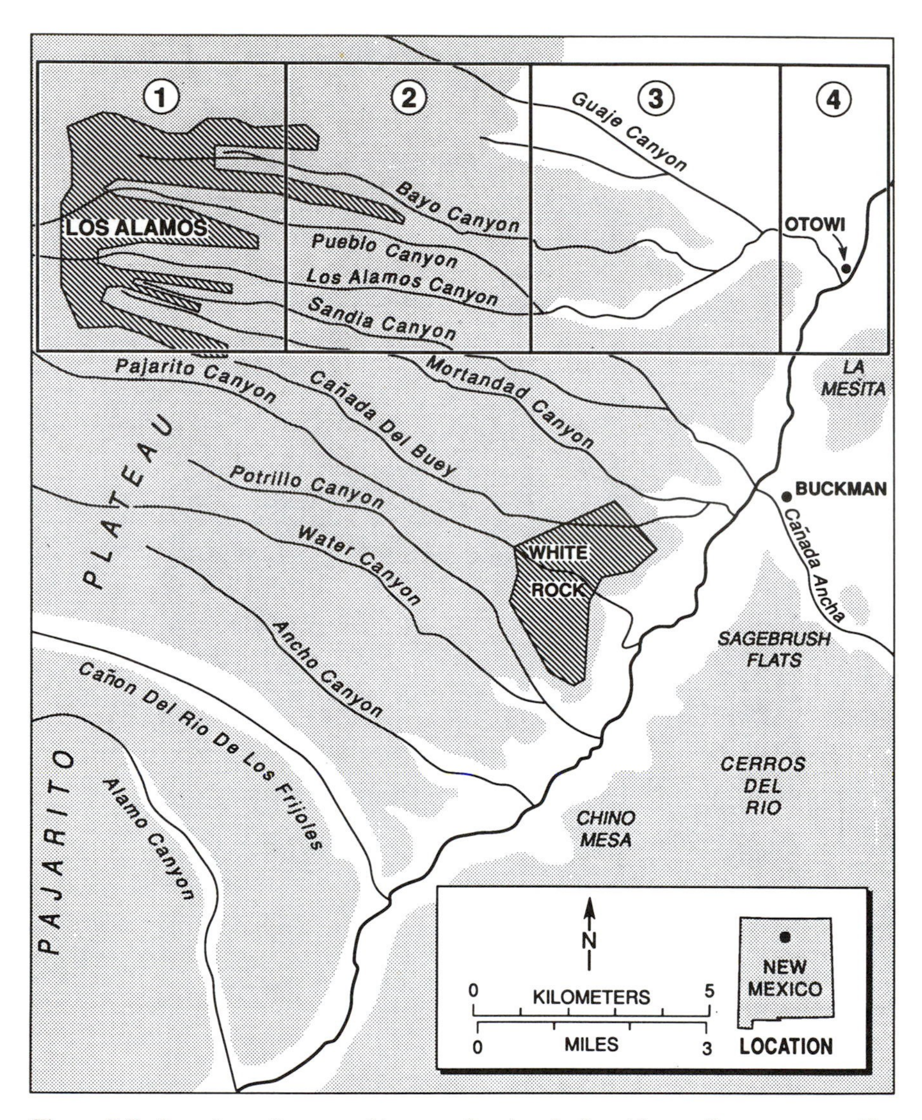

Figure 7.7. Locations of topographic maps showing the Los Alamos Canyon system. The numbered sections correspond to the detail maps: (1) Figure 7.7a, (2) Figure 7.7b, (3) Figure 7.7c, and (4) Figure 7.7d.

Figure 7.7a. Topographic map of the Los Alamos area. (Section of U.S. Geological Survey quadrangle)

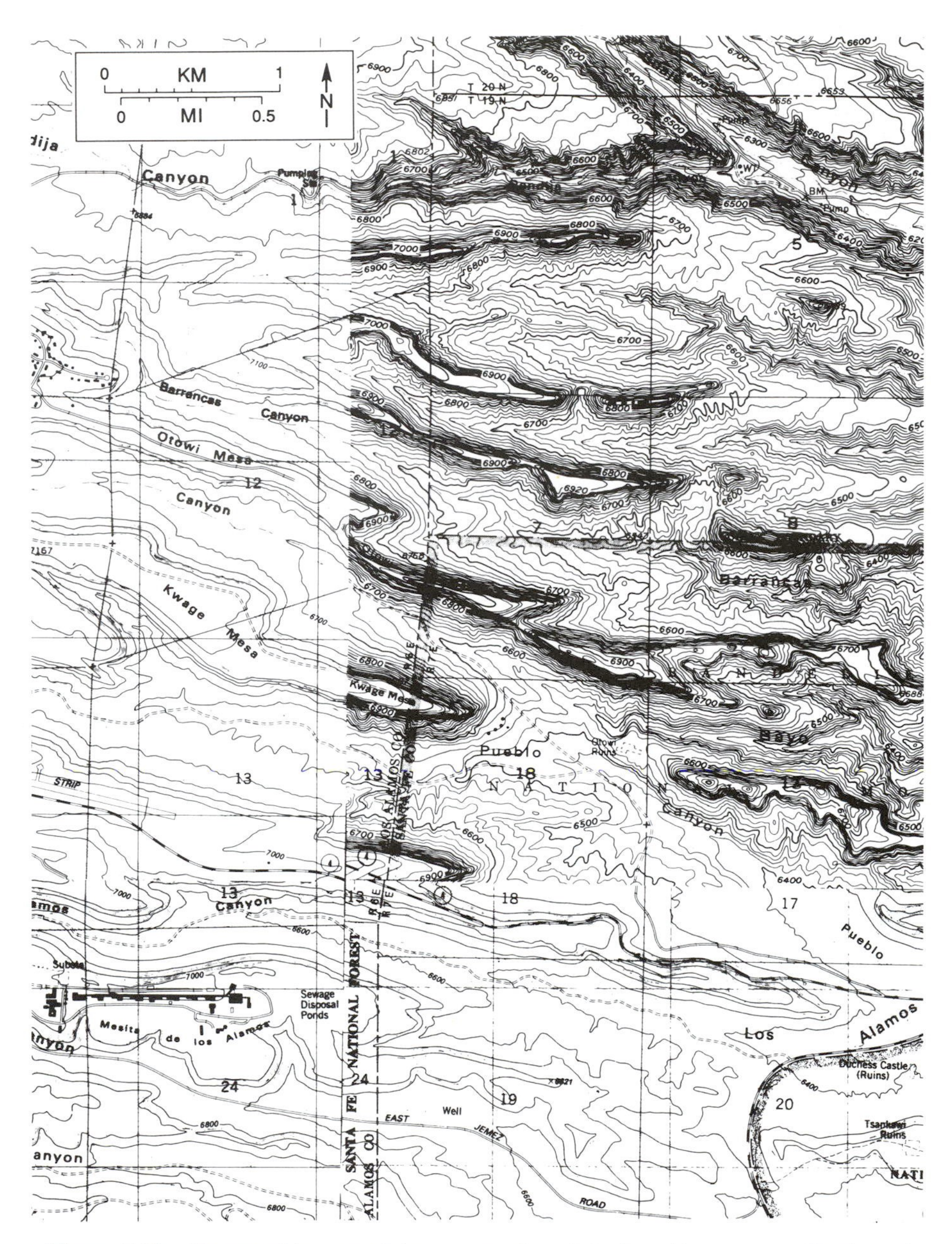

Figure 7.7b. Topographic map of the eastern edge of the Los Alamos area. (Section of U.S. Geological Survey quadrangle)

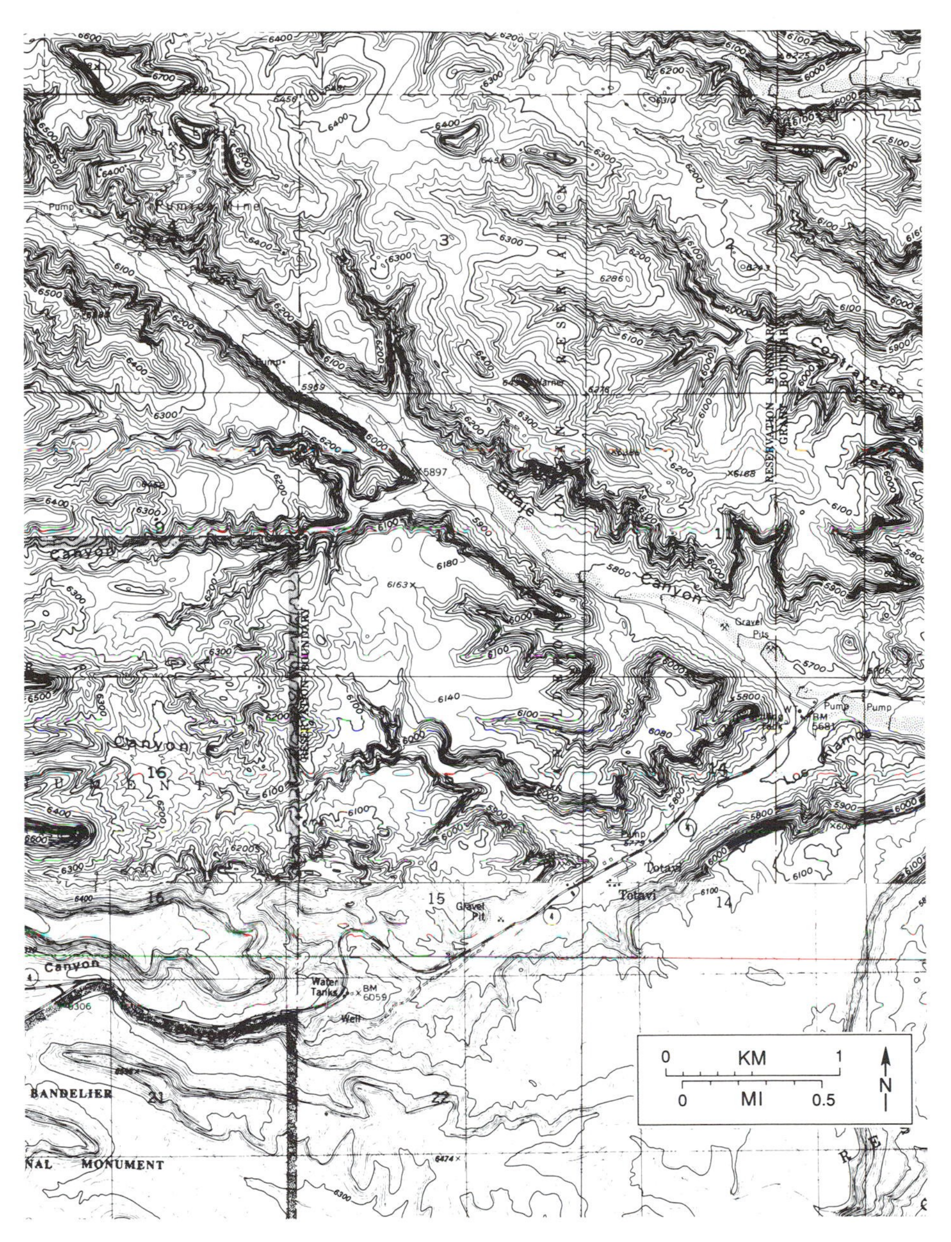

Figure 7.7c. Topographic map of the Totavi area. (Section of U.S. Geological Survey quadrangle)

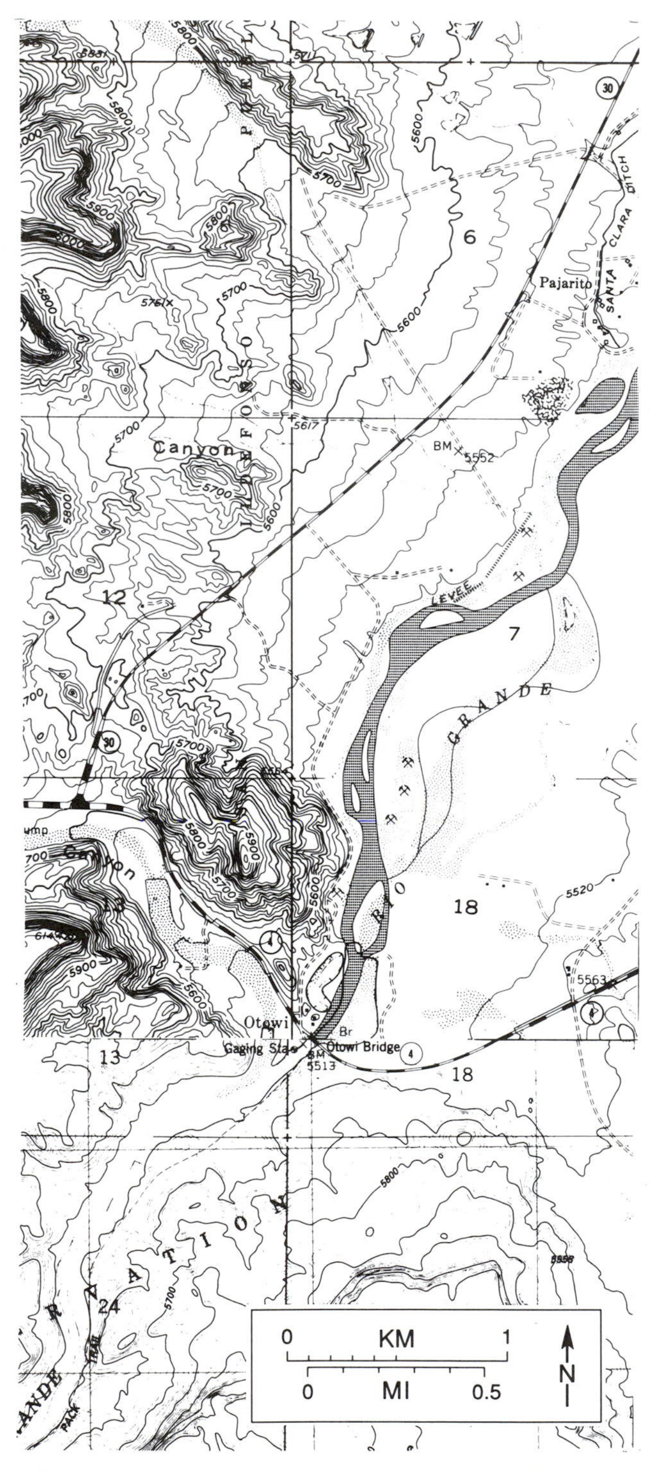

Figure 7.7d. Topographic map of the Otowi area. (Section of U.S. Geological Survey quadrangle)

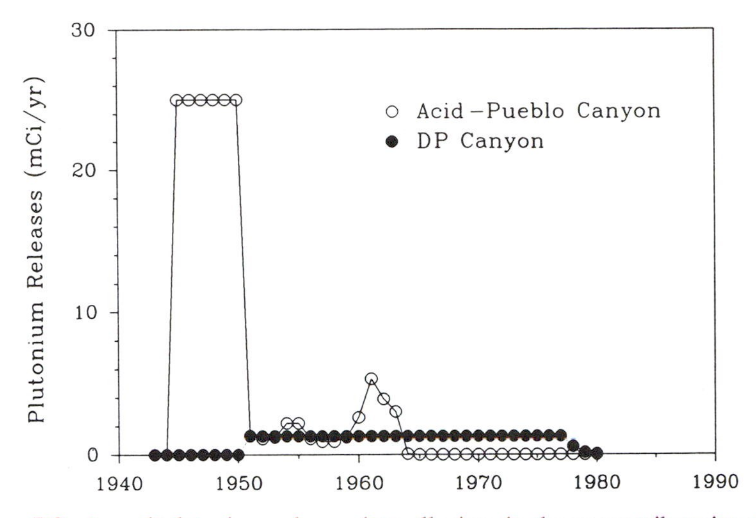

Figure 7.8. Annual plutonium releases into alluvium in the upper tributaries of Los Alamos Canyon. (Data from A. Stoker, A. J. Ahlquist, D. L. Mayfield, W. R. Hansen, A. D. Talley, and W. D. Purtymun, *Radiological Survey of the Site of a Former Radioactive Liquid Waste Treatment Plant (TA-45) and the Effluent Receiving Areas of Acid, Pueblo, and Los Alamos Canyons, Los Alamos, New Mexico*, U.S. Department of Energy and Los Alamos National Laboratory Report LA-8890-ENV, UC-70 [Los Alamos, N.M.: Los Alamos National Laboratory, 1981])

there was no monitoring or treatment. First, Stoker and his associates estimated the total for 1943–1950 period as 150 mCi.[18] Second, Lane and his co-workers used data from the plutonium concentrations in the canyon alluvium as published by Stoker and estimated that an input of 150 mCi was too low to explain the amount measured in the sediments.[19] Lane calculated from the empirical data gathered in the late 1970s that the total amount of plutonium in the canyon was 300 to 900 mCi. Because that amount represents the remains after the removal of some material by natural erosion, there would have been more in the canyon originally. Based on this reasoning, Lane estimated that the 1943–1950 releases could have been as high as 3,000 mCi.[20] For comparison, the amount of radioactive material released in the Three Mile Island, Pennsylvania, incident was about 30 mCi, and the amount of plutonium remaining at the Nevada Test Site, even after cleanup, is about 200,000 mCi.[21]

The variability in the estimates for Los Alamos releases makes impossible a general time series for the release of plutonium into Acid and Pueblo canyons for the 1943–1950 period, although some refinement of the estimates of the timing of the releases is possible. Lane and his associates estimated that in the absence of better information, the minimum estimate of 150 mCi was distributed between 1943 and 1950 at an average rate of 18.75 mCi per year.[22] However, historical accounts of plutonium delivery to Los Alamos (Chapter 2) show that significant quantities did not arrive at the site until mid-1945. Therefore, the minimum estimate of 150 mCi was probably distributed over the 1945–1950 period, with an average rate of 25 mCi per year. If the empirical data that Lane used are correct, the rate could have been three to 30 times higher than this minimum value.

After treatment and monitoring began in 1951, the releases were very low and also well defined. A similar estimation period followed by lower but better-known values characterizes the record of releases for DP and upper Los Alamos canyons. Because the mean values are spread over several years, the time series of releases for both canyons contain periods with no trends. The magnitudes are probably best represented as minimum values, but with the reservation that substantial errors are possible. The general trends for releases over several decades show that the period before 1951 is the one of greatest importance for plutonium loading of the system (Figure 7.8 and Table 7.2). After that time, natural erosion, transportation, and depositional processes moved the contaminated sediments downstream toward the Rio Grande, and the new loadings were relatively small.

The transfer of the contaminated sediments from Los Alamos Canyon into the Rio Grande through natural stream processes in the tributary canyons was inconsistent. Stream flow in the canyons is sporadic, with some years having no flow. In other years, snowmelt created spring discharges, and summer thunderstorms produced flash floods. As a result, the canyon system discharged plutonium-bearing sediments into the Rio Grande several times, with the quantity related to the amount of water available for transporting the materials. In work estimating contaminant transport through the Los Alamos Canyon system, Lane and his associates calculated that runoff would transport the entire plutonium inventory in the canyon system to the Rio Grande sometime before 2000.[23] The exact timing depends on the assumed original loading and the utility of their assumptions about magnitude and frequency of runoff events. Lane and his colleagues calibrated their model using reconstructed records of runoff events between 1943 and 1980 and estimated the amount of plutonium reaching the Rio Grande by combining sediment transport rates, magnitude and frequency of water discharges, and plutonium concentration data.

Previously unpublished data and calculations used in Lane's conclusions show that between 1943 and 1980 there were many years when no plutonium entered the main stream.[24] During four years—1951, 1952, 1957, and 1968—plutonium discharge was at a maximum (Figure 7.9). Also during these four years, floods in Los Alamos Canyon carried sediment and plutonium to the confluence with the Rio Grande, where the materials entered the general transport processes in the main stream. In many cases, the flood peaks on the tributary stream did not coincide with the annual peak discharge in the main stream, and so sediments from the tributary moved only a short distance downstream in the Rio Grande. They often were deposited at the mouth of Los Alamos Canyon, along the north bank within a

Table 7.2. Plutonium releases into canyon sediments at Los Alamos (mCi/yr)

Year	Acid and Pueblo Canyons	DP Canyon	Comments
1943	0	0	Operations begin, no treatment or monitoring
1944	0	0	
1945	25.0	0	Delivery of large quantities of plutonium begins
1946	25.0	0	
1947	25.0	0	
1948	25.0	0	
1949	25.0	0	
1950	25.0	0	
1951	1.3	1.27	Treatment at TA-45, monitoring for Acid and Pueblo canyon
1952	1.1	1.27	
1953	1.2	1.27	Waste from TA-3 and TA-43 added to system
1954	2.2	1.27	
1955	2.2	1.27	
1956	1.1	1.27	
1957	0.9	1.27	
1958	0.9	1.27	
1959	1.2	1.27	
1960	2.6	1.27	
1961	5.3	1.27	
1962	3.9	1.27	
1963	3	1.27	Treatment begins at TA-50
1964	0.04	1.27	
1965	0	1.27	
1966	0	1.27	TA-45 dismantled
1967	0	1.27	
1968	0	1.27	
1969	0	1.27	
1970	0	1.27	
1971	0	1.27	
1972	0	1.27	
1973	0	1.27	
1974	0	1.27	
1975	0	1.27	
1976	0	1.27	
1977	0	1.27	
1978	0	0.54	
1979	0	0.14	
1980	0	0.045	

Note: Data for 1945–1950 for Acid and Pueblo canyons and 1951–1977 for DP Canyon are average values.

Source: Data from A. Stoker, A. J. Ahlquist, D. L. Mayfield, W. R. Hanasen, A. D. Talley, and W. D. Purtymun, *Radiological Survey of the Site of a Former Radioactive Liquid Waste Treatment Plant (TA-45) and the Effluent Receiving Areas of Acid, Pueblo, and Los Alamos Canyons, Los Alamos, New Mexico*, U.S. Department of Energy and Los Alamos National Laboratory Report LA-8890-ENV, UC-70 (Los Alamos, N.M.: Los Alamos National Laboratory, 1981); and L. J. Lane, W. D. Purtymun, and N. M. Becker, *New Estimating Procedures for Surface Runoff, Sediment Yield, and Contaiminant Transport in Los Alamos County, New Mexico*, Los Alamos National Laboratory Report LA-10335-MS, UC-11 (Los Alamos, N.M.: Los Alamos National Laboratory, 1985).

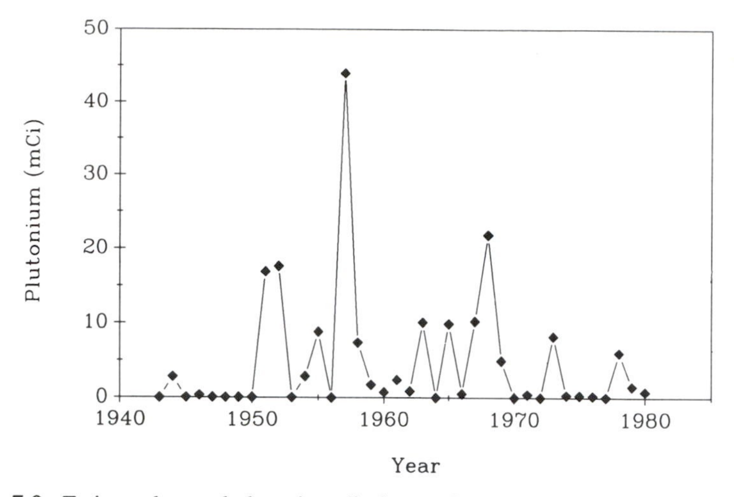

Figure 7.9. Estimated annual plutonium discharges from Los Alamos Canyon to the Rio Grande. (Data from L. J. Lane, Agricultural Research Service, Tucson)

few hundred meters of the mouth, or on bars and in sloughs within a few kilometers downstream along the Rio Grande.

Remobilization from Storage

This limited range of transport for the initial injection of sediments from Los Alamos Canyon is likely because the materials from the tributary are coarse grained and probably are mostly bedload in the energy environment of the Rio Grande except during major floods. Silt and clay particles, with their higher concentrations of plutonium, make up a small amount of the sediment load from the tributary. These lesser amounts of fines would have remained in suspension in the Rio Grande, which would have transported them at least as far south as Peña Blanca. After Cochiti Dam closed in 1973, infusions of suspended sediment would have become reservoir deposits, and virtually none would have passed Cochiti Dam.

The actual amounts of suspended load and bedload injected by natural processes in Los Alamos Canyon into the Rio Grande are not precisely known. The calculations in this study assume that the bedload carries the significant plutonium in the system based on the importance attached to coarse particles by sedimentary studies of the canyon.[25] Measurements at gaging sites along Los Alamos Canyon have produced too few data to reach firm conclusions about the subdivisions of the sediment load. The sparse data from 16 observations show that in Los Alamos Canyon where plutonium is present, bed sediments accounted for an average of about 70 percent of the total sediment load (with a standard deviation of 21 percent and a range of 24 to 95 percent).[26] At the confluence of the canyon with the Rio

Grande, the bedload moving in Los Alamos Canyon is only about 40 percent of the load (with a standard deviation of 21 percent and a range of 11 to 66 percent). Some important materials in the system are therefore probably stored in overbank areas along the lower reaches of the canyon. The fines are probably just as important to transporting the Los Alamos–derived plutonium into the Rio Grande as the bedload is, but further research is required to describe the relationships accurately. The best present estimate is that the plutonium discharges from the tributary into the Rio Grande are about 3 percent by solution, 57 percent by suspended sediments, and 40 percent by bedload.[27]

Sediments in Los Alamos Canyon impregnated with plutonium move down the canyon system in a stepwise fashion, with each step taken as a few meters to a few kilometers during each flood event. Each flood stores the sediments as channel or flood-plain deposits, and each subsequent flood remobilizes them until they reach the Rio Grande. The same temporary storage and remobilization processes probably also occur in the main stream, but because the channel has consistently become smaller and more stable, storage consumes much of the material and relatively little becomes mobile again, at least on a time scale of several decades. After 1973, the remobilized contaminated sediments came to rest in Cochiti Reservoir, and so continued monitoring of the reservoir is critical to an effective surveillance program.

In summary, there are two sources of plutonium in the Rio Grande sediment system, and the maximum input from both sources was during roughly the same time period. Fallout plutonium from atmospheric testing of nuclear weapons and from satellite debris reached a peak in the 1960s, and erosion processes moved a substantial percentage of that fallout from slopes to stream channels during the same period. Contributions from the fallout on the landscape continue at a reduced level. Plutonium releases at Los Alamos were highest from the late 1940s to the early 1950s, but the sediment transport processes in Los Alamos Canyon that injected the peak amounts of the plutonium into the Rio Grande took place in 1951, 1952, 1957, and 1968. Sediments deposited in the Rio Grande immediately downstream from Otowi during the 1950–1970 period are therefore most likely to contain the highest amounts of plutonium in storage.

Plutonium in River Water

Plutonium occurs in the natural environment of the Northern Rio Grande in the atmosphere, water, biologic systems, and surface materials. When plutonium has been present, these systems and materials have undergone some general environmental changes affecting their distribution (Figure 7.10). This book concentrates on the sediments because they are the portion of the total system that contains 99

Figure 7.10. Changes in channel configuration and sediment storage have been minor along the Rio Grande near Embudo, New Mexico. All the views are looking northwest from the highway grade on the east side of the valley: (*a*) 1915 (Museum of New Mexico, photo 53099); (*b*) during the 1920s (Museum of New Mexico, photo 53562); (*c*) 1935 (T. H. Parkhurst photo,

c

d

Museum of New Mexico); (*d*) 1992 (W. L. Graf, photo 107-21). The cottonwood tree on the bank of the side channel in the left center of the view appears in the first three photos but had disappeared by 1992. Although the road grade and embankment have undergone several reconstructions, the channel configuration remained largely the same from 1915 to 1992.

percent of the plutonium in static environments,[28] including the environment of the southwestern United States.[29] In dynamic systems such as rivers, the movement of materials complicates general static impressions. For example, the concentrations in water may be very low, but the amount of water passing through the system is large, and so when considering the entire flux, the quantity may make up for the low concentrations. Therefore, a complete view of the plutonium in sediments depends on an accounting of the amount in river water. Decades-long sampling of river water and sediment in the Northern Rio Grande, combined with the exceptional data for water and sediment discharges in the system, provides a unique opportunity to explore these relationships in a system hundreds of square kilometers in extent (Figure 7.11).

For more than 20 years, Los Alamos National Laboratory has collected systematic and detailed regional river samples of river water for plutonium analysis and has published the results in its annual environmental surveillance reports. There are six major regional sites for which high-quality, published data are available for 1977–1988, with summaries of grouped data published for earlier years. The published values for 1974–1988 are measured plutonium concentrations, whereas in prior years the published values appear as "less than" a given

Figure 7.11. The outlet of untreated contaminated water and sediment into Los Alamos Canyon from the Los Alamos National Laboratory laundry in 1947. Note the sediment-laden water entering from the left side. (Los Alamos National Laboratory photo, document LAMS-516)

concentration. The sample sites on the Rio Grande are at Embudo, Otowi, Cochiti, and Bernalillo. Additional sites are on the Rio Chama and the Jemez River.

The grand mean concentration for plutonium 238 in river water from the six major regional sites (excluding the streams draining the immediate Los Alamos National Laboratory area, which is treated separately) for 166 samples taken in 1974–1988 is nearly zero, and the mean concentration for plutonium 239 and 240 is 0.0041 pCi/l, a value close to the minimum level of detection. Table 7.3 provides measures of the plutonium data, and Appendix E contains the original data. There are no apparent temporal trends in the plutonium concentrations in river water for the 1977–1988 period (when the published data represent individual sample sites rather than averages across all sites), in either the grand mean values or the values for individual sample sites. The highest individual values, 0.090 pCi/l for plutonium 238 and 0.130 pCi/l for plutonium 239 and 240, were from a sample collected from the Rio Grande at Embudo on August 7, 1984. There is no apparent explanation for the occurrence, although the U.S. Geological Survey's gaging records indicate that 1984 was the year of the lowest annual water yield and the lowest flood peak for the river at the sample site during the 1977–1988 period. The association between low flows and higher plutonium concentrations in water is not generally found in the data.

There are no well-defined geographical patterns in the values for plutonium concentrations in river water in the Northern Rio Grande. The concentrations throughout the 1977–1988 period were consistently higher in the Rio Grande at Embudo and the Rio Chama at Chamita than at the other sites. According to a *t*-test of difference of means, the mean for the Rio Grande at Embudo is significantly

Table 7.3. Summary of plutonium concentration data for regional fluvial systems

		Plutonium 238			Plutonium 239 and 240		
Samples	Units	Mean	St. Dev.	N	Mean	St. Dev.	N
River water	pCi/l	−0.0078	0.0521	106	0.0050	0.0246	106
Tributary water	pCi/l	0.0053	0.0191	73	0.0090	0.0178	73
Bedload sediment	pCi/g	−0.0002	0.0003	113	0.0048	0.0086	113
Flood-plain sediment	pCi/g	0.0009	0.0012	8	0.0067	0.0070	8
Reservoir sediment	pCi/g	0.0007	0.0007	60	0.0127	0.0073	60

Source: All data in table from samples collected in 1977–1988 as reported in Los Alamos National Laboratory surveillance reports, except all flood-plain data, which were collected in 1989. For more detailed data for each sample type, see Appendix E.

higher than the mean for the Rio Grande at Otowi (at the 0.05 level). It is not immediately obvious why these relatively higher values do not continue in samples from sites downstream in the Rio Grande, but in any case all the values are small and close to levels of detection.

Los Alamos National Laboratory has also collected data on plutonium concentrations in the water of streams flowing from the Jemez Mountains and the Pajarito Plateau to the Rio Grande in the vicinity of Los Alamos. These local streams have slightly higher concentrations of plutonium than does the main stream (Table 7.3 and Appendix E2), probably because of fallout on the higher mountain watersheds of the tributaries. These local streams include those draining areas directly affected by laboratory operations, as well as those draining unaffected areas.

Extensive sampling during the mid-1980s of water flowing in streams associated with laboratory operations (Los Alamos Canyon, Pajarito Canyon, Pueblo Canyon, and Water Canyon) showed that the streams carry no more plutonium in their water than do other regional or local streams. The plutonium concentrations in the laboratory-affected streams are statistically similar to the concentrations in unaffected streams.

Plutonium in River Sediments

The regional sampling for sediment in rivers for the Northern Rio Grande is also widespread and of long duration compared with that for sediment in similar environments on a global scale. Los Alamos National Laboratory annually samples sediments from the Rio Chama at Chamita, the Jemez River at Jemez Pueblo, and the Rio Grande at Embudo, Otowi, Cochiti Reservoir, and Bernalillo. Additional and less regular sampling of sediments provides data for the Rio Grande at several points in White Rock Canyon between Otowi and Cochiti Reservoir. The laboratory's published environmental surveillance reports include these data as well as the results of occasional sampling in the sediments of Rio Grande, Heron, El Vado, Abiquiu, and Cochiti reservoirs (for locations, see Figure 5.5).

The stream channels of the Northern Rio Grande transport sediment as bedload and suspended load. The coarser bedload rests on the channel floor between flood flows and is the material often sampled for plutonium concentrations and reported in Los Alamos National Laboratory surveillance reports as river "sediment." Investigators collected these materials near the channel margins as grab samples of sediment resting on the channel floor in the lee of cobbles or boulders.[30] Bedload sediment data for streams in the Northern Rio Grande for the 1974–1986 period have been previously published.[31] Appendix E3 summarizes the plutonium concentrations for the Rio Chama at Chamita, the Jemez River at Jemez Pueblo, and sites along the Rio Grande at Embudo, Otowi, four sites in White Rock Canyon, Cochiti, and Bernalillo. The mean concentrations of plutonium 238 are near the minimum level of detection, and those of plutonium 239 and 240 are significantly higher (Table 7.3).

The data show that there is about three orders of magnitude more plutonium in bed sediments than in river water (note that the unit of measure for plutonium in sediments is picocuries per gram and for liquids it is picocuries per liter and that a liter of pure water has a mass of 1,000 g). The values do not show temporal trends at any of the measurement sites. The bedload plutonium concentrations do not show a statistically significant geographic trend along the Rio Grande, although the concentrations at Otowi are the highest ones. The Otowi concentrations probably reflect the combined inputs from fallout and from Los Alamos National Laboratory via Los Alamos Canyon. The contribution from the laboratory must be small in active-bed sediments at the times of measurement because the difference in mean concentrations between Otowi and Embudo (above Los Alamos Canyon and reflecting only fallout contributions) are statistically insignificant. About 5 km downstream from the confluence with Los Alamos Canyon (at the confluence with Sandia Canyon, immediately below Buckman), the plutonium concentrations in bed sediments return to levels so low that they are below the regional mean. Plutonium-bearing bed sediments in the active channel may therefore be deposited between the two sites.

For comparative purposes, plutonium concentrations in bedload sediments in areas directly affected by operations of Los Alamos National Laboratory are considerably higher than the regional values in Table 7.3. The maximum is probably found in an effluent area of Mortandad Canyon where the maximum plutonium 239 and 240 concentration in sediments in 1988 was 33.5 pCi/g. No values approaching this magnitude have been found outside the laboratory's boundaries.

Plutonium concentrations in the finer suspended sediments of regional streams are generally higher than those in the coarser bedload because of the affinity of heavy metals for fine particles. The overall mean concentration of plutonium 239 and 240 in bedload sediments is 0.0086 pCi/g, and in reservoir sediments (derived from settled suspended sediments) the overall mean is 0.0127 pCi/g, a useful figure for budget calculations. The mean from a few direct samples of suspended sediments in the Rio Grande at Otowi was 0.0316 pCi/g (reported in the surveillance report series of Los Alamos National Laboratory), but almost all the samples were from one year (1985) so that the figure is probably not representative. The reservoir data are probably more reliable as long-term indicators. Suspended sediments from small streams directly affected by operations of Los Alamos National Laboratory and located within the laboratory's boundaries have considerably higher concentrations sometimes ranging up to more than 20 pCi/g, but they are usually less than 2 pCi/g. These relatively high values are not found outside the laboratory boundaries.

Almost all the plutonium detected in the bedload and suspended sediments of the active channel of the main river is likely to be derived from fallout. The only time that plutonium from Los Alamos National Laboratory is likely to be in the active channel of the Rio Grande is during and immediately after discharges from the canyon channels draining the laboratory area. Because active sediments in the main stream were rarely sampled during such events, the data describing the active sediments probably represent the activity of the system without direct input from laboratory operations. Only reservoir and flood-plain deposits that integrate the inputs over time are likely to contain plutonium from both sources.

Earlier chapters showed that although there is a great deal of sediment moving through the Northern Rio Grande system, a great deal is also being stored, and so much of the plutonium in the system is also being stored. The storage process for plutonium varied greatly through time because the inputs have varied. The storage process is also geographically diverse, but it focuses on flood plains and reservoirs.

Before this project, there was no systematic sampling of flood-plain deposits and related materials such as abandoned channel areas, filled channels, and flood bars. The preliminary results of sampling the deposits in the Buckman area indicate that the concentrations of plutonium in such materials are highly changeable, reflecting the difference in plutonium concentrations at the various times of sedimentation (Table 7.3 and Appendix E4). The plutonium concentrations in those deposits laid down in years when there were discharges from Los Alamos Canyon are higher than those in the presently active channel. Because the channel of the Rio Grande is generally becoming smaller, straighter, and more stable, the remobilization of this stored plutonium back into the active channel is minimal.

Plutonium in Reservoir Deposits

Unless it is deposited on flood plains or in abandoned channels, sediment moves through the streams of the Northern Rio Grande until it encounters a reservoir. As the stream flow encounters the still waters of the reservoir, its transport capacity drops nearly to zero, and the sediment is deposited on the reservoir floor.[32] The finest materials travel farthest into the middle and lower reaches of the reservoir.[33] Beyond the delta built by the deposited bedload, the reservoir sediments tend to be relatively fine compared with the other deposits, because the reservoir sediments mostly reflect the materials in suspension during transport through the channel.[34] Reservoir sediments from the upper reaches of Cochiti Lake, which I sampled for this study near Frijoles Canyon, contained about 40 to 50 percent silt and clay, a figure that is a likely approximation of materials suspended in the channel portions of the Rio Grande above the lake.

The plutonium concentrations in reservoir sediments differ from one reservoir to another in the Northern Rio Grande system (summarized in Table 7.3, with data in Appendix E5). The concentrations are highest in Cochiti Reservoir, which traps sediment containing contamination from fallout and Los Alamos National

Laboratory. The concentrations in Cochiti Reservoir for plutonium 238 and 239 and 240 are significantly higher (at the 0.01 level in a *t*-test) than in Abiquiu Reservoir, the nearest reservoir upstream, indicating the possibility of significant additions from Los Alamos (Figure 5.5 shows the locations). The plutonium concentrations in the Rio Grande Reservoir (in the mountainous headwaters of the river system) are similar to those in Cochiti Reservoir, but in the Rio Grande Reservoir the only potential source is fallout on alpine slopes in the system's headwaters. Heron Reservoir, in the headwaters of the Rio Chama, has intermediate values, probably because its sediments derive from lower-elevation areas than do those of the Rio Grande Reservoir. El Vado Reservoir, downstream from Heron Reservoir, does not receive sediments from the high landscapes above Heron, and its plutonium concentrations are therefore lower. Abiquiu Reservoir also has low values for the same reason.

Surveillance reports by Los Alamos National Laboratory explain that the plutonium concentrations in Cochiti Reservoir are the result of the fact that Cochiti has finer sediment and more organic material than do the other reservoirs. This explanation may be at least partly correct, but it is not demonstrated by particle-size or organic-content data. The particle-size and organic-material explanation also begs the question concerning the amounts of plutonium entering the reservoirs. If all reservoirs receive the same amount of plutonium and if Cochiti collects more in sediment than the others do, the fate of the plutonium entering the others is still unknown. Water samples do not indicate that discharges from the other reservoirs contain more plutonium than does the discharge from Cochiti. The best explanation is that Cochiti and Rio Grande reservoirs contain more plutonium than the others do because more plutonium enters them as a result of their geographic locations: There is more plutonium in the Rio Grande Reservoir because it receives much fallout from the surrounding mountain slopes, and there is more in Cochiti Reservoir because it receives it from Los Alamos.

This brief summary of plutonium in water and sediments demonstrates the importance of understanding the regional sediment budget in deducing the regional plutonium budget. Almost all the plutonium in the Northern Rio Grande system is associated with sediment. A comparison of the amount of plutonium in transit in the Rio Grande at Otowi shows the overriding significance of sediment-bound plutonium in the regional budgets and fluxes of the contaminant. The total annual movement of plutonium in river water at the site is

$$T = Q * P$$

where T = the total flux of plutonium in water per year, Q = the mean annual water yield from U.S. Geological Survey records, and P = the mean plutonium concentration from Los Alamos National Laboratory surveillance reports. For the period 1977–1988, the total plutonium flux in water (combining plutonium 238,

239, and 240) at Otowi was 0.0199 mCi per year. A similar calculation for sediment (discussed in greater detail in the following section) shows that for the same period, the total plutonium flux in sediment was 29.4390 mCi, or three and a half orders of magnitude greater than in water.

Plutonium in Los Alamos Canyon Sediments

The sampling program of Los Alamos National Laboratory has collected over 400 stream sediment samples in Acid, Pueblo, DP, Los Alamos, Bayo, and Guaje canyons at and near the laboratory. These samples, apparently collected as bedload materials, provide a useful assessment of the level of plutonium concentrations in stream sediments directly affected by the releases associated with laboratory activities in the 1940s and early 1950s. The mean values that reflect this intensive sampling effort show the general degree of plutonium loading in sediments by the laboratory, and a detailed analysis of the data reveals the geographic variability of the concentrations within a few kilometers of the source.

The plutonium values of bedload sediments from streams draining the eastern slopes of the Jemez Mountains and parts of the Pajarito Plateau unaffected by the laboratory activities (Table 7.4) are somewhat higher than the values in the lower-elevation regional streams (Table 7.3). Apparently the fallout contributions to the higher landscapes, with their greater precipitation, account for the difference.

The bedload sediments from streams directly affected by the releases from Los Alamos have the highest concentrations of plutonium of any of the environments considered in this study. The mean concentrations in the bedload sediments of Acid, Pueblo, DP, and upper Los Alamos canyons (above the confluence with Pueblo Canyon; for general locations, see the maps in Chapter 9) are two and a

Table 7.4. Summary of plutonium concentration data for Los Alamos Canyon systems (pCi/g)

	Plutonium 238			Plutonium 239 and 240		
Samples	Mean	St. Dev.	N	Mean	St. Dev.	N
Background streams*	0.00100	0.00140	44	0.00210	0.00210	44
Acid and Pueblo canyons	0.03409	0.11876	132	3.30054	7.11496	132
DP and upper Los Alamos canyons	0.17077	0.65337	165	0.56531	1.17599	165
Lower Los Alamos Canyon	0.00930	0.01277	49	0.19624	0.22290	49

*Includes sites in the Los Alamos Canyon system not directly affected by laboratory activities: sites in Bayo and Guaje canyons, at the highway bridge and reservoir sites in Upper Los Alamos Canyon, and sites upstream from the points of plutonium releases in DP and Acid canyons.

half to three orders of magnitude higher than the mean values of similar sediments in the regional river system (Table 7.4). In lower Los Alamos Canyon (downstream from the confluence with Pueblo Canyon), plutonium concentrations are only a fraction of the values in those areas directly affected by Los Alamos National Laboratory releases, but they are still many times higher than the values in the regional river system.

The mean values of plutonium concentrations in the bedload sediments of Los Alamos Canyon system are useful general indicators of contamination levels, but the mean values mask important temporal and geographical variation in the data. In many places, repetitive sampling over a period of decades has revealed substantial variation from year to year, with each site producing elevated values of plutonium concentrations as well as values approaching regional background levels (Figures 7.12 and 7.13). This temporal fluctuation is the product of sediment transport processes, and in some years the direct inputs to downstream sample sites from discharge events transport the contaminated sediments directly from the release sites upstream. The result is relatively high concentrations of plutonium at the sample site. In other years, the release sites may have not experienced a significant discharge, but other uncontaminated portions of the system

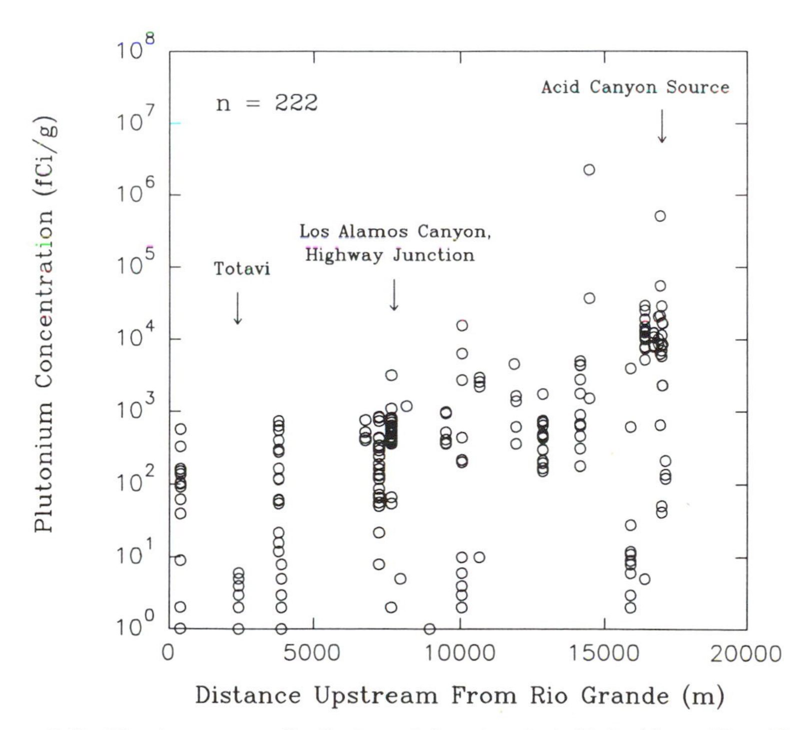

Figure 7.12. The along-stream distribution of plutonium in Acid, Pueblo, and Los Alamos canyons. For general locations, see Figure 7.7. (Data from Los Alamos National Laboratory)

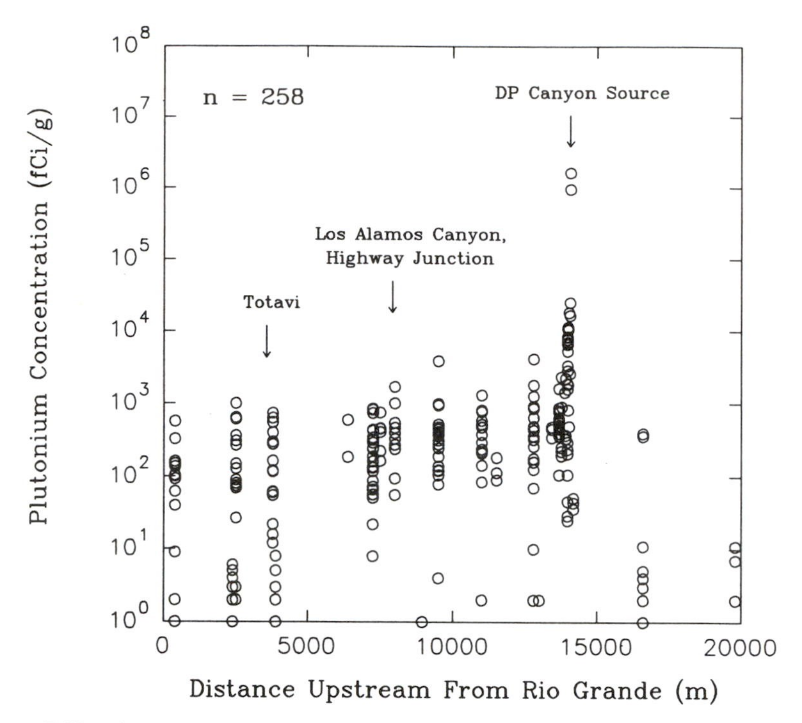

Figure 7.13. The along-stream distribution of plutonium in DP and Los Alamos canyons. For general locations, see Figure 7.7. (Data from Los Alamos National Laboratory)

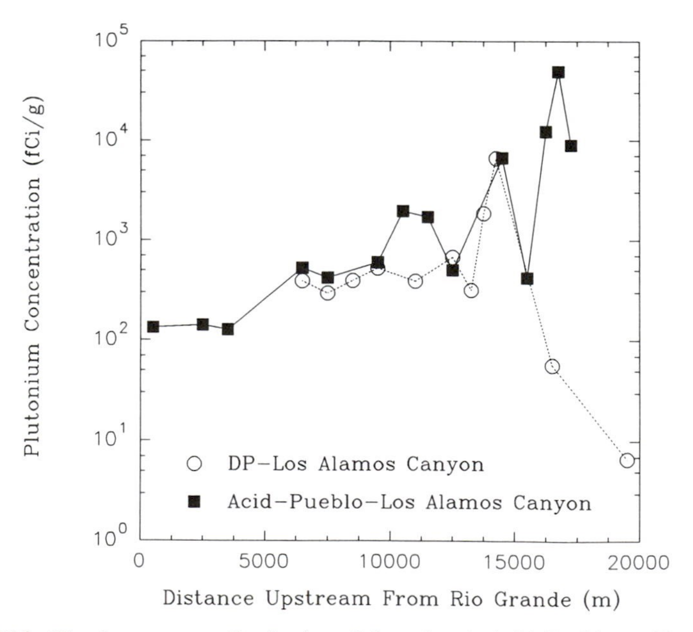

Figure 7.14. The along-stream distribution of plutonium in Acid, Pueblo, and Los Alamos canyons and in the DP and Los Alamos canyons, plotted as average values within individual 500-m reaches. Reaches without samples are not plotted. For general locations, see Figure 7.7. (Data from Los Alamos National Laboratory)

may have been flooded, contributing relatively "clean" sediments to the downstream sample site and thus producing relatively low plutonium concentrations when sampled.

As distance increases downstream away from the source of plutonium in Acid and DP canyons, the mean values within limited reaches decline (Figure 7.14). The mixing of relatively uncontaminated sediment with polluted materials, the deposition of contaminated materials in flood plains where they are not sampled as part of the bedload sediments, and the dispersion resulting from stream-flow processes contribute to a downstream drop in concentrations. Figure 7.14 shows that the highest concentrations are in Acid Canyon, with a general downstream decline to the junction with the Rio Grande. The mean concentrations in the DP and upper Los Alamos Canyon system are low in the uppermost portions because these reaches are above the injections of plutonium from DP Canyon.

This chapter provides a backdrop for consideration of the movement of plutonium through the system. This complicated issue occupies much of the remainder of this book. The most important lesson of mean values is that although they offer a valuable insight into the relative roles of various environmental compartments and geographical locations, the temporal and spatial variability of the concentrations must temper any detailed views of the system. Chapter 8 attempts to assess, from a geographical perspective, the annual movement of plutonium through the Northern Rio Grande.

8

Annual Plutonium Budget for the Rio Grande

Sources of Data

A mean annual plutonium budget for the Northern Rio Grande provides an accounting of the amounts of plutonium moving into and out of various reaches of the river during a typical year. Such a budget is a basis for assessing the rates of plutonium transport and the location of storage along the river. The budget presented in the following pages is for bedload and suspended sediments. It does not include plutonium in water because water-borne plutonium is such a small portion of the total in the system (as discussed in Chapter 7). The budget as calculated here requires data concerning sediment and plutonium concentrations in the sediment. The sediment discharge data that are available from U.S. Geological Survey gaging sites (Chapter 4) define the overall framework for budget construction. A reasonably detailed picture is possible for the river system from the Rio Grande at Embudo and the Rio Chama at Chamita southward to the Rio Grande at San Marcial (for locations, see Figure 3.9) where the river empties into Elephant Butte Reservoir.

Data collected by Los Alamos National Laboratory and published in the annual surveillance reports by the laboratory's Environmental Studies Group and later by the Environmental Surveillance Group provide plutonium concentrations for bedload and suspended sediments. The calculations for each site in this study used mean values of plutonium concentrations from all measurements at or near the site. Table 8.1 reviews the sources of plutonium concentration data for each of the sediment-gaging sites in the regional budget calculations. Unfortunately, the sites for collecting the plutonium data were not always colocated with the gaging sites that produced the sediment discharge data (Figure 8.1). In addition, most of the plutonium concentration data are for bedload sediments because of the manner in which the workers collected samples. In some cases, the best estimates of plutonium concentrations in suspended load for gaging sites are from concentrations found in sediments of the nearest reservoir downstream because those sediments are likely to have been in suspension before their emplacement on reservoir floors. The assumption that the mean concentration is a useful representative value seems reasonable given that in those reaches with relatively large amounts of data, concentration values do not show temporal or geographic trends.

There are no direct plutonium concentration data for sediments in the Rio Puerco and Rio Salado. The budget calculations in this study assumed that in the absence of more reliable estimates, the data from Frijoles Canyon were also representative of sediments in the two southern tributaries. The Frijoles drains a portion of the Jemez Mountains and therefore might be expected to produce excessively high estimates of plutonium concentrations for the other two streams. This overestimation may be balanced by the fact that the Rio Salado and Rio Puerco have more erodible soils that probably shed more of their burden of fallout plutonium than do the soils of the Frijoles system. Whatever shortcomings are imposed by the small number of samples, expensive chemical processing and analysis have always precluded the acquisition of additonal data.

The budget calculations used summary values for total plutonium transport in Los Alamos Canyon (rather than summing the bedload and suspended transport), because only the total values were available.[1] In the case of the Rio Grande at

Table 8.1. Sources of mean values for plutonium concentrations in budget

Site	Bedload	Suspended Load	Total Load
Rio Chama at Chamita	15 values, 1974–1986	5 values, Abiquiu Reservoir	Sum
Rio Grande at Embudo	16 values, 1974–1986	Difference between Chama and Otowi	Sum
Rio Grande at Otowi	16 values, 1974–1986	3 values, 1981–1987	Sum
Los Alamos Canyon	Not used in calculations	Not used in calculations	*
Jemez River below dam	15 values at Jemez Pueblo, 1974–1986	None	Sum
Rio Grande at Albuquerque	14 values at Bernalillo, 1974–1986	5 values, Cochiti Reservoir	Sum
Rio Puerco near Bernardo	3 values, Frijoles Canyon	3 values, Frijoles Canyon	Sum
Rio Salado near San Acacia	3 values, Frijoles Canyon	3 values, Frijoles Canyon	Sum
Rio Grande at San Marcial	14 values at Bernalillo, 1974–1986	5 values, Cochiti Reservoir	Sum

Sources: All data from Los Alamos National Laboratory Surveillance Reports, except *, which is from unpublished data from L. J. Lane in support of L. J. Lane, W. D. Purtymun, and N.M. Becker, *New Estimating Procedures for Surface Runoff, Sediment Yield, and Contaminant Transport in Los Alamos County, New Mexico*, Los Alamos National Laboratory Report LA-10335-MS, UC-11 (Los Alamos, N.M.: Alamos National Laboratory, 1985).

Figure 8.1. The State Line Bridge on the Rio Grande near the boundary between New Mexico and Colorado is the site of a U.S. Geological Survey stream gage that records water and sediment exiting the San Luis Valley and flowing downstream into New Mexico. (*a*) Looking north on the day of the dedication of the bridge in 1911. (O. T. Davis, photo 1137, Denver Public Library, Western History Collection) (*b*) The same view in 1989 when the bridge is no longer in use: few changes are evident in the banks and channel of the river when the lower water level in the later view is taken into account. The positions of the boulders at the lower edge of the view remained unchanged over the 78-year period. (W. L. Graf, photo 76-1)

Embudo, plutonium concentration data for the bedload sediments are available, but not for suspended sediments. Therefore, the total budget for the Rio Grande at Otowi minus the total budget for the Rio Chama at Chamita provided the estimate for the Rio Grande at Embudo.

Methods of Calculation

The basic approach to calculating the plutonium budget in this work was to determine the mean annual flux (expressed as a rate of movement) of all plutonium (238, 239, and 240 isotopes) at each of nine sites in the Northern Rio Grande. The total mass of annual sediment discharge (bedload and suspended) times the plutonium concentration in each sediment type for plutonium 238 and plutonium 239 and 240 produced the data for total flux at each site. Algebraically,

$$\frac{\partial(Pu)}{\partial(t)} = C_b^{236}\frac{\partial(S_b)}{\partial(t)} + C_s^{238}\frac{\partial(S_s)}{\partial(t)} + C_b^{238,240}\frac{\partial(S_b)}{\partial(t)} + C_s^{238,240}\frac{\partial(S_s)}{\partial(t)}$$

where

$\partial(Pu)/\partial(t)$ = total plutonium flux per unit time (Ci/yr)

C_{b}^{238} = mean concentration of plutonium 238 in bedload (Ci/Mg)

$\partial(S_b)\partial(t)$ = mean bedload sediment discharge (Mg/yr)

C_{s}^{238} = mean concentration of plutonium 238 in suspended load (Ci/Mg)

$\partial(S_s)/\partial(t)$ = mean suspended sediment discharge (Mg/yr)

$C_{\mathrm{b}}^{239,240}$ = mean bedload concentration of plutonium 239 and 240 (Ci/Mg)

$C_{\mathrm{b}}^{239,240}$ = mean concentration of plutonium 239 and 240 in suspended sediment (Ci/Mg)

The budget function for each of the sites in the river system given in Table 8.1 could be solved for two scenarios: a general best estimate budget and a specific budget for the time period 1970–1979. The general best estimate or composite budget used the mean sediment discharge values from the entire record at each of the sites. The length of these records and their dates of initiation and cessation varied from one site to another, so that the resulting general budget represents a statistical composite and does not reflect the values for any particular year. On a several-decade basis, however, this budget is likely to be the most accurate, given the nature of the available data.

A second set of budget calculations using the sediment discharge data for the 1970–1979 period is useful for comparison with the best estimate budget as an indication of variability and as a view of one representative time period. The decade of the 1970s was also particularly useful for sediment discharge data because it is the 10-year period for which the most data are available—only the

Jemez River below Jemez Dam lacked data for this period, with all other stations providing data for the majority of years. In addition, the plutonium data included values from the mid- and late 1970s, obviating the need to assume that mean values have not changed over the time period of this analysis.

Magnitude of Error

The annual plutonium budgets for the Northern Rio Grande calculated in this book are only first approximations. Although the calculations use the best available data from an environment that is rich in data in comparison with most other areas, there may be substantial errors in the results. The sources of error include measurement, sampling, and estimation errors. Measurement errors occur when laboratory analysts measure plutonium concentrations in sediment and when field workers measure the sediment concentrations in river flows. Information published in the annual surveillance reports of Los Alamos National Laboratory indicates that one standard deviation for the measurements of plutonium concentrations in sediments is usually in the range of 0.0002 to 0.002 pCi/g for plutonium 238 and 0.0002 to 0.01 pCi/g for plutonium 239 and 240. The larger mean values are associated with larger standard deviations, and generally the mean and standard deviation are approximately equal.

The measurement errors for suspended sediment are relatively small compared with the measurement errors for plutonium. The reported mean values for suspended sediment include errors resulting from perturbations in the water flow caused by the sampling device and errors in weighing the samples. The measurement errors for suspended sediment are probably standard deviations equal to about 0.1 times the mean value.

The estimation errors for bedload are relatively large compared with the other sources of errors. Direct bedload measurements are not available for the system except on a limited experimental basis, and so it is necessary to estimate them from limited amounts of research, as mentioned earlier. Uncertainties associated with bedload data result in errors that are standard deviations equal to the means.[2] Because the quality of the data for the Northern Rio Grande is superior to that for most other environments, a reasonable estimate is that the one standard deviation error is probably 0.5 times the mean value.

Sampling errors result from the method of using a small sample to represent a much larger population. Because the samples of sediment taken for plutonium analysis or for the calculation of sediment discharge represent data from only a few years of a much larger and more variable population of yearly data, sampling error must be considered when assessing the usefulness of the final estimates for the total annual plutonium budget. The statistical measure of this error is the "standard error," defined by the standard deviation divided by the square root of the sample size. For most of the stations in our calculations, the standard error is less than 0.5 times the mean value. For the sediment discharges, the standard error is about 0.18 times the mean value for all the stations taken together. The sediment discharges for Los Alamos Canyon with a value of 0.28 times the mean and the Rio Salado with a value of 0.30 times the mean have high standard errors because of the variability of their processes and, in the case of the Rio Salado, a short record.

The sum of these errors indicates that the composite annual plutonium budget probably has a standard deviation of about half an order of magnitude. The error in the 10-year budget for 1970–1979 is probably somewhat smaller because the data for the restricted time period are less variable. The budgets produced from the currently available data cannot therefore be used to predict precisely the amounts of plutonium in a particular place at a particular time. They can show the geographic variation in regional transport and storage processes, and they can provide a context for analyzing one particular source in relation to all other sources.

Annual Plutonium Budgets

The composite annual plutonium budget represents a view of the system constructed with grand general mean values for plutonium concentrations and sediment discharges. It represents a picture made with the most stable mean values from the available data and does not represent any one particular year. It is a useful general guide to the nature of the system from the mid-1940s to the mid-1980s. Figure 8.2 shows the composite annual budget in graphic form, with the width of the flow arrows related to the magnitude of the plutonium's movement. Table 8.2 reviews the basic plutonium data for the calculation and provides more detailed numbers for the budget.

The budget shows that generally, fallout plutonium enters the Northern Rio Grande sediment system in relatively large amounts from the Rio Grande as it flows past Embudo and from the Rio Chama. The Rio Chama carries more sediment (Figures 4.4 and 4.5), but the Rio Grande carries more plutonium because it occurs there in higher concentrations. Because reservoirs on the Rio Chama trap most of the sediment and fallout plutonium from upper elevations in the watershed, the sediments in the lower Rio Chama are derived from low-elevation landscapes that received less fallout than did the higher terrain that drains into the Rio Grande.

The Rio Puerco empties huge quantities of sediment into the Rio Grande system, but that sediment is likely to contain only low concentrations of fallout plutonium because of the relatively low elevation and the dry landscapes that the tributary drains. The resulting influx of plutonium is nonetheless large with respect to the amount entering from other tributaries because so much sediment is present in the Rio Puerco. The Jemez River and the Rio Salado contribute mostly bedload materials, which have low concentrations of plutonium, and so they contribute relatively small quantities of the element.

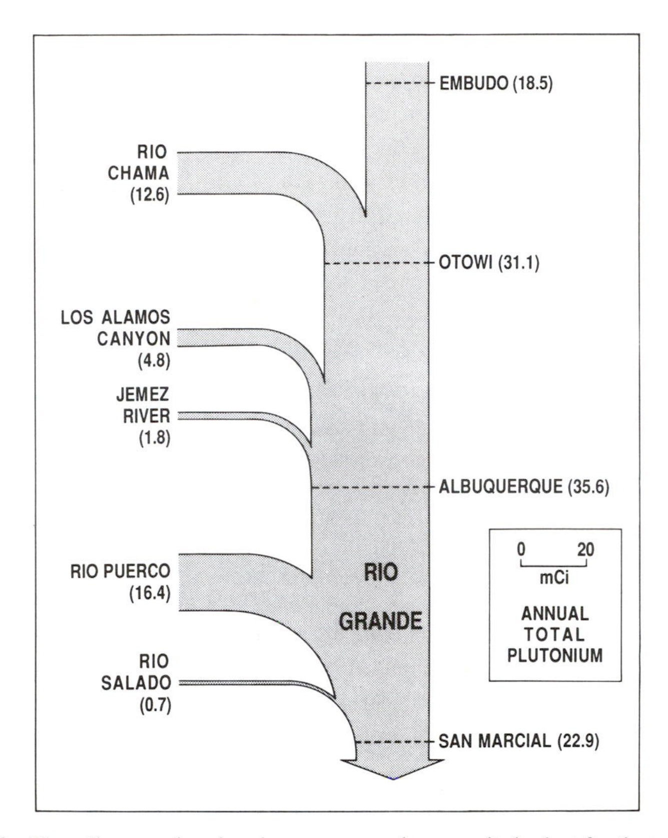

Figure 8.2. Flow diagram showing the mean annual composite budget for the total amount of plutonium in the Northern Rio Grande.

Los Alamos Canyon contributes only tiny amounts of sediment to the total system, but because the concentrations of plutonium are higher in its sediment than in sediments elsewhere in the system, the influx from the tributary is proportionally larger than its sediment alone might suggest. But the amount is still small in comparison with the amount coming down the Rio Grande. On a mean annual basis, the amount contributed by Los Alamos Canyon, mostly from the operations of Los Alamos National Laboratory, is about 15 percent of the amount contributed by fallout plutonium from the Upper Rio Grande and the Rio Chama. Thus if the laboratory did not exist, the amount of plutonium in flux and storage would be still be about 85 percent of the present total inventory in the vicinity of Otowi.

The composite budget indicates the general location of plutonium storage in the system. A small amount, about 2.1 mCi per year, was stored along the river between Otowi and Albuquerque. After 1973, Cochiti Reservoir stored most of this amount, but before that year storage was likely to have been more widespread.

Table 8.2. Budget calculations for annual plutonium budget

Site	In Bedload				In Suspended Load				Total Load		
	Pu 238		Pu 239 and 240		Pu 238		Pu 239 and 240		Pu 238	Pu 239 and 240	Total Pu
	Conc. (pCi/g)	Mass (mCi)	Conc. (pCi/g)	Mass (mCi)	Conc. (pCi/g)	Mass (mCi)	Conc. (pCi/g)	Mass (mCi)	(mCi)	(mCi)	(mCi)
Rio Chama at Chamita	0.000133	0.029	0.001533	0.337	0.000500	0.686	0.008400	11.534	0.715	11.871	12.586
Rio Grande at Embudo	–	–	–	–	–	–	–	–	2.733	15.802	18.535
Rio Grande at Otowi	0.000357	0.112	0.005000	1.570	0.001700	3.336	0.013300	26.103	3.448	27.673	31.121
Los Alamos Canyon	–	–	–	–	–	–	–	–	–	–	4.788
Jemez River	0.000467	0.204	0.00400	1.749	–	–	–	–	0.000	1.749	1.749
Rio Grande at Albuquerque	0.000000	0.000	0.005000	1.406	0.001100	1.932	0.018300	32.157	1.932	32.157	34.089
Rio Puerco near Bernardo	0.000000	0.000	0.002500	4.756	0.000000	0.000	0.002500	11.600	0.000	16.356	16.356
Rio Salado near San Acacia	0.000000	0.000	0.002500	0.323	0.000000	0.000	0.002500	0.323	0.000	0.647	0.647
Rio Grande at San Marcial	0.000357	0.203	0.005000	2.889	0.002500	1.290	0.017400	62.853	1.496	21.385	22.881

The system stored a large amount of plutonium, about 29.8 mCi per year, between Albuquerque and San Marcial on flood plains and in abandoned channels. Less than half of the plutonium that entered the Albuquerque–San Marcial reach exited the reach to Elephant Butte Reservoir. For the entire system, fluvial deposition stored more plutonium internally (about 32 mCi per year) than left the system and was stored in Elephant Butte Reservoir (about 23 mCi per year).

For this study, the downstream end of the Northern Rio Grande system is Elephant Butte Reservoir, where the river deposits its sediment and associated plutonium. Very little of either passes Elephant Butte Dam to continue the journey down the Rio Grande. This arrangement explains the extremely low sediment and plutonium yields from the Rio Grande system indicated in Foster and Hakonson's calculations.[3]

A comparison of the composite annual budget using all the available data (representing mostly the period 1948–1985 because of the nature of the sediment records) with the 1970–1979 mean annual budget shows the temporal variability of the annual plutonium budget (Table 8.3). During the 1970s, the Rio Chama annually contributed only about one-third as much plutonium as it did annually during the longer period. The sediment discharge of the Rio Chama during the 1970s was relatively low because of the closure of Abiquiu Dam in 1963 and because of relatively low water yields and small annual floods. As a result, the plutonium flux was also relatively low. Los Alamos Canyon also discharged less water because of climatic conditions, and it therefore transported relatively small quantities of plutonium to the Rio Grande during the 1970s.

The processes on the main stream were the reverse of these tributary processes. Sediment discharges on the Rio Grande at Albuquerque and San Marcial were large because of channel adjustments resulting from the erosion downstream

Table 8.3. Comparison between the composite budget and the 1970–1979 budget

Site	Composite Budget (mCi/jr)	1970–1979 (mCi/yr)
Rio Chama at Chamita	12.586	4.257
Rio Grande at Embudo	18.535	24.802
Rio Grande at Otowi	31.121	29.059
Los Alamos Canyon	4.800	1.702
Jemez River below Jemez Dam	1.749	–
Rio Grande at Albuquerque	34.084	50.546
Rio Puerco near Bernardo	16.356	14.169
Rio Salado near San Acacia	0.647	1.111
Rio Grande at San Marcial	22.881	79.968
Total inputs	54.661	46.041
Storage (+) and output (−)		
Otowi–Albuquerque	3.574	−19.785
Albuquerque–San Marcial	28.206	−14.142
Total system	31.780	−33.927

from Cochiti Dam (closed in 1973) and flash floods on smaller tributaries. The construction of levees and pilot channels near San Marcial contributed additional inputs. The result was a 3.5-fold increase in plutonium flux on the Rio Grande at San Marcial.

Because internal erosion dominated the 1970–1979 processes, there was a net loss of plutonium stored in the river system during the decade, and more was flushed into Elephant Butte Reservoir than was stored along the channel. About 46 mCi of plutonium per year entered the river system, but 79 mCi per year exited to Elephant Butte Reservoir. This arrangement was not typical of the entire 1948–1985 period, which was mostly a period of river deposition, but it illustrates the variability of the internal processes.

The Relationship Between Laboratory and Fallout Contributions

The composite budget provides a useful framework for assessing of the relative inputs of plutonium from fallout sources and from Los Alamos. The data in the composite budget are empirical and are specific to the Northern Rio Grande rather than being general predictions based on assumptions about the latitudinal distribution of fallout. Despite the limitations imposed by measurement, sampling, and estimation errors, the composite budget is likely to be more accurate than others previously published, because it depends on local empirical data rather than on general estimates from global approximations.

The best estimate of inputs to the system during the 1948–1985 period (Figure 8.2 and Table 8.2) indicates that fallout sources introduced about 55 mCi of plutonium per year and that Los Alamos National Laboratory introduced through Los Alamos Canyon about 4.8 mCi per year. During the 1970–1979 period, fallout produced about 46 mCi per year, and Los Alamos, about 1.7 mCi per year. In any case, the contribution of Los Alamos to the annual plutonium flux in the entire Northern Rio Grande system (that is, the portion above Elephant Butte Reservoir) is relatively small, accounting for about 9 percent of the total. Fallout occurs in very small concentrations, but because relatively large amounts erode from the landscape and move through the river system, it accounts for more than 90 percent of all the plutonium in the river sediments. In the more restricted vicinity of Otowi (an area not including inputs from the Rio Puerco and the Rio Salado), the inputs from Los Alamos accounted for about 15 percent of the plutonium in the composite annual budget and about 6 percent during the 1970s.

The significance of these annual fluxes is that large amounts of plutonium remain on the landscape of the general system and in Los Alamos Canyon. If the general estimate by Harley and others is correct for the mean fallout of plutonium for the latitudinal belt of the Northern Rio Grande (1.8 mCi/sq km),[4] then a total of about 66,800 mCi have entered the landscape of the general basin above Otowi. Assuming a middle-range estimate of 1,500 mCi for the total plutonium loading of Los Alamos Canyon,[5] the Los Alamos contribution to the total plutonium inventory of the basin upstream from the confluence of the Rio Grande and Los Alamos Canyon at Otowi is about 2.2 percent. The calculation that I have used in this study suggests that the discharges of plutonium from Los Alamos Canyon have evacuated only about 10 percent of the total stored in the canyon, so that 90 percent of the amount released at Los Alamos is still in the tributary canyon. In the general basin above Otowi, erosion and fluvial transport have removed less than 2 percent of the fallout plutonium inventory. At the current rates of flux, as estimated in the composite annual budget, more than 2,100 years will be required to remove the fallout plutonium stored upstream from Otowi.

Particular Cases of Plutonium from Los Alamos

Plutonium from the fallout and from Los Alamos National Laboratory is not equally distributed in either time or space. Fallout plutonium is mostly associated with fine-grained materials and suspended sediment, whereas the inputs from Los Alamos are mostly associated with coarse particles and bedload. These differences imply that in a restricted time period, fallout plutonium moves more readily over long distances in the river system, and plutonium from Los Alamos moves over shorter distances. Also, fallout plutonium enters the system more gradually, and the Los Alamos plutonium enters the main stream only sporadically as a result of infrequent flash floods in Los Alamos Canyon. During many years, no plutonium from Los Alamos enters the system. For these reasons, an analysis of Los Alamos National Laboratory's contribution requires us to focus on bedload budget processes for the few selected years when the contributions from Los Alamos Canyon were relatively large: 1951, 1952, 1957, and 1968.

The calculations for the bedload plutonium budget in the Northern Rio Grande near Los Alamos for these particular years required sediment discharge data from the U.S. Geological Survey's gaging stations at Chamita on the Rio Chama and at Otowi on the Rio Grande, unpublished data on plutonium discharges from Los Alamos Canyon by L. J. Lane,[6] and plutonium concentration data published in surveillance reports by Los Alamos National Laboratory. There were two important assumptions in these calculations. The first was that the plutonium from Los Alamos Canyon as calculated by Lane was mostly contained in bedload. This assumption seems reasonable because the original contamination was in coarse alluvium, the plutonium in transit in the late 1940s was in coarse materials along the stream channels, as shown in the photographs taken at that time,[7] and subsequent findings were that most of the contamination was associated with coarse particles.[8]

The second assumption was that the mean plutonium concentration data regarding the bedload in the main stream were consistent over time. Because there are no identifiable trends in the concentration data collected at any of the stations in the data set, this assumption seems safe for the mid-1950s and thereafter when the amount of plutonium available from fallout was substantial. It is likely that there was less plutonium from fallout in the system for the 1951 and 1952 calculations than this assumption implies, because there had been relatively few atmospheric tests at that time. Therefore, for 1951 and 1952 the amounts of fallout plutonium calculated as entering the system through the Rio Chama and Rio Grande are maximum estimates. The magnitudes of the numbers are relatively small in relation to the contributions from Los Alamos, however, because the main stream carried small amounts of sediment and bedload during those two years. The errors are also probably small for budget considerations.

The annual budgets for plutonium in bedload for 1951, 1952, 1957, and 1968 for the Rio Grande in the vicinity of Los Alamos show the overwhelming importance of inputs from Los Alamos Canyon in bedload (Figure 8.3). The inputs from Los Alamos were clearly larger for these four years than the mean condition, but the inputs from fallout were also larger than average. The increase in fallout contributions was not as great as that in the more variable contributions

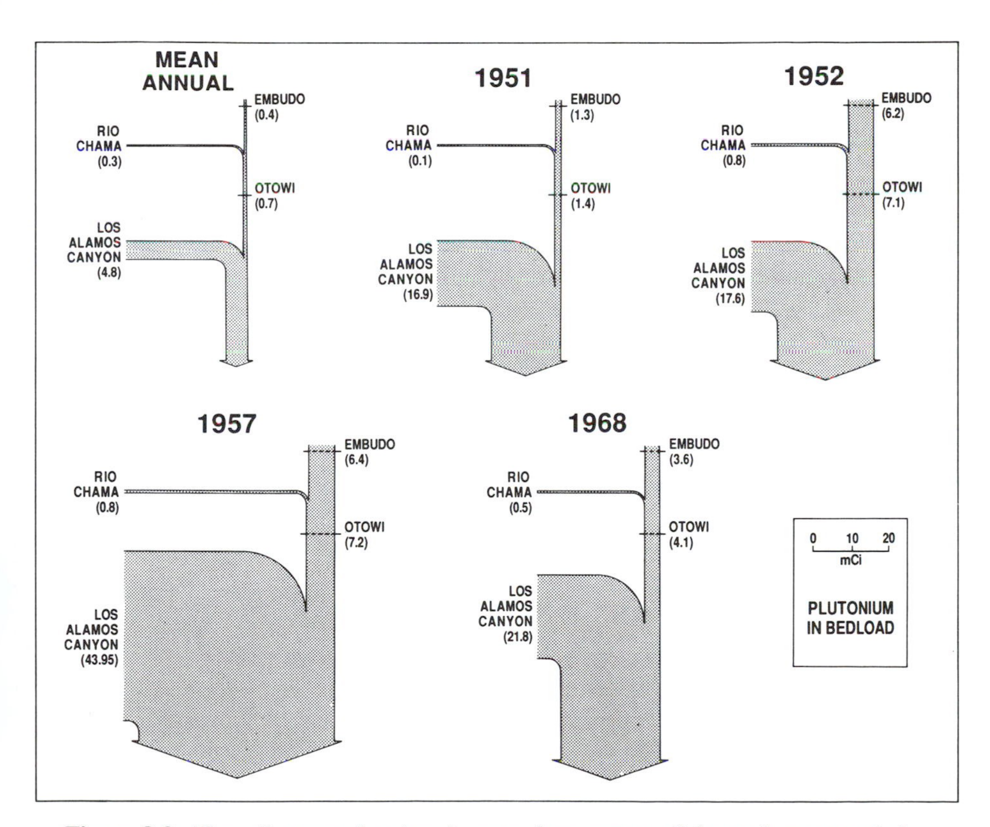

Figure 8.3. Flow diagrams showing the annual movement of the total amount of plutonium in bedload sediments in the Northern Rio Grande in the vicinity of Los Alamos.

from Los Alamos Canyon. The result was that the role of the Los Alamos contributions was exaggerated and that 71 percent (in 1952) to 86 percent (in 1957) of the bedload plutonium below Otowi was from Los Alamos.

For the four years of interest, water yield and annual floods on the Rio Grande were not especially large, and so flash-flood sediments (coarse bedload particles) from Los Alamos Canyon entering the main river were not likely to have traveled far downstream. Evidence developed in Chapter 9 shows that these materials probably contributed to deposits immediately below the confluence of Los Alamos Canyon and the Rio Grande and to accumulations at Buckman, about 5 km downstream. The configuration of the Rio Grande downstream from Otowi during these four years was a channel constricted by the walls of White Rock Canyon, with wide portions suited for deposition only at Buckman, a few pockets in White Rock Canyon, and the flood plains near Cochiti Pueblo and Peña Blanca. It seems unlikely that the sediments would have traveled farther downstream than the Peña Blanca reach, but further research using hydraulic simulations could test this hypothesis.

Given these considerations, along with the fact that the general river system was aggrading (storing sediment) throughout the 1950s and 1960s, the 1951, 1952, 1957, and 1968 contributions of plutonium-bearing sediments from Los Alamos are likely to have formed deposits along the channel in flood-plain deposits, channel fills, and bars. It is possible to estimate the probable concentrations of plutonium in the deposits by combining the sediment discharge data for the Rio Grande developed in this book with the data on plutonium inputs from Los Alamos Canyon developed by Lane and others.[9] The estimates given in Table 8.4 indicate that the concentrations of total plutonium (238, 239, and 240) in flood-plain and channel deposits are probably between about 0.01 and 0.1 pCi/g, values

Table 8.4. Estimated plutonium concentrations in bedload deposits for specific years

	Bedload			Plutonium			
Year	Otowi (Mg)	Los Alamos (Mg)	Combined Total (Mg)	Otowi (mCi)	Los Alamos (mCi)	Combined Total (mCi)	Concentration (pCi/g)
1951	130,744	8,077	138,821	1.42	16.90	18.32	0.13197
1952	649,323	5,730	655,053	7.06	17.61	24.67	0.03766
1957	661,508	14,942	676,450	7.19	43.95	51.14	0.07560
1968	373,576	12,809	386,385	4.06	21.82	25.88	0.06698
Total mean	2,163,396	2,241	2,165,637	1.57	4.79	6.36	0.00293
Active bedload, 1974–1986*	–	–	–	–	–	–	0.01056

*Data from W. D. Purtymun, R. J. Peters, T. E. Buhl, W. N. Maes, and F. H. Brown, *Background Concentrations of Radionuclides in Soils and River Sediments in Northern New Mexico, 1974–1986*, Los Alamos National Laboratory Report LA-11134-MS, UC-11 (Los Alamos, N.M.: Los Alamos National Laboratory, 1987).

confirmed by analyses of samples recently taken from the deposits (Table 8.4). The deposits from different years may reasonably be expected to exhibit different concentrations, depending largely on the magnitude of inputs from Los Alamos Canyon.

The bedload sediments moving through the Rio Grande during the 1970s and 1980s in the vicinity of Otowi had mean plutonium concentrations of about 0.01 pCi/g, a value at the lower end of the range of predicted values for the deposits. The concentration values are an order of magnitude higher in the flood-plain deposits near Buckman, probably representing levels in the bedload sediments in the past, but they are not high enough to be hazardous. The United States does not have standards for sediment quality with regard to plutonium, but experience on the Puerco River in northwestern New Mexico (not to be confused with the Rio Puerco, a tributary of the Rio Grande) provides a value for comparison. The Environmental Protection Agency recommended a safe maximum limit of 30 pCi/g for thorium 230 after an accidental spill of contaminated sediments into the Puerco River.[10] This limit is three to four times greater than values associated with any data for plutonium in the Rio Grande system outside the boundaries of Los Alamos National Laboratory. The contributions from the laboratory are therefore probably not hazardous given our current understanding of the element, but they are in recognizable amounts.

In summary, the development of a regional plutonium budget for the Northern Rio Grande produces the following generalizations:

1. Plutonium enters the system from fallout and from Los Alamos National Laboratory.
2. In the best estimate of processes for the 1948–1985 period, about half the plutonium that entered the system was stored along the river. The remainder moved into storage in Elephant Butte Reservoir.
3. In the total budget, fallout accounts for more than 90 percent of the plutonium in the system, and Los Alamos National Laboratory accounts for slightly less than 10 percent.
4. The contribution of Los Alamos National Laboratory is mostly in bedload sediments, in which for four particular years the contribution was dominant.
5. Most of Los Alamos National Laboratory's contributions remain in storage along the river between Otowi and Peña Blanca; since 1973, their downstream progress has ended in Cochiti Reservoir.

9

Sediment and Plutonium Storage Upstream from Cochiti

Representative Reaches

The foregoing chapters demonstrated that large amounts of sediment and much of the plutonium entering the Northern Rio Grande have been stored along the river channel. A composite budget analysis gives the quantities of materials involved annually, but except in very broad terms it does not describe where the materials are stored. It is a matter of scale: The budget indicates the overall quantities of sediment and plutonium stored in the system but does not reveal on a local scale where one might search for the materials. The next chapters show that the storage process has particular geographic characteristics and that in representative reaches it is possible to map those sediments that were deposited during the years of maximum input of plutonium into the system. These critical deposits are likely to contain more plutonium than are similar deposits of other years. In this way, the evidence of environmental change along the river provides a guide for determining the fate of plutonium in the system (Figure 9.1).

A sampling program for assessing the storage of plutonium along the Northern Rio Grande depends on the development of the connections among vegetation communities, fluvial landforms, sedimentary deposits, and plutonium contents. Although it is not possible here to map and interpret completely the entire 313 km of river from Española to San Marcial, limited reaches can serve as representatives of larger portions of the whole. Eleven representative reaches, each about 3 to 6 km long, provide information on the entire study area because each representative reach exemplifies the conditions that obtain over a much larger portion of the total length of the river. My selection of the representative reaches began by reviewing the entire river by aerial photography and then directly in the field. The river divides itself into sections based on the geomorphologic conditions as modified by engineering works. Each representative reach illustrates the

Figure 9.1. Considerable changes have occurred in the channel of the Rio Grande between Otowi Bridge and Black Mesa near San Idelfonso Pueblo. (*a*) Looking north upstream in 1937, showing numerous unstable sandbars and a braided channel. (F. Broeske, photo 120125, Museum of New Mexico) (*b*) The view in 1991 in the same direction but from a slightly higher perspective, showing a single, narrower, and more stable channel than in the earlier view. (W. L. Graf, photo 104-13)

conditions within one larger section. For example, the Frijoles representative reach is similar to other relatively short reaches throughout White Rock Canyon. The 11 representative reaches with the geographic coordinates of their approximate center points are as follows (for general locations shown with corresponding numbers, see Figure 9.2):

1. *Santa Clara* (35° 54′ N, 106° 06′ W). A control reach for comparison with those farther downstream, Santa Clara is located between Española and a channel constriction caused by basalt flows. The Rio Grande flows through a broad valley unaffected by dam construction at Cochiti and by processes in Los Alamos Canyon.
2. *Otowi Bridge* (35° 52′ N, 106° 08′ W). This reach, the origin of contaminants from Los Alamos National Laboratory in the Northern Rio Grande system, is immediately upstream from Otowi Bridge and the confluence of Los Alamos Canyon with the Rio Grande.
3. *Buckman* (35° 50′ N, 106° 09′ W). This deposition area in White Rock Canyon is about 2 km downstream from Otowi and is a representative canyon site where some sediments are stored.
4. *Frijoles Canyon* (35° 45′ N, 106° 15′ W). Frijoles Canyon, the headwaters area of Cochiti Reservoir at the edge of Bandelier National Monument, is a representative canyon site where little sediment is stored except for reservoir backwater deposition.
5. *Peña Blanca* (35° 35′ N, 106° 20′ W). This reach from near Peña Blanca to Gallisteo Creek, is representative of unstable reaches downstream from Cochiti Dam in a broad alluvial valley without extensive urban development.
6. *Coronado* (35° 20′ N, 106° 32′ W). Coronado is north of Bernalillo at Ranchito, representing a partly confined channel with a levee on one side and strong tributary influences from the Jemez River and smaller streams.
7. *Los Griegos* (35° 17′ N, 106° 36′ W). In northern Albuquerque at the Alameda Bridge, Los Griegos is a channel confined by levees on both sides in an urban area typical of reaches in the Albuquerque area.
8. *Los Lunas* (34° 48′ N, 106° 43′ W). At the New Mexico Route 49 Bridge, Los Lunas is a confined channel with a narrow space between the levees and an abandoned meander.
9. *San Geronimo* (34° 22′ N, 106° 50′ W). About 10 km south of Bernardo, San Geronimo is a complex backswamp zone with the influence of an along-channel railroad embankment, downstream from the confluence with the Rio Puerco and its massive sediment load.

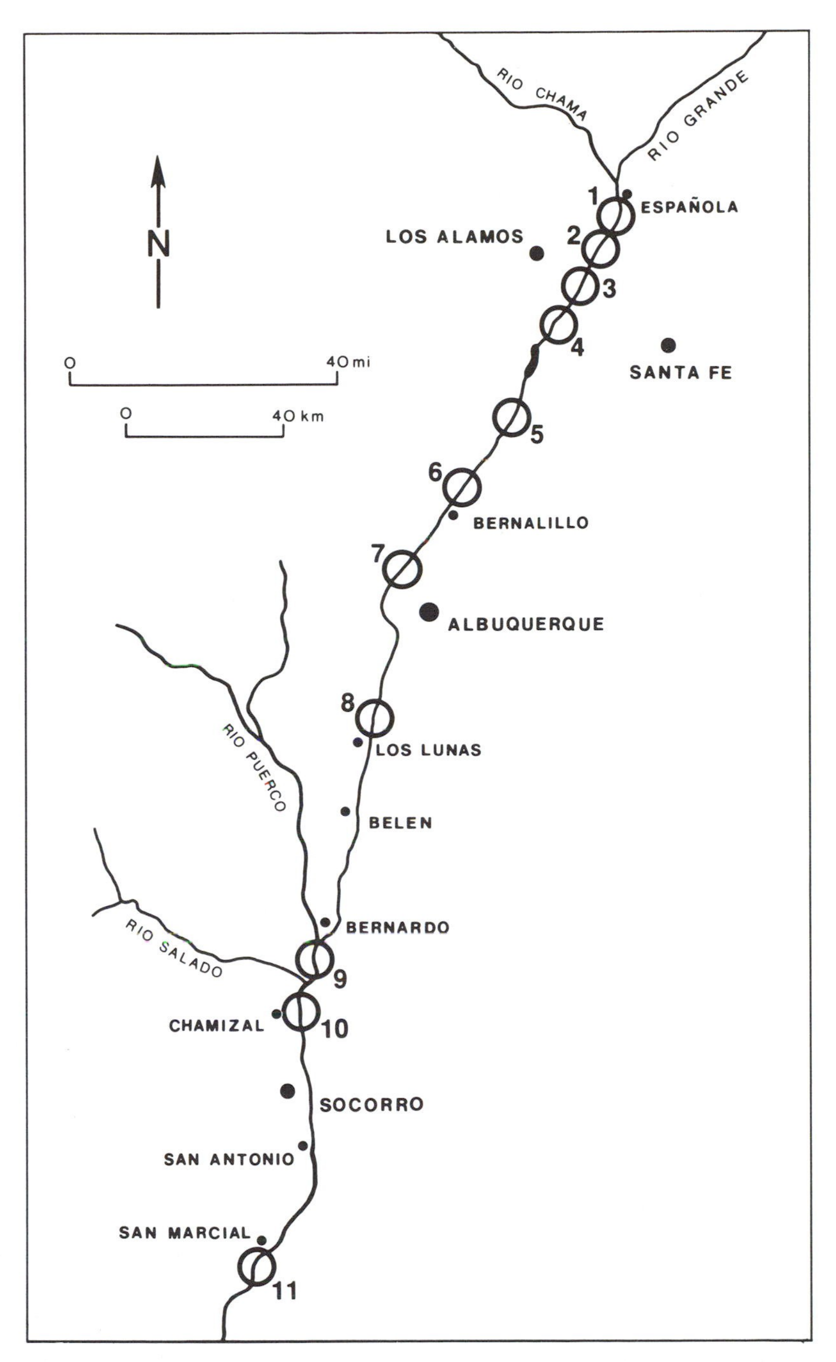

Figure 9.2. Locations of representive reaches along the Northern Rio Grande. See the text for descriptions keyed to numbers on the map: (1) Santa Clara, (2) Otowi, (3) Buckman, (4) Frijoles, (5) Peña Blanca, (6) Coronado, (7) Los Griegos, (8) Los Lunas, (9) San Geronimo, (10) Chamizal, and (11) San Marcial.

10. *Chamizal* (34° 14′ N, 106° 55′ W). Downstream from the confluence with the Rio Salado and San Acacia Diversion Dam, Chamizal is a wide channel area with a pilot channel and a settling basin, in an agricultural area.
11. *San Marcial* (33° 45′ N, 106° 53′ W). San Marcial is in Bosque del Apache, near Black Mesa in the backwater and delta area of Elephant Butte Reservoir, and has dense phreatophytes.

The purpose of examining limited, specific reaches was to identify sites for sample collection, evaluate the changes in the river channels, and aid in the development of environmental monitoring and surveillance programs. All these objectives must take into account the geographic and temporal variability of the fluvial environment where the plutonium is likely to be in transit or stored. Plutonium has been sampled in deposits along the first three reaches, but the plutonium in the deposits of the other reaches has not yet been investigated. Plutonium concentrations in active-bed sediments have been examined in several reaches, but because the time of the sampling did not coincide with any of the critical years, these data show relatively low concentrations reflecting only small amounts of fallout plutonium in transit.

The base maps used for the illustrations are sections of the U.S. Geological Survey's topographic quadrangles (for a complete list, see Appendix F). Often the original topographic map shows conditions during the 1970s or early 1980s (depending on the date of the aerial photography that provided the initial mapping information), but because of the unstable channels and construction, some details of the landscape had changed by the late 1980s. The locations of mid-channel sand bars and channel-side bars on the maps were correct for the year of the photography on which the maps were based, but by the late 1980s, the bar configurations and locations often were different from the forms depicted by the maps.

The vegetation maps are sections from a major ecological survey completed for the U.S. Army Corps of Engineers.[1] The basic data for the vegetation communities came from aerial-photographic interpretation and field surveys by Hink and Ohmart and from my confirmatory field checks. The U.S. Geological Survey's topographic outlines served as base maps for the data on vegetation communities. The maps use the abbreviations for vegetation communities and community structure types that are defined in the original ecological survey and summarized in Table 9.1.[2]

The vegetation maps and numerous coverages of aerial photographs from a variety of dates made it possible to create the geomorphologic and sedimentologic maps that also used the U.S. Geological Survey's topographic outlines as base maps. Extensive field investigations, mapping, and sampling supplemented the vegetation maps for the final interpretations of the geomorphholology and sedimentology. The geomorphology and sedimentology maps represent conditions between 1986 and 1990.

Table 9.1. Vegetation community and community structure types used for mapping representative reaches

Map Symbol	Explanation
Vegetation community type	
C	Cottonwood (*Populus fremontii* var. *wislinzenii*)
RO	Russian olive (*Elaeagnus angustifolia*)
SC	Salt cedar (tamarisk, *Tamarix chinesis*)
CW	Coyote willow (*Salix exigus*)
TW	Tree willow (*Salix goodingii*, *Salix Amygdaloides*)
SE	Siberian elm (*ulmus pumila*)
NMO	New Mexico olive (*Foresteria neomexicana*)
I	Indigo bush (*Amopha fruticosa*)
SW	Seepwillow (*Baccharis salicina*)
J	Juniper (*Juniperus monosperma*)
Wb	Wolfberry (*Lycium andersonii*)
Rb	Rabbit brush (*Chrysothamnus nauseosus*)
Community structure type	
1	Vegetation in all foliage layers, mature trees up to 20-m high; mixed-age forest stands
2	Vegetation mostly in the canopy layer, mature trees up to 20-m high, sparse and patchy understory
3	Vegetation mostly in a dense layer below 10 m; intermediate-age trees, with some growth extenting above 30 m
4	Open stands of intermediate-age trees; most vegetation between 10- and 15-m high; widely spaced shrubs
5	Vegetation increasing in density approaching ground level, most below 20-m high; thick layer of grasses and annuals
6	Low and sparse herbaceaous or shrub vegetation; most foliage below 2-m high

Santa Clara Reach

The Santa Clara reach represents the Northern Rio Grande in an area not affected by the closure of Cochiti Dam or by potential inputs of plutonium from Los Alamos National Laboratory. The Santa Clara reach extends approximately 6 km downstream from the vicinity of the Santa Clara Pueblo to the vicinity of Black Mesa (Figures 9.3 and 9.4). The river flows through a broad alluvial valley that is up to 2 km wide and similar to valleys farther south. During the 1920s and 1930s, the river in the Santa Clara reach was a braided channel with unstable margins and a wide, shallow, sandy channel. Those places where the flow was concentrated had larger particles, of the cobble size. Maps made from 1930s aerial photographs show a braided channel several times wider than the 1990s channel.[3] A 1930s photograph of boys from the Los Alamos Ranch School swimming in the river near Black Mesa shows a flood bar of cobbles along one side of an otherwise sandy channel. The 1941 flood coursed through a broad, shallow channel, but thereafter low flows occupied a series of multiple threads across what once was a wider braided channel. The small single threads were gradually abandoned, either through slight entrenchment of a

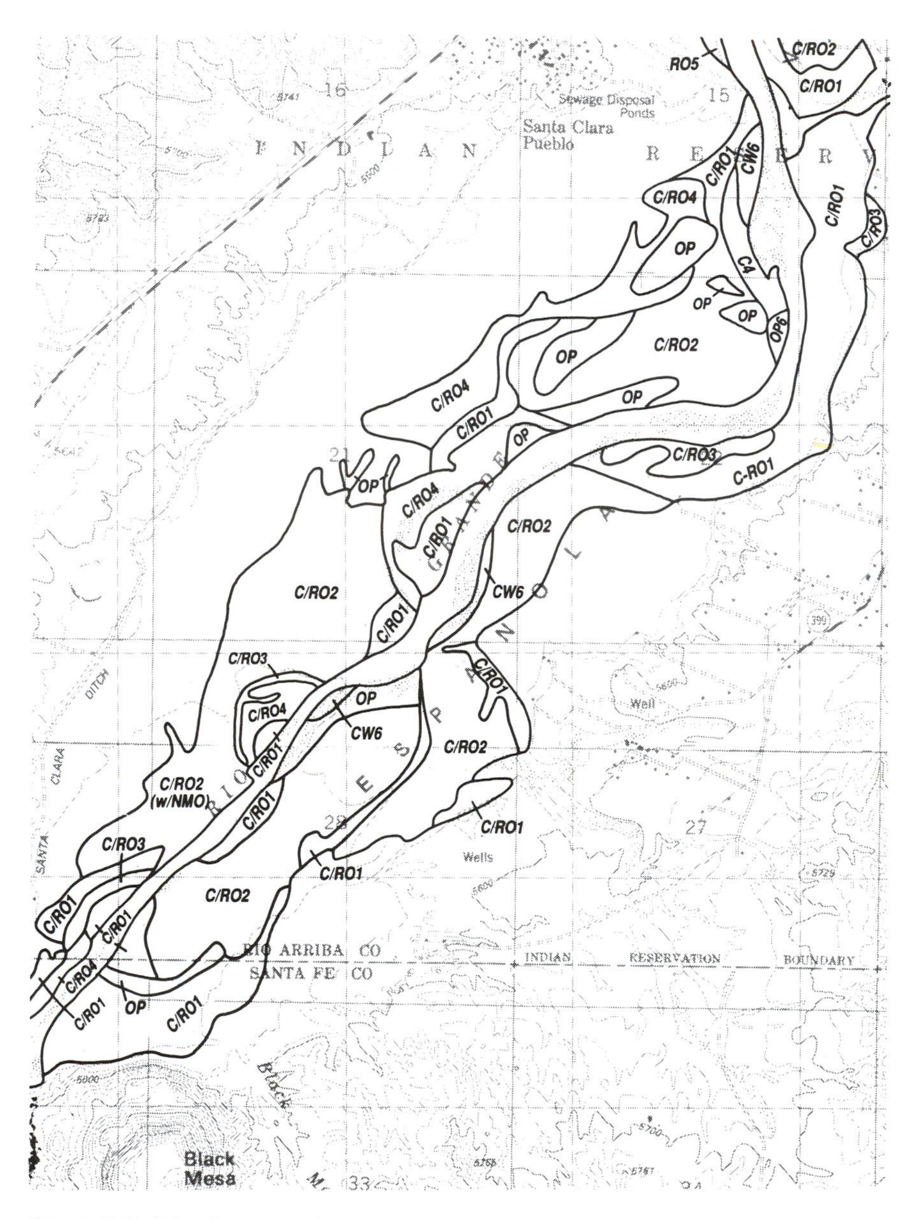

Figure 9.3. Riparian vegetation communities along the Santa Clara representative reach. See number 1 in Figure 9.2 for the location and Table 9.1 for symbols. (Modified from V. C. Hink and R. D. Ohmart, *Middle Rio Grande Biological Survey: Final Report*, U.S. Army Corps of Engineers Contract Report DACW47-81-C-0015 [Albuquerque: U.S. Army Corps of Engineers, 1984])

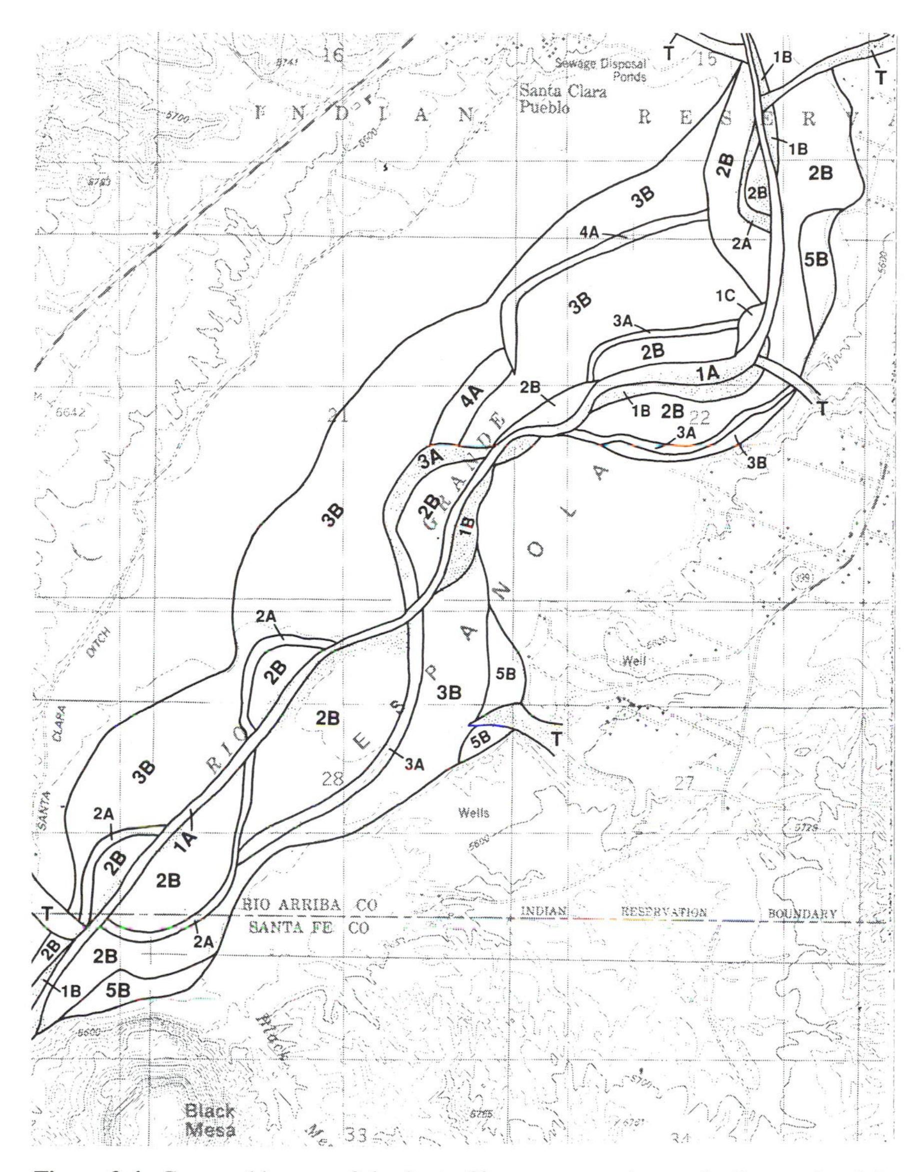

Figure 9.4. Geomorphic map of the Santa Clara representative reach. See number 1 in Figure 9.2 for the location. Labeled areas on the map are (1A) active channel, 1990; (1B) active flood plain, 1990; (1C) abandoned flood plain, active during the 1970s; (2A) abandoned channel, active 1941–1968; (2B) abandoned flood plain, active 1941–1968; (3A) abandoned braided channel, active before 1941; (3B) abandoned flood plain, active before 1941; (4A) abandoned channel, active in the 1930s or earlier, older than the unit mapped as 3A and 3B; (5B) abandoned flood plain, active in the 1920s or earlier; and (T) tributary alluvial fan materials and channel fills.

single channel that eventually would be the dominant one or by filling of the small secondary channels. The 1968 flood followed the dominant channel left from this shrinkage process and defined a single channel with a pattern that was relatively straight compared with previous configurations. After 1968, the changes were minor and represented only slight modifications of the newly established, relatively straight, single channel. Flash floods in tributaries contributed small deposits on top of the flood plains and abandoned channels of the main river.

The boundaries of the vegetation communities in the Santa Clara reach often approximate the locations of the abandoned braided channels of the 1930s, the abandoned threads of the 1941–1968 system, and the flood plains of various episodes. Cottonwood and russian olive dominate most of the reach, but the community structure varies from one place to another, depending on the history of the location. For example, open canopies of cottonwood and russian olive are common on the flood plains that were active before 1941, whereas the same species occur in very dense stands with substantial growth close to the surface in abandoned braided and single thread channels.

The geomorphic map of the Santa Clara representative reach shows that the area contains deposits from several different time periods and of several different types (Figure 9.4). The tributary deposits are coarse sands and gravels, forming fans radiating outward from the tributary mouths. The abandoned braided channels are sandy, and the abandoned single-thread channels are clearly defined in the field as linear depressions in the otherwise relatively flat flood plain. Sometimes the depressions are floored with a veneer of fine-grained materials. Those channels filled with fine materials and abandoned during the 1941–1968 period are likely to contain more fallout plutonium than are other deposits in the reach.

Recently collected samples from the Santa Clara reach illustrate the relationship between geomorphic history and plutonium concentrations in flood-plain sediments. The plutonium concentration in sediment from a flood plain that was active during the 1960s (2B at the top of Figure 9.4), a time of peak fallout loading, was 0.0266 pCi/g. The channel that was active at that time yielded sediments with a concentration of 0.0081 pCi/g (a tributary of 2A at the top of Figure 9.4). Both samples were from a portion of the near-channel environment commonly referred to as "flood plain," but clearly the landform is complex, as shown in Figure 9.4. Both these values exceed the concentration in sediments from the presently active channel nearby, where the value was only 0.0038 pCi/g. All the plutonium in this reach is from fallout derived from erosion upstream and none is from Los Alamos National Laboratory.

Otowi Reach

The Otowi representative reach extends 4 km downstream on the Rio Grande into White Rock Canyon (Figure 9.5) and includes the lower 2 km of Los Alamos

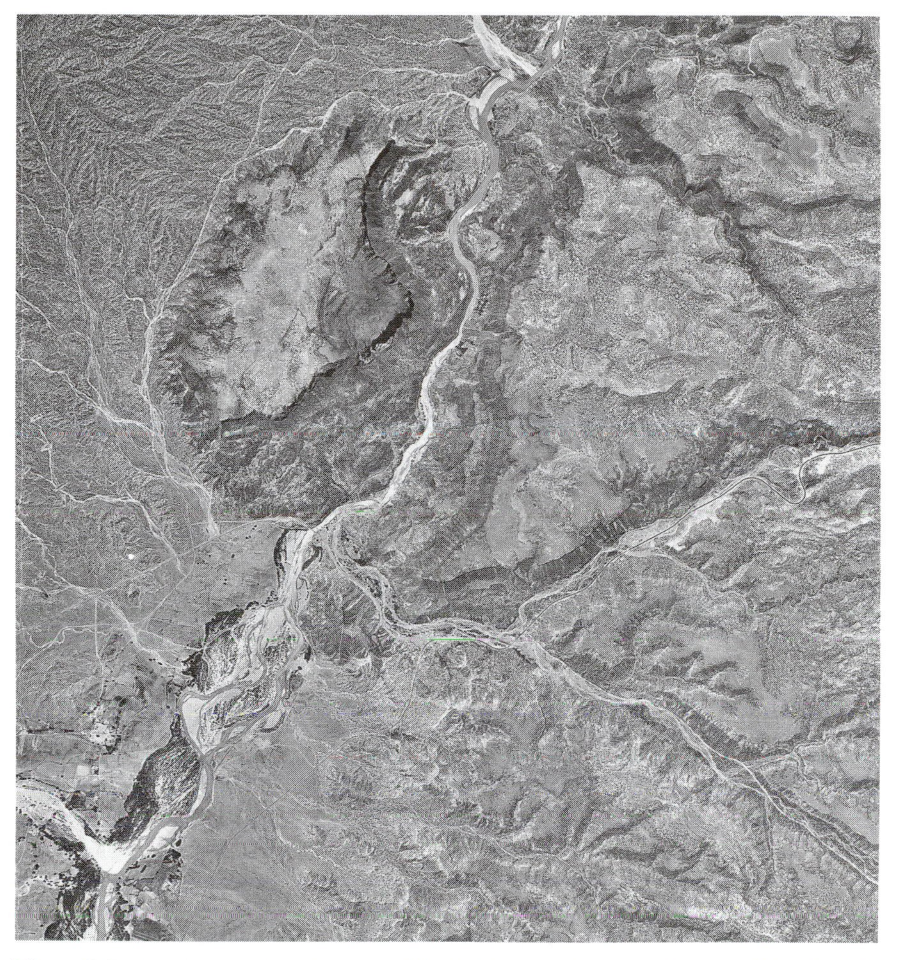

Figure 9.5. Vertical aerial photograph of the Rio Grande from near the Santa Clara Pueblo (upper border of the photo) to Buckman (lower border), a distance of about 10 km. Los Alamos Canyon joins the main stream at Otowi in the center of the view. The dark mass is the collapsing basalt cap of La Mesita. (U.S. Army Map Service photograph, EROS Data Center, photo VV BE M79 AMS 120, frame 11121, May 27, 1954)

Canyon. The reach is significant because it includes the point of introduction of plutonium into the main stream from Los Alamos National Laboratory. The confluence of Los Alamos Canyon and the Rio Grande was unstable during the 1945–1988 period. Historical ground photographs show that during the early and mid-1940s the tributary stream included a distributary channel that entered the main river north of the general location of the bridges. The highway (New Mexico Route 4) passed over a small suspension bridge that spanned the Rio Grande and then crossed the distributary on a water-level crossing. The completion of a larger truss bridge across the Rio Grande in 1947 included filling in the distributary to provide an embankment for the roadway.

The general alignment seen in the late 1980s dates from 1947, although there have been changes in its detail. For example, because the Los Alamos stream had a bend that encroached on the highway, engineering works stabilized the channel by introducing reveted banks and a cleared, artificially formed channel. In 1988 a deck bridge replaced the truss structure, although the original small suspension bridge remains as an unused crossing. These engineering activities, supplemented with levee construction and channelization on the tributary, established and then maintained the current confluence location southwest of the bridge area, although its exact configuration changed during several floods.

The arrangement of the channel of the Rio Grande in this reach changed relatively little between the mid-1940s and the late 1980s (Figures 9.6 and 9.7). Writers give impressionistic descriptions of the early-1940s channel in the vicinity of the house and garden of Edith Warner, who lived immediately north of the bridges' present location.[4] Sandy banks lined with cottonwoods were common. Historical ground photographs indicate that in the 1930s and early 1940s the channel had more gravel and boulders than it did in the late 1980s. There has been some narrowing of the channel with the deposition of sand along the margins. Significant deposition began in the 1930s along the north bank immediately downstream from the confluence with Los Alamos Canyon, and the subsequent colonization of the deposits by cottonwood has resulted in three groups of mature trees (Figure 9.6). Newer sediments deposited during the 1958 and 1967 floods buried the root collars of the trees, which date from the 1930s. Flood sediments from Los Alamos Canyon that were swept partly into the main channel and flood plain provide additional recognizable deposits that have mostly russian olive as a vegetation cover, differentiating them from the older deposits (Figure 9.7). The area is a typical slack-water depositional zone immediately downstream from the alluvial fan created by the Los Alamos stream.

The geomorphic map showing the conditions in the late 1980s suggests that the primary potential storage sites for plutonium from Los Alamos National Laboratory include the flood plain and terrace of Los Alamos Canyon and the slack-water depositional site on the north bank of the Rio Grande immediately downstream from the confluence (T2A, 1B, and 2B on Figure 9.7). These environments are most likely to have received materials from tributary floods. The meandering stream in Los Alamos Canyon has produced a series of abandoned channel courses and flood plains that are recognizable in the field from their vegetation and geomorphological indicators. Sediments along the main channel at the north end of the bridge area are not be likely to yield useful plutonium information, because construction has disrupted the entire area several times. Deposits along the channel margin north of the bridge are likely to contain only fallout plutonium, since road building closed the distributary channel from Los Alamos Canyon.

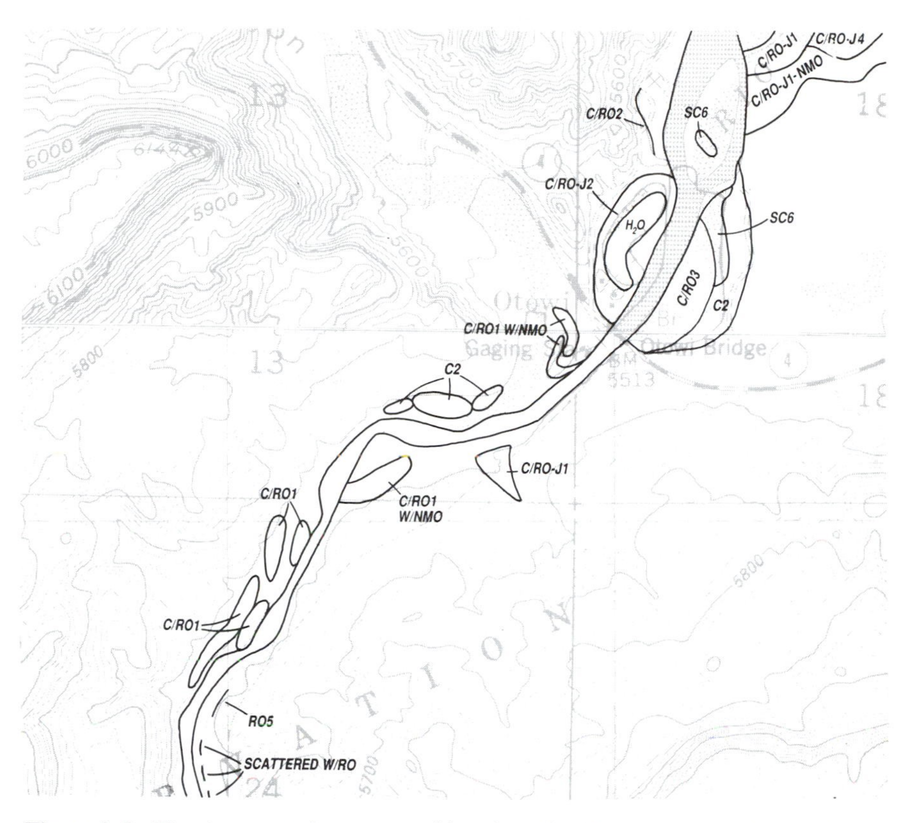

Figure 9.6. Riparian vegetation communities along the Otowi representative reach. See number 2 in Figure 9.2 for the location and Table 9.1 for symbols. (Modified from V. C. Hink and R. D. Ohmart, *Middle Rio Grande Biological Survey: Final Report*, U.S. Army Corps of Engineers Contract Report DACW47-81-C-0015 [Albuquerque: U.S. Army Corps of Engineers, 1984])

Because the Otowi reach is the entry point into the main stream for the plutonium that originates at Los Alamos National Laboratory, the laboratory has extensively sampled the sediment over a period of several years to assess its plutonium concentrations. A dozen samples from almost as many years taken from the active bedload in the Los Alamos stream near Otowi revealed plutonium concentrations ranging from 0.001 to 0.528 pCi/g, with a mean value of 0.127 pCi/g. Active bedload sediments in the Rio Grande near Otowi produced a range of 0.002 to 0.065 pCi/g from a similar number of samples. The mean concentration in the sediments from the Rio Grande was 0.011 pCi/g. A difference-of-means test shows that there is a significantly higher level of plutonium in the Los Alamos Canyon stream than in the main river. The main stream must quickly dilute the concentrations in sediments contributed by Los Alamos National Laboratory, however, because the mean concentrations of plutonium in sediments in Cochiti

Figure 9.7. Geomorphic map of the Otowi representative reach. See number 2 in Figure 9.2 for the location. Labeled areas on the map are (1A) 1981 channel of the Rio Grande; (1B) 1981 active flood-plain and bar surfaces of the Rio Grande; (2B) flood plain deposited in either the 1958 or the 1967 flood; (T1A) 1981 active channel of the tributary stream in Los Alamos Canyon; (T1B) 1981 active flood plain and overflow channels of the tributary stream, active since at least 1954 and probably since 1947 when the northeast outlet of the tributary was sealed by construction related to the truss bridge; (T2A) younger pre-1954 active channel; (T3B) younger pre-1954 flood plain; (T4B) older pre-1954 active channel; and (T5B) older pre-1954 flood plain.

Reservoir (probably including bedload and suspended load from the Rio Grande) are only 0.0134 pCi/g, similar to the sediments of the Rio Grande without the infusions of plutonium from Los Alamos National Laboratory.

Buckman Reach

The Buckman reach represents the portions of the Rio Grande that flow in canyon segments with floors up to 1 km wide. The reach extends about 5 km from Buckman Mesa, past the alluvial fan of Cañada Wash, to a bedrock restriction in White Rock Canyon (Figures 9.5, 9.8, and 9.9). The reach is located about 5 km downstream

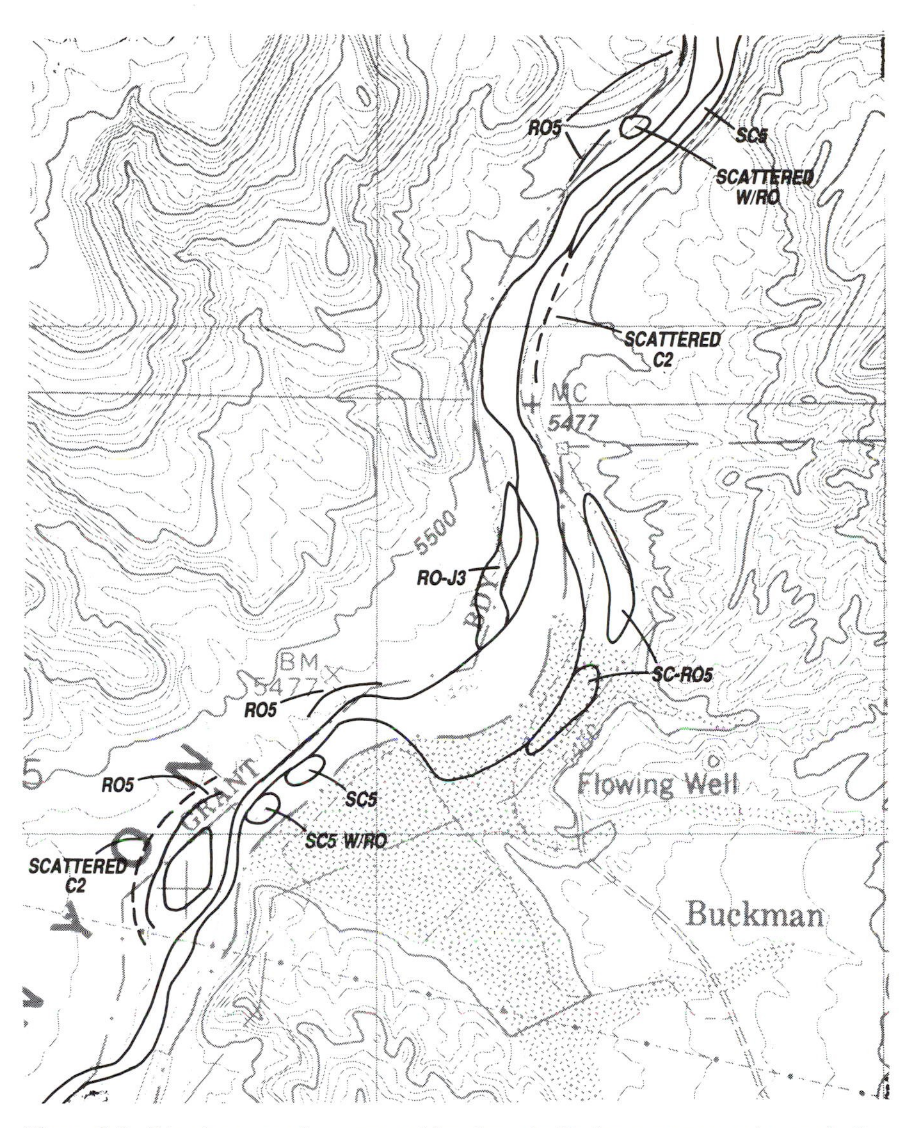

Figure 9.8. Riparian vegetation communities along the Buckman representative reach. See number 3 in Figure 9.2 for the location and Table 9.1 for symbols. (Modified from V. C. Hink and R. D. Ohmart, *Middle Rio Grande Biological Survey: Final Report*, U.S. Army Corps of Engineers Contract Report DACW47-81-C-0015 [Albuquerque: U.S. Army Corps of Engineers, 1984])

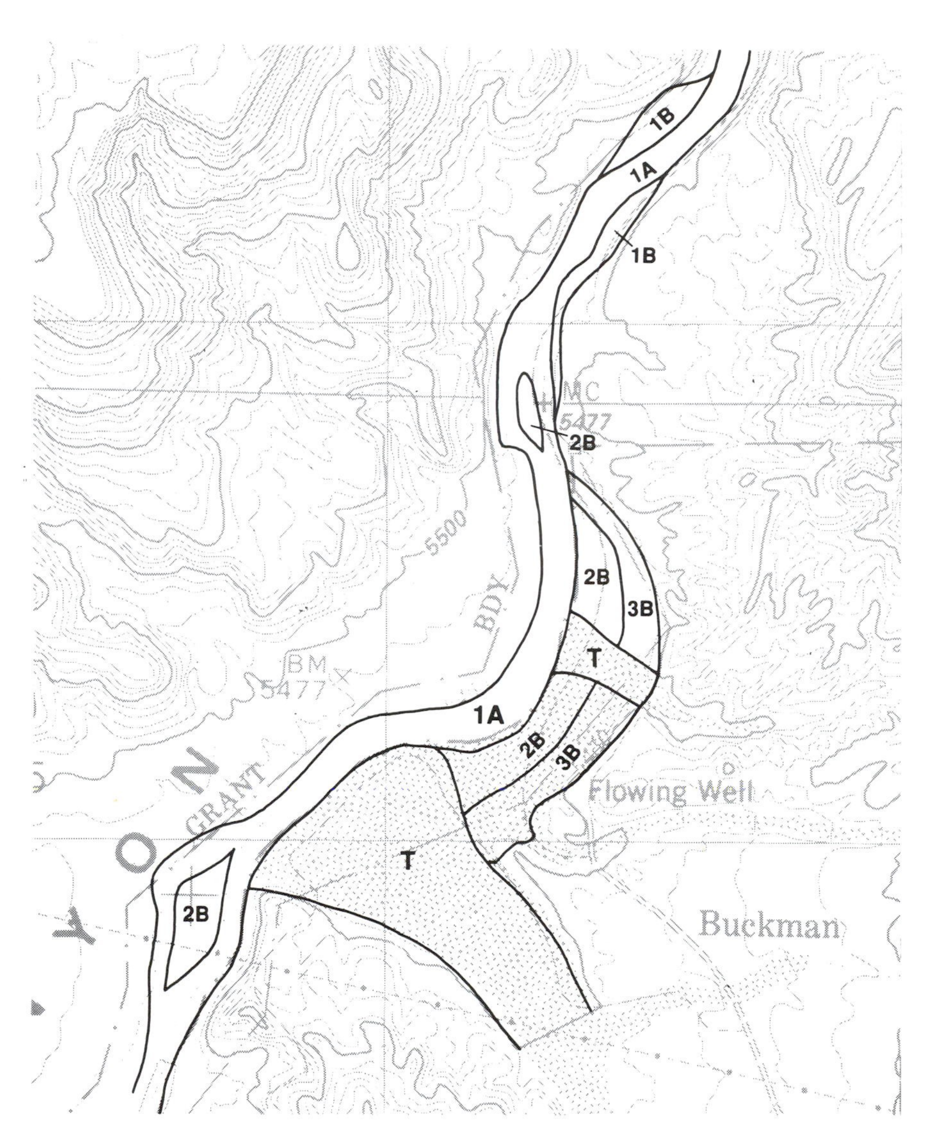

Figure 9.9. Geomorphic map of the Buckman representative reach. See number 3 in Figure 9.2 for the location. Labeled areas on the map are (1A) 1981 active channel; (1B) 1981 active flood plain; (2B) two mid-channel islands: the southernmost island has existed since before the 1940s, and the northernmost island is a bar that detached from the east channel side between 1976 and 1981 – the 2B area at Buckman is a channel overflow area that was a channel in 1912, was widened in 1913 flood, and was a frequent overflow area until isolated by the 1958 or 1967 flood; (3B) pre-1912 channel location maintained as a decreasing channel until the 1958 or 1967 flood – during 1940–1958 or 1967 it was a slough; and (T) tributary alluvium.

from the confluence of Los Alamos Canyon and the Rio Grande, and so sediments in the Buckman area combine inputs from the main river, Los Alamos Canyon, and two nearby tributaries. In the vicinity of Buckman, Cañada Wash (from the south) and Sandia Canyon (from the north) empty large amounts of sandy sediment into the main stream. The Rio Grande channel reflects the influence of these fans by transcribing a curving course around their bases. Buckman was a sawmill town established in 1899 on a Pleistocene terrace on the southeast side of the channel.[5] It later served as a railroad stop on the Chili Line, the narrow-gauge route along the Rio Grande, with its well supplying water for the steam engines.[6] The area is now uninhabited, but an active well field remains, supplying water to Santa Fe.

The Rio Grande channel in the vicinity of Buckman has consistently become narrower since at least 1912. Historical photography in the Museum of New Mexico at Santa Fe shows that before 1920, the channel was braided and spanned the 0.5-km valley floor from one terrace to the other. Sedimentation in the area caused the progressive abandonment and filling of the channels and the attachment of bars to the channel banks. In the bend of the river at Buckman, the early-twentieth-century channel was braided. A large mid-channel bar developed in the reach in the mid- and late 1920s and produced two smaller channels by splitting the main stream. The southeastern channel outside the bend, next to Buckman, developed into a slough, a relatively inactive channel except during floods. During the 1940–1958 period, the slough gradually filled with sediment, and after the 1967 flood, slack water had completely filled it with sediment. Thereafter, the Rio Grande had a single narrow channel northwest of the bar, and the bend in the river's single-thread course is now less pronounced than it was previously.

The trees in the Buckman area are mostly scattered individuals of cottonwood, russian olive, and juniper (Figure 9.8), but the ecological data from the early 1980s does not show the extensive tamarisk revealed in my work.[7] Although tamarisk now dominates (in percentage) the trees found on the two mapping units representing the old bar and slough, the tree count is radically different between the two. One unit, the abandoned slough, contains two to three times as many trees as the bar area does. The slough also has sediments with about twice as much fine material as the bar has.

Historical information, aerial-photographic evidence, and field-derived data for the vegetation and sediment make it possible to map distinct geomorphic features and associated sediment deposits in the Buckman reach (Figure 9.9). The geomorphic map shows the slough and bar areas with sandy sediments from tributaries draped over the older Rio Grande deposits. The slough (3B on Figure 9.9) is a primary site to sample for plutonium storage because it is a short distance (6–7 km) downstream from Los Alamos Canyon and because its arrangement formed an ideal trap for fine sediments during periods when the river was likely to have been transporting plutonium from Los Alamos and from fallout. The bar, with its coarser sediments and earlier stabilization, probably stores less plutonium.

Recent measurements of plutonium concentrations confirm that the concentration (0.017 pCi/g) is highest in a slough area south of the tributary (3B in Figure 9.9). This area stored materials during the early releases from Los Alamos Canyon and from the period of maximum fallout. The mean concentrations of plutonium are about 50 percent higher in the slough than in the bar area (2B in Figure 9.9). The concentrations in currently active channel sediments (0.0027 pCi/g) are lower than those in the historical deposits, and sediments from the tributary are the lowest of all (0.0008 pCi/g). The concentrations in the tributary sediments are one-fourth those in the present Rio Grande and one-eighth those of historical sediments stored in the reach.

The sedimentary deposits of the Rio Grande can be distinguished from those derived from local tributaries not influenced by Los Alamos National Laboratory on the basis of plutonium concentrations. However, the data from deposits at Santa Clara (unaffected by Los Alamos) and Buckman (likely to have both Los Alamos and fallout contributions) are not statistically different from each other. Thus in this case, fallout pollution and laboratory pollution cannot be distinguished from each other on the basis of plutonium concentrations alone. The isotopic ratios may serve this discriminating function. Data from Los Alamos Canyon indicate that the ratio of plutonium 239 and 240 to plutonium 238 in alluvium affected by laboratory releases is about 100. The same isotopic ratio for global fallout is less than 30. Therefore, deposits derived from fallout might be expected to have isotopic ratios lower than the ratios for deposits of pure Los Alamos plutonium or of plutonium from both sources. Data from the Santa Clara and Buckman deposits suggest that this expected trend is found in the Rio Grande. The mean ratio of plutonium 239 and 240 to plutonium 238 from Santa Clara is 4.3, and at Buckman, influenced by infusions from Los Alamos, the ratio is 27.1, with one sample showing a ratio of 72.3.

A special study of isotopic ratios using the geomorphic data generated by my work as a sampling guide also shows that fallout plutonium and Los Alamos plutonium are mixing in flood-plain sediments. D. B. Curtis, R. E. Perrin, and D. W. Efurd investigated the ratio of plutonium 240 to plutonium 239 from three flood-plain deposits. In the flood plain at Santa Clara, which presumably contains only fallout products, the 240:239 ratio is 0.120, which compares favorably with the global fallout value of 0.130. In a sample from Acid Canyon, a site directly affected by inputs from Los Alamos, the ratio was one-tenth as much, 0.013, consistent with laboratory records for material processed during the late 1940s to mid-1950s. In a sample from the slough area at Buckman, presumably showing the mixing of plutonium from both fallout and laboratory sources, the 240:239 ratio was 0.077, midway between the values of samples affected by only one source. A reasonable conclusion is that the plutonium in flood plains downstream from Los Alamos Canyon in White Rock Canyon include contributions from both fallout and the laboratory.

Frijoles Canyon Reach

The Frijoles Canyon representative reach is typical of the confluences of large tributaries with the Rio Grande in White Rock Canyon. It is also typical of narrow canyon reaches and of those few junctions with relatively short tributaries draining plateau and canyon country west of the main stream (Figure 9.10). Frijoles Canyon contains a stream draining the southeastern flank of the Jemez Mountains that flows through a defile cut into the Pajarito Plateau. Quaternary Bandelier Tuff, particularly the rhyolitic Tshirege Member, constitutes the rocks of the tributary canyon walls and contributes sandy sediments to the stream.[8] In regard to their minerology, these materials are distinctly different from the sediments of the Rio Grande. The tributary materials have accumulated in a large alluvial fan at the junction of the tributary and the main stream. The Rio Grande has a single-thread channel largely confined by the resistant older basalt of the Santa Fe Group in the walls of White Rock Canyon. Fluvial deposits are not voluminous along the channel and consist only of scattered terrace remnants and limited active channel deposits.

The Frijoles Canyon reach also represents the upstream slack-water area of Cochiti Reservoir. Although Cochiti Dam closed in 1973, its high runoff periods and operations did not produce a full reservoir until the early and mid-1980s. By 1985, high lake levels intruded into the Frijoles Canyon reach for long periods and with water depths of about 15 m in the area, so the Rio Grande's flows deposited materials in the reach by means of lacustrine processes. Sedimentation during the high-lake-water periods draped 1.0 to 1.5 m of lake sediment over the fluvial materials from the tributary's main stream and alluvial fan. In the late 1980s, the lake level fell and abandoned the reach to channel processes. The reactivated Rio Grande channel eroded a course through the lake sediments and returned to its original single-thread configuration. In 1988, the channel margins exposed vertical sections of lake and fluvial sediments.

Before being inundated by Cochiti Lake, the riparian vegetation in the Frijoles Canyon reach was mostly on the alluvial fan. The area included a cottonwood forest and a considerable number of tamarisks. The 1985 inundation was deep enough and lasted long enough to kill all the trees on the fan, and by the summer of 1988 their regrowth was minimal. Vegetation elsewhere in the reach was mostly upland species that returned after the lake waters receded. The distribution of riparian vegetation reflects the arrangement of prelake sediments and does not say anything about the lacustrine materials that now dominate the surfaces of the reach.

The sediments in the reach are of three groups: the materials from the active Rio Grande channel, the sediments from the tributary, and the sediments from the lake. The main stream contains bed sediments that are relatively coarse, with less than 5 percent of their bulk as silt and clay. At the time the samples were collected

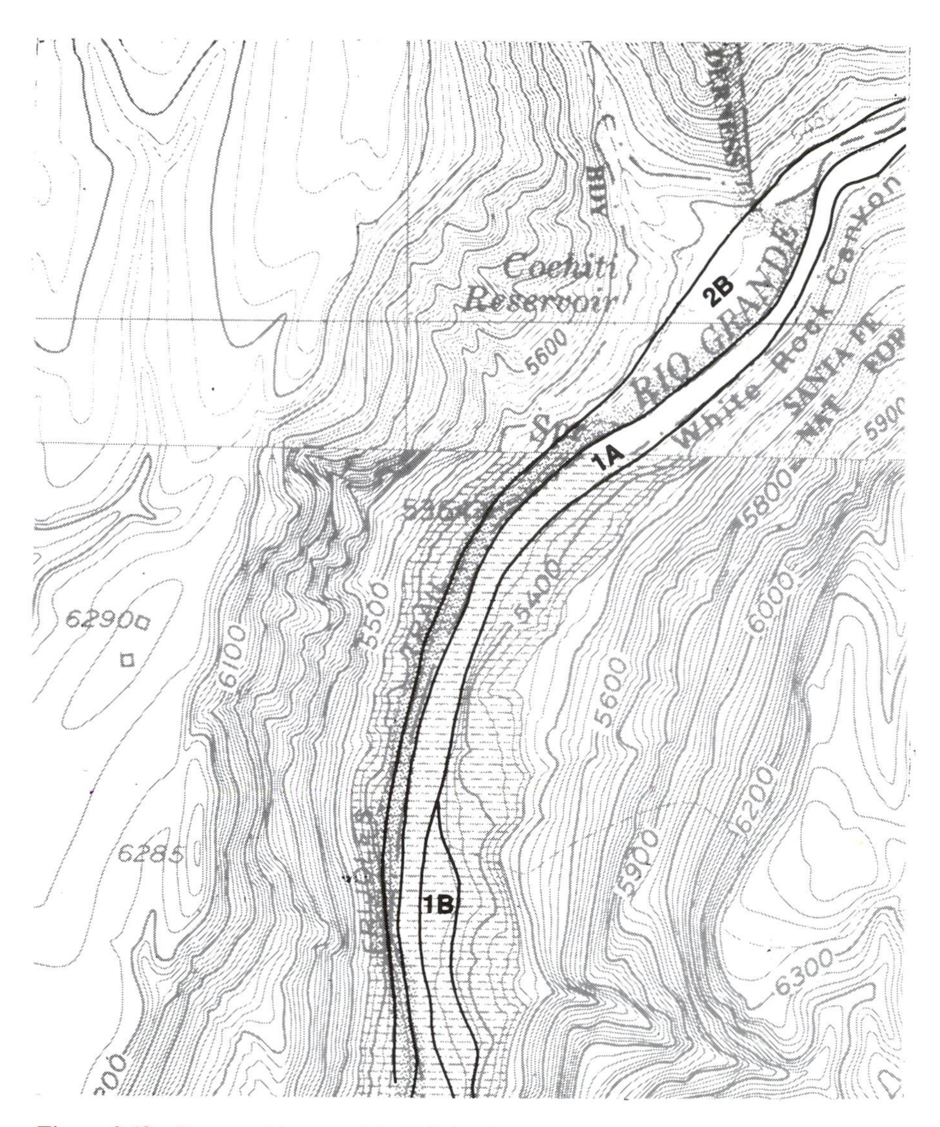

Figure 9.10. Geomorphic map of the Frijoles Canyon representative reach. See number 4 in Figure 9.2 for the location. Labeled areas on the map are (1A) 1981 active channel; (1B) younger, pre-1950 flood plain, active deposition and erosion, gradual net erosion through 1981, with field evidence in 1988 indicating partial burial by reservoir sedimentation in mid-1980s; and (2B) older, pre-1950 flood plain and alluvial fan deposits from Frijoles Canyon, with field evidence in 1988 indicating partial burial by reservoir sedimentation in mid-1980s.

(summer 1988), the channel in the reach was still excavating its course into the underlying sediments, and its gradient was steeper than normal as it cut its way through the accumulated materials at the head of the lake. As a result, the fine materials were transported farther downstream, leaving the coarser sediments behind. The sediments from Frijoles Canyon in the alluvial fan are also relatively coarse (fine materials about 7 percent). The lake sediments are very fine (40–50 percent silt and clay), representing materials deposited from the suspended load of the main stream as it entered the lake.

The storage of plutonium in the sediments of the reach is highly variable. The total volume of stored sediments in this restricted canyon reach is small, and the tributary sediments from Frijoles Canyon are coarse and therefore unlikely to contain large amounts of fallout. The lacustrine sediments may represent a significant storage location for plutonium from fallout and possibly from Los Alamos National Laboratory. Recent samples of fine sediments from the middle and lower portions of Cochiti Reservoir show concentrations of plutonium 238 and plutonium 239 and 240 (combined concentraton of 0.0134 pCi/g) that are higher than those in most of the other soil and sediment samples.[9] The sediments exposed along the river in the Frijoles Canyon reach probably also contain higher-than-average levels. The ratio of plutonium 239 and 240 to plutonium 238 is 20–25, similar to the ratio for global fallout, indicating that the plutonium in the sediments is likely to have come from the Rio Grande watershed. The only exception is a sample collected in the upper reservoir area in 1987 when high water extended into the Frijoles Canyon reach. The plutonium ratio for the sample was 85,[10] similar to ratios recorded earlier in lower Los Alamos Canyon.[11] The plutonium in sediments deposited near the head of the delta during the high-lake stand was therefore derived from runoff from the Los Alamos area.

10

Sediment and Plutonium Storage Downstream from Cochiti

Downstream from White Rock Canyon and the reaches discussed in Chapter 9, the Rio Grande takes on a different character because of the presence of Cochiti Dam at the lower end of the canyon. From that point downstream, the river's present appearance and behavior reflect the influence of the dam, which was closed in 1973 (Figure 10.1). Although the channel has become narrower throughout the length of the Rio Grande since the 1930s, this change is most pronounced south of Cochiti Dam. Downstream from the Los Lunas representative reach (which ends near Bernardo), the character of the Rio Grande changes radically. Immediately below Bernardo, the Rio Puerco joins the main river, bringing with it a huge load of sediment. The Rio Grande Valley becomes much wider below Bernardo, and the twentieth-century narrowing of the channel, aided by engineering works, is even more pronounced than in upstream areas, and the vegetation community is dominated by tamarisk. The final three representative reaches discussed in this chapter share the features of great valley width, extensive channel changes, and widespread impacts of engineering works.

Peña Blanca Reach

The Peña Blanca reach, a 5-km channel section, represents conditions common along 40 km of the Northern Rio Grande between Cochiti Pueblo (site of Cochiti Dam) and the confluence with the Jemez River. The river passes Peña Blanca, a settlement based on irrigated agriculture dating from the early nineteenth century. The reach is typical of the conditions in a portion of the river where the flood plain is several times the width of the channel (Figures 10.2 and 10.3) and where the channel has been exceedingly unstable. The reach is also instructive concerning the results of levee construction (in 1953) and dam closure (in 1973).

Figure 10.1. The Rio Grande has become progressively narrower in the vicinity of Albuquerque during the past century. (*a*) The view looking east across the stream at the Old Town Bridge about 1895 showing a broad channel in flood. (Photographer unknown, photo J-78041-4623, Colorado State Historical Society) (*b*) The same view showing a modern bridge over a considerably smaller channel in 1989. (W. L. Graf, photo 77-13)

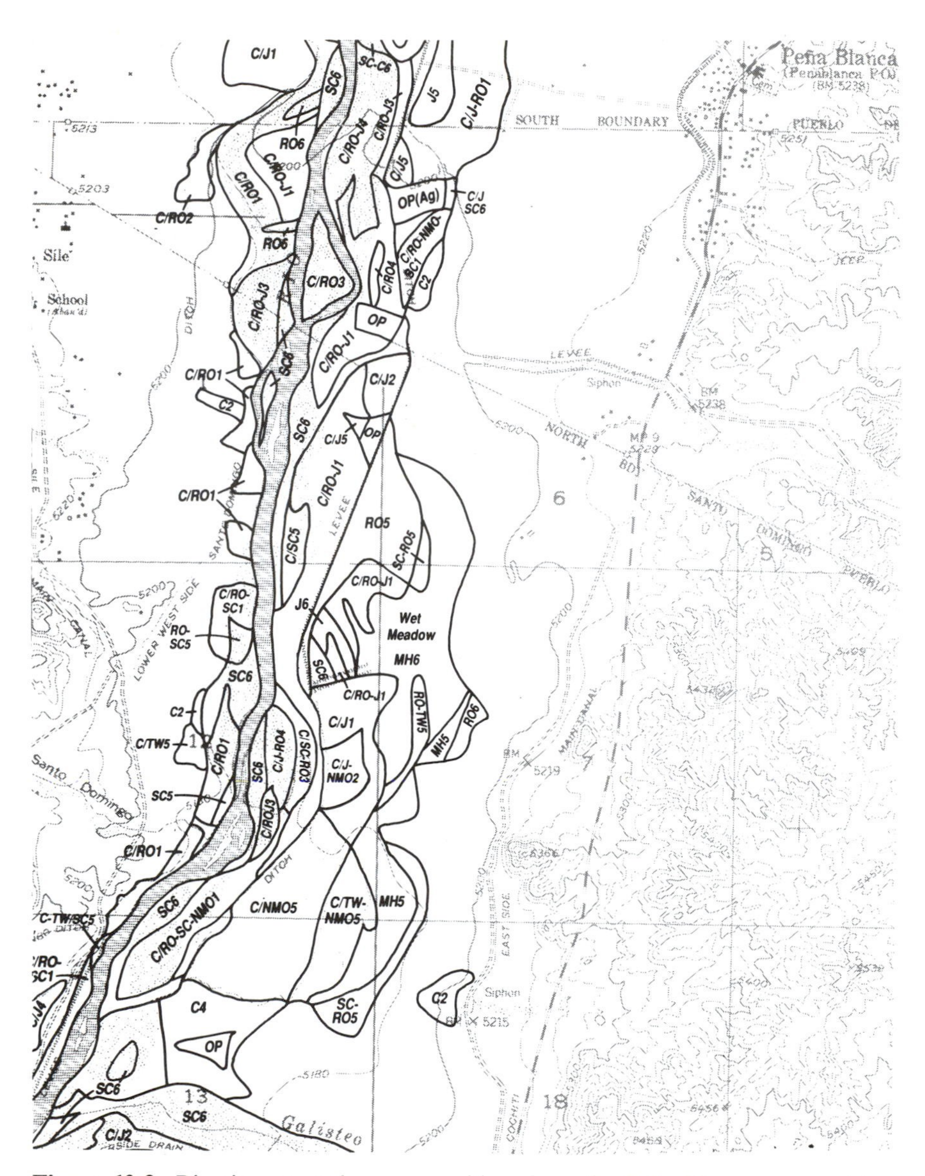

Figure 10.2. Riparian vegetation communities along the Peña Blanca representative reach. See number 5 in Figure 9.2 for the location and Table 9.1 for symbols. (Modified from V. C. Hink and R. D. Ohmart, *Middle Rio Grande Biological Survey: Final Report*, U.S. Army Corps of Engineers Contract Report DACW47-81-C-0015 [Albuquerque: U.S. Army Corps of Engineers, 1984])

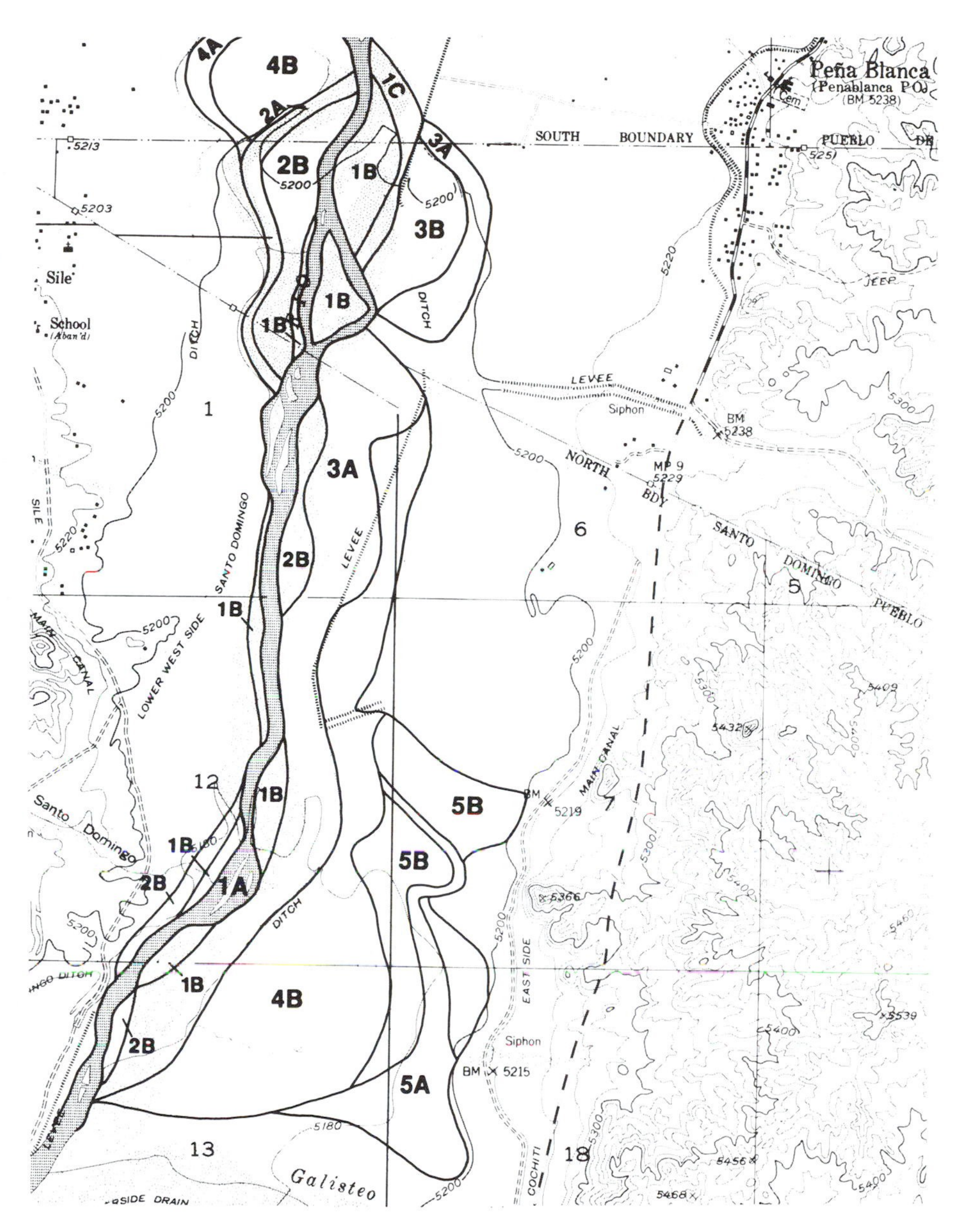

Figure 10.3. Geomorphic map of the Peña Blanca representative reach. See number 5 in Figure 9.2 for the location. Labeled areas on the map are (1A) 1982 active channel; (1B) 1982 active flood plain; (1C) 1975–1980(?) channel; (2A) 1954–1967 channel, closed by 1967 flood; (2B) 1954–1967 flood plain; (3A) mid-1940s–1955 channel; (3B) mid-1940s–1955 flood plain; (4A) 1941–1955 channel; (4B) 1941–1955 flood plain (includes older area cut off by dike on the east side of the channel system); (5A) pre-1940 channel; and (5B) pre-1940 flood plain.

The behavior of the channel in the Peña Blanca reach between the early 1940s and about 1990 has consistently included locational instability and progressive adjustment from a broad-braided configuration to a narrow, straighter alignment. In the 1940s, the channel was wide and unstable, with numerous major and minor threads, but the gradual reduction in water yield and radical reduction in the annual flood peaks resulted in the progressive isolation and closure of secondary channels. Certainly the installation of Cochiti Dam accelerated these changes, but they were already well established before the dam was built. The product of the changes is a single remaining channel with many abandoned channels and flood-plain deposits on both sides. The present vegetation patterns are complex reflections of this complicated history (Figure 10.3).

The most critical time in the channel's evolution was the low-flood period in about 1955 that caused the closure and abandonment of many of the secondary channels in the reach (for the flood history, see Figure 3.13). Immediately west of Peña Blanca, for example, a sinuous channel and associated flood plain displayed a prominent westward loop (4A in Figure 10.3). But after 1955, this loop was no longer active, and overbank flows from the active main channel filled the inactive loop with fine sediment. There are similar, wider segments of abandoned braided channel along the eastern side of the present channel throughout the reach (3A on Figure 10.3).

Between about 1955 and the 1967 spring flood, the channel pattern resembled later arrangements except that it was wider and occupied a different alignment. After the 1967 flood, the alignment remained relatively unchanging, though the channel became narrower, partly as a result of reduced outflows from Cochiti Dam. Erosion of the channel bed by relatively sediment-free flows from the dam after 1973 has probably contributed to its locational stability.

The riparian vegetation in the Peña Blanca reach is primarily cottonwood (Figure 10.2). The flood plains once associated with the abandoned channel areas are distinctive because they contain significant amounts of tamarisk (up to 74 percent of the trees), which favors the fine-grained soils on these forms. The vegetation communities and clear topographic expression of the abandoned channels as minor troughs 1.0 to 1.5 m deep permits field differentiation of the channels and the planar surfaces of flood plains. The sediment characteristics of the two forms vary. The amount of fine sediments in the abandoned channels and flood plains that date from the 1941-1955 period is similar, about 20 to 25 percent silt and clay. Pre-1940 flood-plain sediments are much finer, with about 40 to 60 percent silt and clay.

Plutonium in the Peña Blanca reach is likely to be associated with the abandoned flood plains and channels that were active during the 1941-1967 period and that coincided with three of the four years of maximum plutonium discharge from Los Alamos Canyon (1951, 1952, and 1967). These features (3A, 3B, 4A, and 4B on Figure 10.3) differ in terms of their content of fine materials, and further field sampling will be required to determine the locations of concentrated fine materials. Those flood plains that received overbank flows and the abandoned sloughs that were active during the 1960s (such as the loop west of Peña Blanca) are likely locations for future exploratory sampling.

Coronado Reach

The Coronado reach is an example of the Northern Rio Grande significantly influenced by a major tributary (Figures 10.4 and 10.5). In this reach, inputs from the Jemez River contribute large amounts of sediment to the main stream, producing a pronounced widening of the flood plain below the confluence. The

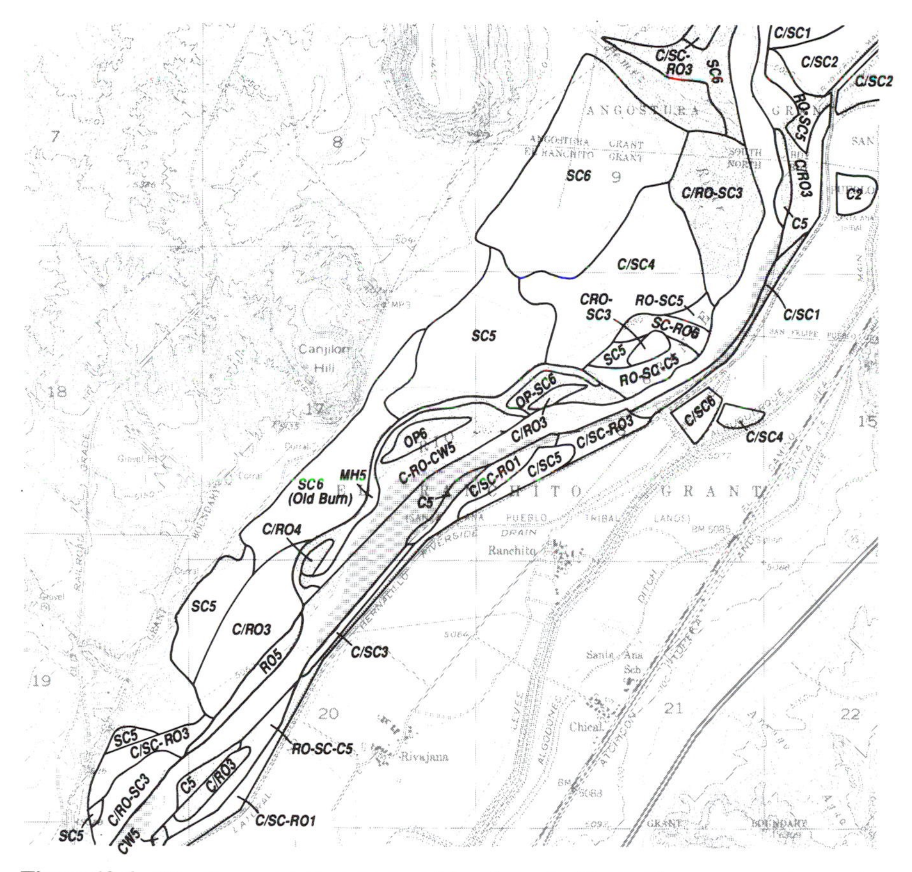

Figure 10.4. Riparian vegetation communities along the Coronado representative reach. See number 6 in Figure 9.2 for the location and Table 9.1 for symbols. (Modified from V. C. Hink and R. D. Ohmart, *Middle Rio Grande Biological Survey: Final Report*, U.S. Army Corps of Engineers Contract Report DACW47-81-C-0015 [Albuquerque: U.S. Army Corps of Engineers, 1984])

Figure 10.5. Geomorphic map of the Coronado representative reach. See number 6 in Figure 9.2 for the location. Labeled areas on the map are (1A) 1982 active channel; (2A) mid-1940s–1961 channel, east of Canjilon Hill, dates of activity are mid-1940s–1952; (2B) mid-1940s–1961 flood plain, east of Canjilon Hill, dates are mid-1940s–1952, active lateral accretion in early 1980s near Rivajana; (3A) mid-1940s–1961 channel, at Jemez River, flood plain that was active in late 1940s, perhaps as a result of flooding on the Jemez; (3B) pre-1940s–early 1940s braided channel, at Jemez River active in late 1940s; (4A) 1955–1961 overflow channel; and (4B) 1955–1961 braided channel, site of flood control fences established shortly before 1961.

modern active channel is only about 100 to 125 m wide, but the flood plain spans more than 2 km between terraces on either side. The reach also exemplifies thecomplicating factor of levee construction, which substantially restricts the available space for the channel's relocations. Conditions in the 5-km reach are typical of the 25-km section of the main stream from the Jemez River through Bernalillo to Albuquerque. Near Bernalillo, the river passes Coronado State Monument, the location of an outpost established by Coronado in 1540 among the ruins of Kuau, the northernmost Tiwa village.

The main channel of the Rio Grande in the Coronado reach has responded to flood events on the main stream and in the watershed of the Jemez River. From the mid-1940s until the early 1960s, its configuration was intricately braided and highly unstable because of periodic inputs of water and sediment from the tributary. During this period, the channel in the Peña Blanca reach immediately upstream gradually narrowed in the Coronado reach, but large floods in 1945, 1946, 1950, 1951, and 1958 in the Jemez River interrupted the general trend. Between 1955 and 1962, there was a narrow looping overflow channel on the west side of the river south of Canjilon Hill, and between the loop and the channel there was a braided-channel area. Both forms were products of greatly fluctuating discharges and high sediment loads.

The 1952 closure of Jemez Dam on the tributary contributed to the stability of the reach by reducing flood magnitudes and sediment input to the main stream. The U.S. Army Corps of Engineers installed an extensive series of Kellner jack systems in the braided-channel areas of the reach and established a narrow, relatively straight pilot channel through the reach. The agency also completed an extensive levee on the east side of the river in late 1961 that restricted the channel's mobility and isolated some of the flood plain from active-channel processes.

Cottonwood and willow are common on older flood plains of the reach, but the dominant riparian vegetation in the area is tamarisk (Figure 10.4). Dense thickets of tamarisk (up to 77 percent of the trees) now occupy the areas of braided channel where the Kellner jacks stabilize the once-active braided channel. Russian olive outlines the margins of presently active channels and some that were recently abandoned. The present vegetation distribution strongly reflects the geomorphologic changes, with vegetation boundaries clearly revealing the abandoned channel system. The abandoned braided channels have dense tamarisk growth, whereas the single-thread channels have almost no tamarisk but substantial amounts of willow. In 1953, clearing operations removed the tamarisk from large areas, but the species quickly reestablished itself, and so the clearing left no lasting impact on the vegetation communities.

Sediments in the Coronado reach are not as variable as they are in some other reaches. The active-channel sediments contain about 36 percent silt and clay, a typical value for most of the Northern Rio Grande. The abandoned braided channels (3B and 4B in Figure 10.5) have less fine material (about 22 percent silt and clay), a tendency observed on the surface and at various depths below the surface down to 90 cm. The flood-plain materials (2B in Figure 10.5) are different: At the surface they contain more fines than does the channel (about 52 percent). Subsurface investigations show a typical "fining upward sequence" in which the materials become progressively finer approaching the surface.[1] In the Coronado reach, those samples collected 90 cm below one example flood-plain surface contained only 5 percent silt and clay, but at the surface in the same location the fines were 44 percent of the bulk sample (see the data in Appendix C1).

The geomorphologic map of the Coronado reach—constructed with evidence from aerial photography and field data for vegetation, sediment, and surface forms—shows an active channel of variable width with a narrow strip of abandoned surfaces on the southeast side and a much wider zone of abandoned deposits on the northwest side (Figure 10.5). There may be plutonium in the abandoned overflow channel that formed a loop south of Canjilon Hill (4A on Figure 10.5), because the loop was active during the late 1950s and early 1960s. But the gradual deposition of and plugging with sediments may have included plutonium from fallout and from Los Alamos National Laboratory. However, sediment from the Jemez River are likely to have diluted plutonium concentrations in the main stream because of the large amounts of material contributed by the tributary.

On the southeast side of the river, immediately north of the settlement of Ranchito, the construction of a levee truncated a meander of a braided channel active between the mid-1940s and 1961. A portion of the channel, an extension of 3A in Figure 10.5 north of Ranchito, now lies outside the levee and is isolated from current river processes. The site contains preserved sediments that represent conditions in the late 1950s without subsequent interference from floods.

Los Griegos Reach

The Los Griegos reach is representative of the state of the Northern Rio Grande in the Albuquerque area (Figures 10.6 and 10.7). Los Griegos (Spanish for a family name, Greeks) is a northern suburb of Albuquerque. The 5-km representative reach is similar to the 25 km of river through the city where levees on both sides constrain the active fluvial zone to a narrow strip 0.5 km or less in width. The reach is heavily engineered with bridges, Kellner jack fields, pilot channels, and designed levees. As a result, fluvial processes in the reach have features that are different from those in the other reaches.

From the early 1940s to the late 1980s, the channel narrowed somewhat and was converted from a broad, slightly braided channel to a single-thread version with some meanders. Although the braided channel had a meandering configuration, the extensive engineering efforts completed in 1960 produced a partially designed channel with a relatively straight alignment. Subsequent floods altered the planned arrangement, however, and by the late 1980s the single thread meandered in a course that crisscrossed previous paths and deposits.

The most common riparian vegetation in the Los Griegos reach is cottonwood, which grows in mature stands on flood-plain areas (Figure 10.6). The abandoned channels contain less mature cottonwood than other species, including russian olive and tamarisk. The distribution of the communities is directly related to the fluvial history of the strip between the two levees. On the east side, south of the Alameda Bridge, for example, a narrow band of mature cottonwood forest (with cottonwood constituting 66 percent of the trees; C2 on Figure 10.6) exactly

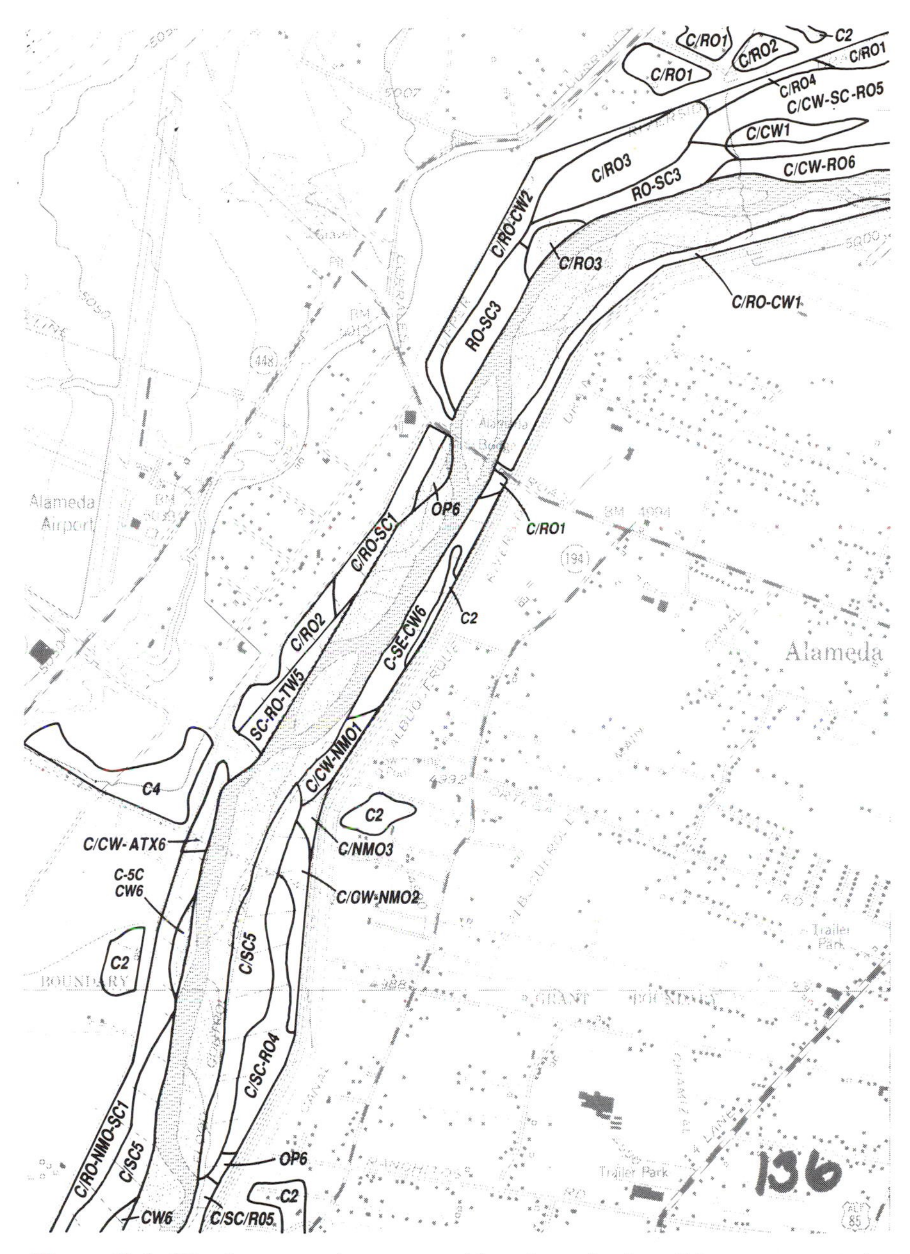

Figure 10.6. Riparian vegetation communities along the Los Griegos representative reach. See number 7 in Figure 9.2 for the location and Table 9.1 for symbols. (Modified from V. C. Hink and R. D. Ohmart, *Middle Rio Grande Biological Survey: Final Report*, U.S. Army Corps of Engineers Contract Report DACW47-81-C-0015 [Albuquerque: U.S. Army Corps of Engineers, 1984])

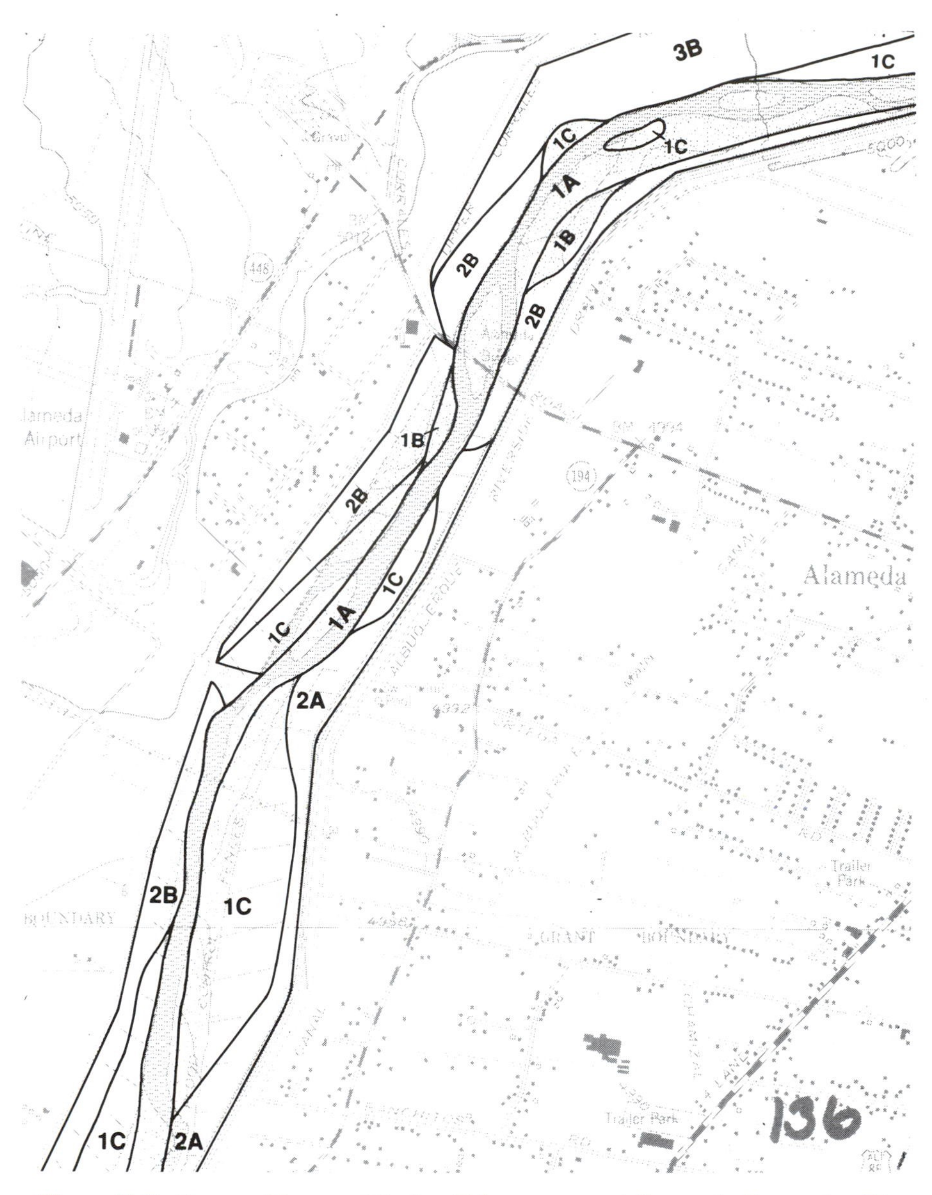

Figure 10.7. Geomorphic map of the Los Griegos representative reach. See number 7 in Figure 9.2 for the location. Labeled areas on the map are (1A) 1982 active channel; (1B) 1982 active flood plain; (1C) braided channel and overflow area, pre-mid-1940s–1960; (2A) pre-mid-1940s–1960 braided channel area; (2B) pre-mid-1940s–1960 flood plain, except northwest of Alameda Bridge where it was a channel overflow and slough area, pre-mid-1940s–1954; and (3B) pre-mid-1940s–1954 flood plain.

outlines the remnant of a flood plain several decades old. A zone of juvenile cottonwood, siberian elm, willow, and tamarisk surrounds the older forest and occupies the surface of a braided-channel area that was active until 1960 (C-SE-CW6 on Figure 10.6). The composition of the communities' species varies dramatically from one surface type to another. One braided channel active from the early 1940s until 1954 and another active until 1960 contain a variety of species not found on the old flood plain (51 percent of the trees are tamarisk; 11 percent, cottonwood; 11 percent, willow; and 26 percent, russian olive).

The sediments reflect the differentiation between the flood plain and the abandoned braided channels in the reach. The older flood-plain surface materials contain about 30 percent silt and clay, with a fining upward sequence from a depth of 60 cm. The braided channel abandoned in 1960 has differing amounts of fine particles, with the mean value of 37 percent reflecting the overbank deposition in the area after abandonment. There is no overbank deposition in the pre-1954 braided channel northwest of the Alameda Bridge, and so it exhibits the relative lack of fine materials (3.8 percent) typical of a braided channel.

The geomorphologic map of the Los Griegos reach (Figure 10.7) shows a suite of landforms and deposits unlikely to contain large quantities of sediment or plutonium. The constriction of the river by levees reduced the depositional areas, and although sediments did collect in the area during the past several decades (resulting in local flooding), the total quantity of material is fairly small. Fine materials, possibly containing some plutonium, appear mostly on a few flood plains.

Los Lunas Reach

The 5-km Los Lunas reach is representative of about 70 km of the Northern Rio Grande in the transition from urban to rural land use along the river from southern Albuquerque to Bernardo (Figures 10.8 and 10.9). Los Lunas is an agricultural settlement south of Albuquerque named after the Luna family, prominent early Hispanic settlers in Rio Grande Valley.[2] In this general segment, engineering works on both sides continue to restrict the course of the river, but the distance between the levees is greater than in the urban area. In the Albuquerque area, the levees' alignments are suited to the needs of the city, but downstream the levees conform to the late-1950s arrangement of the channel. At that time, the channel was braided and somewhat meandering, and included wide bends that still appear as the inherited positions of the levees. Since the late 1950s, however, the channel has shrunk because of the withdrawal of irrigation, less precipitation, and the effects of upstream dams. The current channel is straighter than its predecessor, so that it crosses the meandering deposits left by the earlier, more complex channel (Figure 10.9).

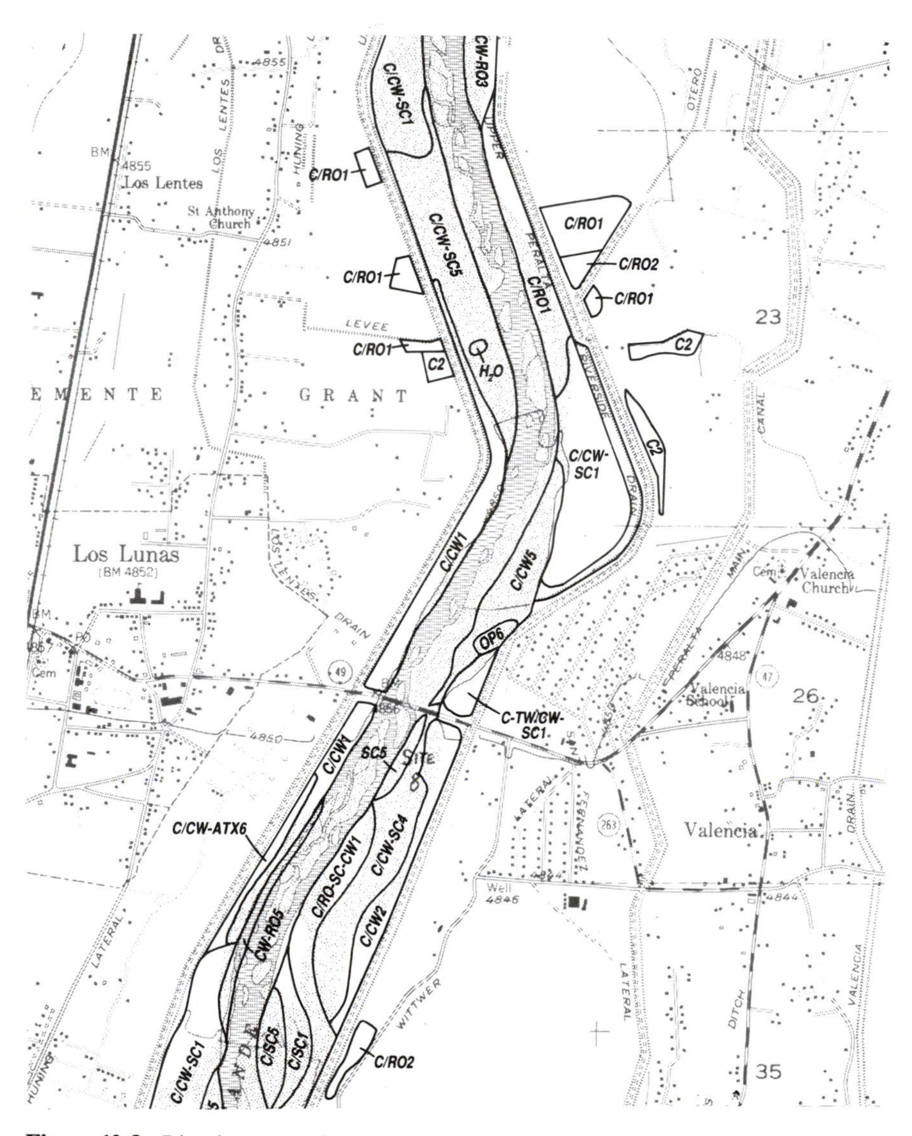

Figure 10.8. Riparian vegetation communities along the Los Lunas representative reach. See number 8 in Figure 9.2 for the location and Table 9.1 for symbols. (Modified from V. C. Hink and R. D. Ohmart, *Middle Rio Grande Biological Survey: Final Report*, U.S. Army Corps of Engineers Contract Report DACW47-81-C-0015 [Albuquerque: U.S. Army Corps of Engineers, 1984])

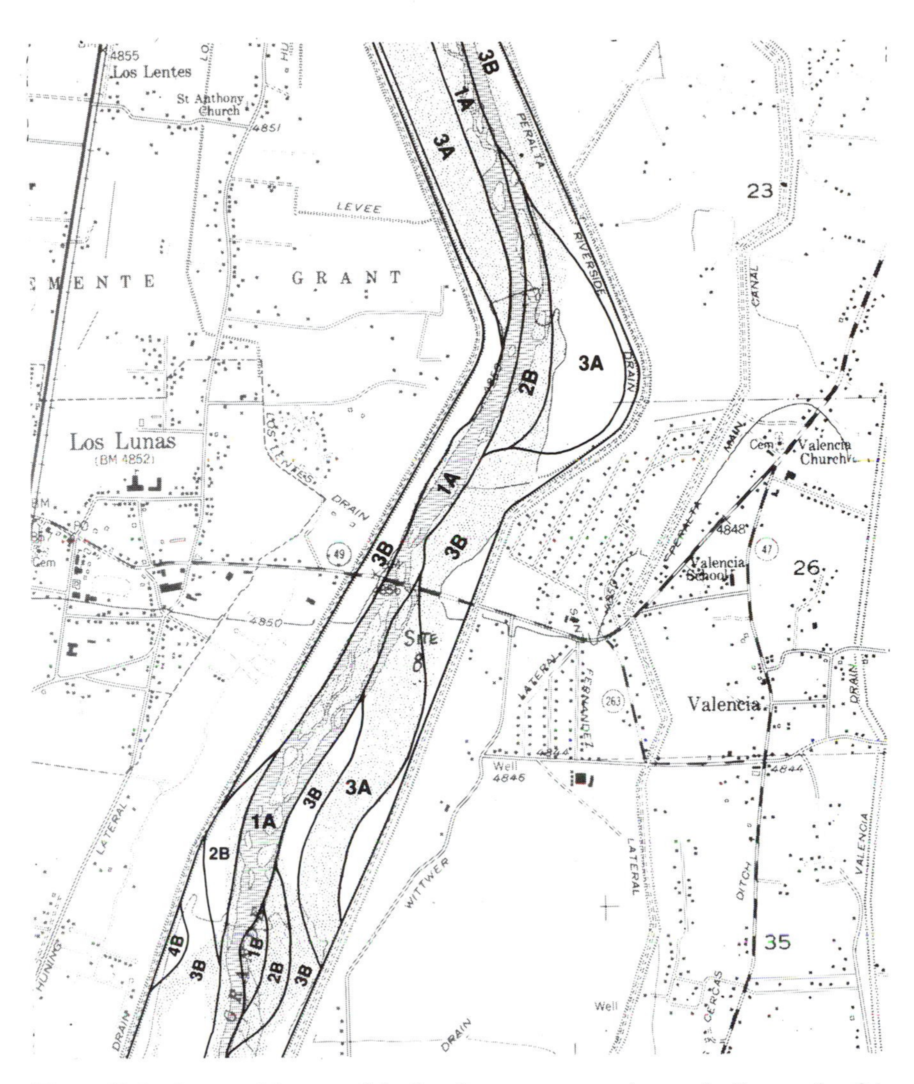

Figure 10.9. Geomorphic map of the Los Lunas representative reach. See number 8 in Figure 9.2 for the location. Labeled areas on the map are (1A) 1984 active channel; (1B) channel side bar established between 1963 and 1971; (2B) pre-1941–1960 channel area; (3A) pre-1941–1941 braided channel; (3B) pre-1941–1941 flood plain; in a small segment northeast of the bridge, pre-1941–1960 channel; and (4B) pre-1941 overflow channel.

The consequences of this history appear northeast of Los Lunas Bridge, which carries New Mexico Route 49 across the Rio Grande, where the levee configuration includes a prominent elbow-shaped area. This bulge in the levee system accommodates a bend in the pre-1960 braided-channel system. The subsequent narrowing and simplification of the channel into a single thread have left the elbow as an anomalous feature that then behaved as a sediment trap during floods (3A on Figure 10.9).

Mature cottonwood forest is common in the Los Lunas reach, occupying the older flood-plain surfaces and braided channels that were abandoned before about 1965. On those meandering segments of abandoned channel that are not part of the present active channel, the composition of species differs from that of the flood plains in its inclusion of tamarisk and willow (Figure 10.8). The newer deposits also contain fewer mature community structures. Willow is clearly associated with the finer materials of the filled channels, accounting for as much as 74 percent of the trees in such areas.

The sediments in the Los Lunas reach vary both vertically and horizontally. In abandoned braided-channel zones, the sediments are coarse, with 5 percent or less as silt and clay. In abandoned flood-plain zones, the fines constitute an average of 50 percent of the materials. In the elbow area just described, the surface sediments are fine (54 percent silt and clay) because of the slack-water overbank deposition in the area during flood periods in the last 20 years. Subsurface samples, however, show that these fine materials are less than 45 cm deep, because a sample from 45 cm below the surface revealed materials with only 2 percent fines, sediments typical of the original abandoned braided-channel deposits.

The geomorphologic interpretation of the Los Lunas reach indicates that it is unlikely to harbor significant quantities of plutonium, although no direct measurements are available. Fine materials deposited during the 1950s and early 1960s are relatively scarce, occurring on a few small flood-plain segments partly destroyed by the modern channel. The fine deposits trapped in the elbow area northeast of Los Lunas Bridge date from the last 20 years, a period during which Cochiti Dam prevented plutonium from entering the lower river from mountain fallout sources and from Los Alamos. If there is any plutonium in the deposit, it is from remobilized materials that had been previously deposited elsewhere upstream and then entrained a second time. Mixing and dilution by tributary sediments have probably been extensive.

The remaining reaches discussed in this chapter are located downstream from the massive sediment influxes of the Rio Puerco and Rio Salado, with the Rio Grande responding to these inputs and wider valley configurations. Significant channel narrowing, extensive engineering works, and the development of dense riparian vegetation are common in the reaches (Figure 10.10).

Figure 10.10. Engineering works produced a partly artificial channel that is more stable and narrow than the original natural channel in the reach immediately below the San Acacia Dam. (*a*) Looking south in 1905, showing a broad channel constrained only by the railroad embankment on the near side. Note the nearly barren sand dune on the left side. (R. H. Chapman, photo 37, U.S. Geological Survey Photography and Field Records Library, Denver) (*b*) The same view in 1989, showing a narrow channel constrained by levees and mature riparian trees. Note the vegetated sand dune on the left. (W. L. Graf, photo 77-6)

San Geronimo Reach

The 5-km San Geronimo reach is similar to the 25-km segment of the Northern Rio Grande between the highway bridge at Bernardo and the San Acacia Diversion Dam (Figures 10.11 and 10.12). The reach has a levee and drain system on the east side, and the embankment of the Southern Pacific Railroad strongly influences the west side. San Geronimo was once a stop on the railroad, but it is now abandoned. The embankment constrains the lateral movement of the active channel and isolates a backswamp and lake area on the west side. The flood plain and abandoned channel area is more than 2.5 km wide and is bounded by sharply defined Pleistocene terraces of sand and gravel.

The changes in the channel in the San Geronimo reach between the early 1940s and about 1990 were an adjustment from a winding braided channel to a meandering single thread and finally to a relatively straight single thread. Through the 1940s, the channel was as much as 1 km wide and nearly filled the available valley floor between La Joya (situated on an east-side terrace) and the railroad embankment on the west side. Near San Geronimo, the railroad embankment confined the channel, but farther south the river broadened again. A remnant of this 1940s channel (4B on Figure 10.12) still exists north of La Joya.

Declining water yields and smaller flood peaks after the early 1940s led to the establishment of a narrower, meandering single channel in the early 1950s (see Figure 3.12 for the annual water yield history and Figure 3.14 for the flood history). Moderate floods in 1957 and 1958 brought minor adjustments, and another moderate flood in 1967 straightened the alignment. The abandoned, sinuous channel and its associated flood plains appear as remnants on the valley floor throughout the reach (2A and 2B on Figure 10.12). After 1967, there were only minor adjustments, usually the addition of channel-side bars.

The riparian vegetation in the San Geronimo reach is not as varied as it is in the reaches upstream (Figure 10.11). Data and historical maps indicate that by the mid-1930s, tamarisk was already established in dense thickets.[3] Tamarisk is by far the most common species, providing a continuous monospecies cover in many areas. Changes in the community structure, however, reflect the variation in underlying deposits. Mature forest with canopies several tens of meters above the surface are rare, but there is considerable variation in structures with vegetation closer to the ground. Less dense growth occupies the younger deposits. Two anomalies are the cottonwood forest growing along the abandoned 1940s channel north of La Joya (C/CS1 in Figure 10.11), and the willow lining the modern active channel.

Unlike the vegetation, the composition of the sediment in the San Geronimo reach is distinctly different from place to place and shows strong connections to fluvial history. Sediments from an abandoned braided channel that was active between 1950 and 1967 (2A in Figure 10.12) contain modest amounts of fine

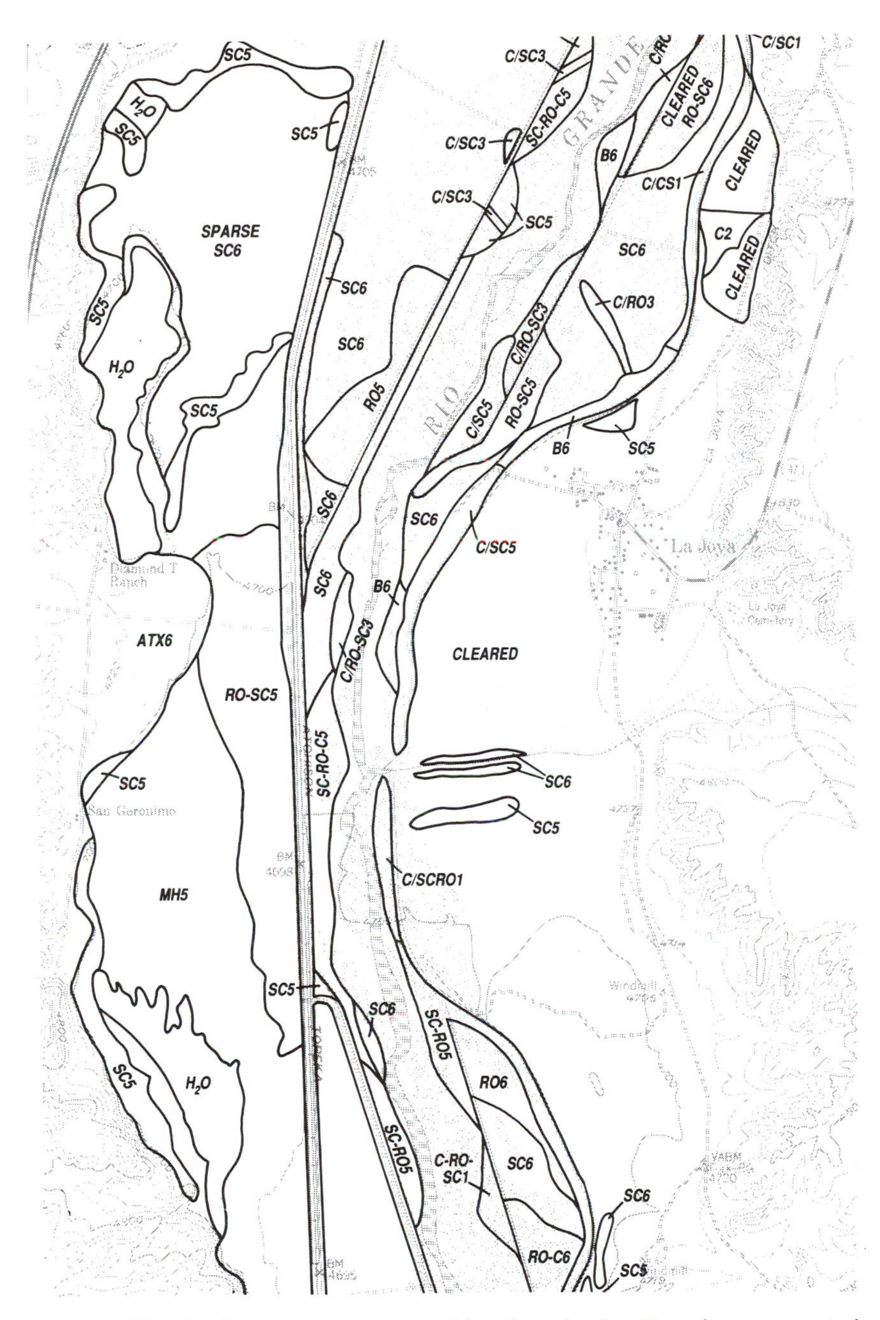

Figure 10.11. Riparian vegetation communities along the San Geronimo representative reach. See number 9 in Figure 9.2 for the location and Table 9.1 for symbols. (Modified from V. C. Hink and R. D. Ohmart, *Middle Rio Grande Biological Survey: Final Report*, U.S. Army Corps of Engineers Contract Report DACW47-81-C-0015 [Albuquerque: U.S. Army Corps of Engineers, 1984])

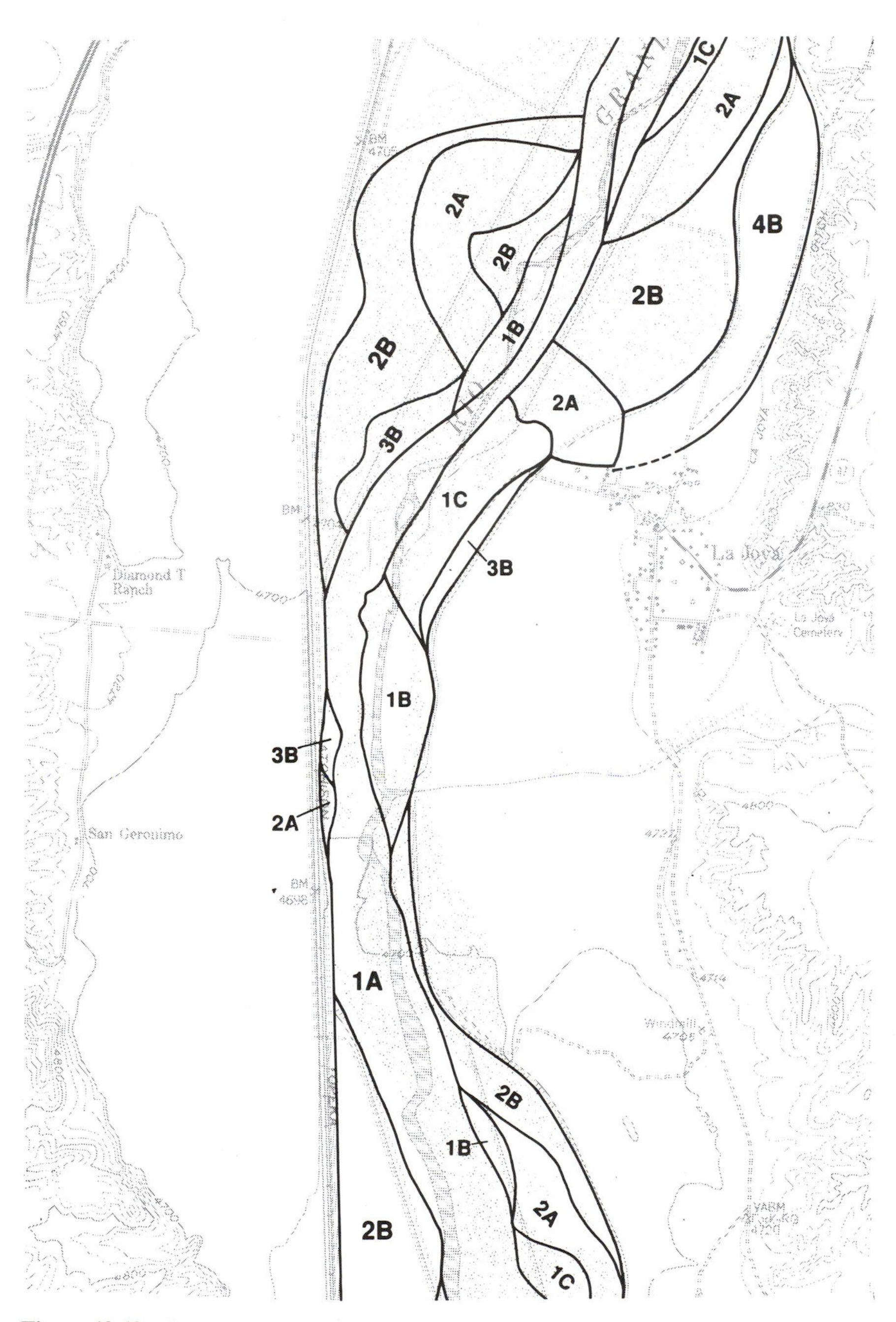

Figure 10.12. Geomorphic map of the San Geronimo representative reach. See number 9 in Figure 9.2 for the location. Labeled areas on the map are (1A) 1984 active channel; (1B) 1984 flood plain; (1C) 1984 overflow area, part of a braided-channel and bar system; (2A) 1950–1967 channel, pre-mid-1940s–1950 a braided channel; (2B) 1950–1967 flood plain, pre-mid-1940s–1950 a braided channel; (3B) 1950–1971 flood plain; and (4B) 1940s–1950 flood plain and braided channel.

material (about 28 percent), but abandoned flood plains active during the same period (2B in Figure 10.12) have significantly more (45 percent). A pre-1940s flood-plain surface has a similarly high content of fine particles. Subsurface samples from depths down to 90 cm show a fining upward sequence in all the features (data in Appendix C1).

The geomorphic map shows the contrast between the winding course of the river's deposits before 1967 and the straighter course of the present channel. The railroad embankment forms a resistant boundary on the west side of the reach, and since the 1940s the channel has stored little material along its length near San Geronimo because the bend in the river forces flows against the barrier. Agricultural activities have obscured the surfaces of the deposits southwest of La Joya, but their forms and materials remain relatively undisturbed in the northern portion of the representative reach.

The flood-plain deposits associated with the 1948–1967 channel systems are the most likely areas for plutonium storage. These areas contain fine-grained materials deposited during the period of maximum plutonium input into the system from fallout and from Los Alamos. The plutonium concentrations have probably been diluted, however, because the San Geronimo reach is immediately downstream from the confluence with the Rio Puerco. The tributary contributes huge quantities of sediment with low concentrations of plutonium that mix with those of the main stream, reducing the concentrations of the contaminant carried by the Rio Grande. The influx of sediment also contributes to massive flood-plain deposits extending up to 3.5 km across the valley in the reach.

Chamizal Reach

The Chamizal reach contains 5 of the 40 km between the San Acacia Diversion Dam and the San Antonio Bridge (Figure 10.13). Because the reach lies below the confluences of the Northern Rio Grande with the Rio Puerco and Rio Salado, the system is flooded with sediments from tributaries, resulting in a flood plain that is 2.5 to 3.5 km wide. During the 1940s, the channel was as much as 1 km wide, with a gently winding course that lacked the extreme wandering seen in the channel near La Joya in the San Geronimo reach. There is only one pronounced bend or elbow in the Chamizal reach near the settlement of Chamizal. The broad-braided channel was abandoned in stages, with some parts abandoned under natural processes by the late 1950s (3A and 3B in Figure 10.13) and others abandoned as a result of engineering activities (2A and 2B in Figure 10.13).

A significant feature of the reach is the elbow in the vicinity of Chamizal. Activities of the Middle Rio Grande Conservancy District in the 1920s produced the Lemitar Riverside Drain and its levee on the outside edge of the elbow. During the construction associated with the Middle Rio Grande Project, engineers installed a pilot channel consisting of straight segments (Figure 10.13). They built a

Figure 10.13. Geomorphic map of the Chamizal representative reach. See number 10 in Figure 9.2 for the location. Labeled areas on the map are (1A) 1984 active channel; (1B) pre-mid-1940s–1984 flood plain; (1C) pre-mid-1940s–late 1970s flood-plain and overflow area for braided channel; (2A) pre-mid-1940s–1962 braided channel; (2B) pre-mid-1940s–1962 flood plain; (3A) pre-mid-1940s–late 1950s braided channel; (3B) pre-mid-1940s–late 1950s flood plain; and (4B) cleared area on unit 2B.

levee that cut off the Chamizal bend in 1959. The U.S. Bureau of Reclamation christened the newly isolated area between the old and new levees the San Lorenzo Settling Basin because it trapped sediments from the San Lorenzo Arroyo, a tributary to the main channel.

Floods in the Chamizal reach result from activities of the Rio Grande, Rio Puerco, and Rio Salado, and major events in 1967, 1972, and the early 1980s caused channel adjustments. Although the engineering works completed in 1959 produced a straight channel, this configuration did not represent an equilibrium geometry. The subsequent channel adjustments reproduced the original channel geometry of a gently meandering course, somewhat constrained by the west-side levee.

Although earlier workers did not map the vegetation in detail in the Chamizal reach (or the San Marcial reach immediately downstream), my field investigations uncovered some useful data. Tamarisk is the most common riparian species in the Chamizal reach. On the narrow strip that is the active flood plain, its coverage approaches 100 percent, and in many places there are no other tree species. In channel and flood-plain areas abandoned by active processes after 1959, cottonwood is beginning to be a recognizable component of the riparian community. In the San Lorenzo Settling Basin, cottonwood is flourishing, but in those areas still subject to overbank flows from the modern channel, tamarisk maintains its dominance. Because tamarisk is so common and other species are not especially variable from place to place, the riparian vegetation is not a reliable indicator of the distribution of fluvial forms and sediments in the reach.

Sedimentary characteristics are useful tools in differentiating deposits in the Chamizal reach. Flood plains have consistently finer materials than do abandoned channels, and flood plains of different ages contain different amounts of fine sediments. The sediments in the major abandoned braided channel in the area (2A in Figure 10.13), excluding the settling basin, are less than 1 percent silt and clay. Among the flood-plain sediments, the more recent the feature is, the more fine material it will contain. The modern active flood-plain sediments have a mean of 53 percent silt and clay; the flood plain abandoned in the late 1950s (2B in Figure 10.13) has 43 percent; and the one abandoned somewhat earlier (3B in Figure 10.13) has 29 percent. As is often the case with braided-stream deposits, there is a vertical variation in particle size below the surface.

The geomorphic map of the Chamizal reach shows mostly post-1941 features, all related to gently curving channels that have changed from a braided to a single-thread configuration. Given the location of the reach far downstream from the sources of plutonium (almost 250 km south of Española) and downstream from the major sources of sediment (the Rio Puerco and Rio Salado), sediments in the reach are not likely to contain high concentrations of plutonium from mountain fallout areas or from Los Alamos. Sediments from depths below 30 cm probably reflect the materials in transport through the system in 1959.

Comparisons with modern active sediments would reveal any temporal trends in plutonium loading of the stream, but currently there are no radionuclide data available on the reach.

San Marcial Reach

The 8-km San Marcial reach represents channel conditions between the San Antonio Bridge and the San Marcial Bridge, site of a major U.S. Geological Survey stream gage and the downstream end of the study area (Figure 10.14). San Marcial was founded as a mid-nineteenth-century trading town and was destroyed by floods in 1866 and 1929. Sedimentation in the reach has produced a flood plain up to 4 km wide, but the active channel is relatively narrow, less than 60 m wide, because of transmission losses, irrigation withdrawals, and the maintenance of a conveyance channel that diverts low flows into a canal separate from the natural channel. Some of the reach's characteristics are similar to those of the San Geronimo reach in that for more than century, the Southern Pacific Railroad embankment has isolated a part of the previously active flood plain from the active channel. This isolated area at San Marcial was a backswamp under natural conditions and included San Marcial Lake. After the construction of the railroad, the lake filled with sediment and organic debris, and so it no longer is a body of water.

The changes in the channel in the San Marcial reach have a complex history. Before 1941, the channel in the reach was a winding, braided system that narrowed as it passed near Mesa del Contadero. This braided arrangement (3A and 3B in Figure 10.14) survived in various forms through a series of floods through the late 1930s, with the largest flows on record in 1905 and 1929 (for the flood history of the Rio Grande at San Marcial, see Figure 3.15). The arrangement included a flood-plain zone along most of the reach (3C in Figure 10.14). During the 1930s, the flow was confined to a channel on the west side of the valley.[4] Floods in the late 1930s and in 1941 established a new configuration that included a broad overflow area or flood plain immediately north of Black Mesa (2D in Figure 10.14). The low flows bifurcated, with one branch flowing along the west side of the valley (2A in Figure 10.14) and the other along the east side.

The Middle Rio Grande Project, completed in 1953, resulted in the abandonment of the west branch and the direction of flows into the east branch. The broad overflow surface is now isolated from the active channel by means of levees. Subsequent engineering efforts produced a relatively straight channel that cut off some meanders along the single thread eastern channel (1B in Figure 10.14).

The riparian vegetation in the San Marcial reach is predominantly tamarisk. The bedrock constriction offered by Black Mesa and Mesa Peak forces groundwater to the surface in the reach, producing ideal conditions for the growth of

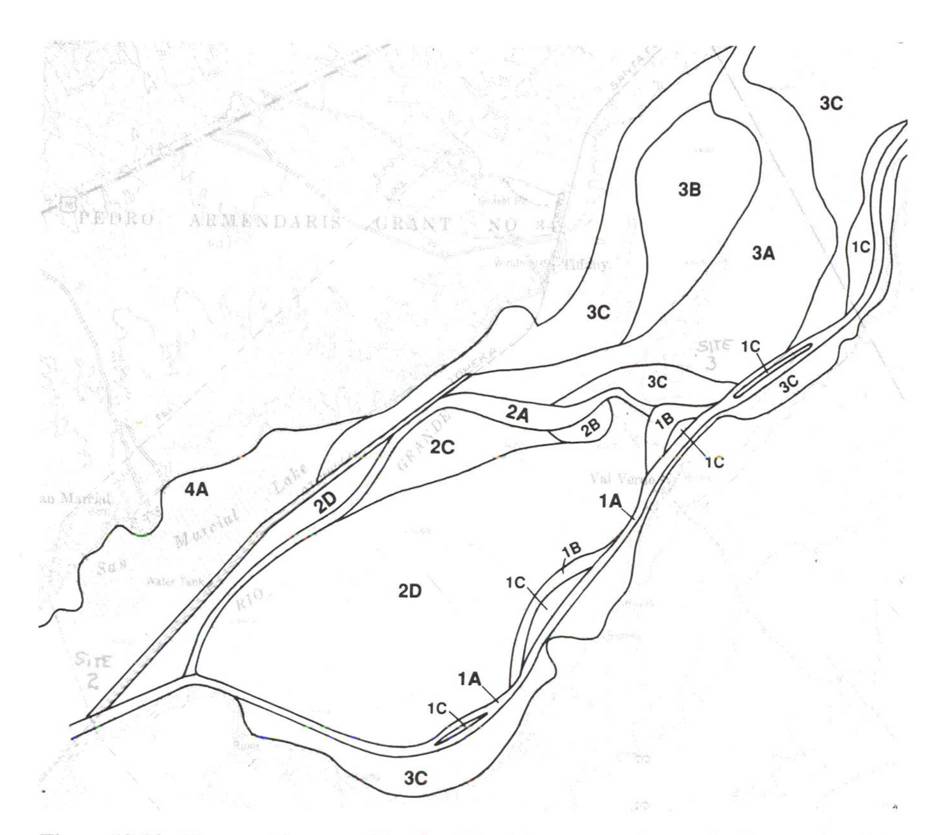

Figure 10.14. Geomorphic map of the San Marcial representative reach. See number 11 in Figure 9.2 for the location. Labeled areas on the map are (1A) 1984 active channel; (1B) 1984 overflow channel, active channel during the 1970s and early 1980s; (1C) 1984 flood plain, active at least from the early 1970s to present; (2A) pre-mid-1940s–1960s active channel; (2B) pre-mid-1940s–1960s flood plain; (2C) pre-mid-1940s–1960s flood plain; (2D) pre-mid-1940s–1960s flood plain and overflow area, probably overflowed in 1941 and established geometry, with flood flow bifurcated with one split on each valley side; (3A) pre-1941 channel; (3B) pre-1941 flood plain; (3C) pre-1941 mid-channel island and channel side bar; and (4A) pre-1941 lake and backswamp area isolated by railroad embankment, sedimented in by 1970s and probably not a deposition site for main channel materials after construction of the railroad.

phreatophytes. In one test plot in my study, the density of tamarisk is more than 500,000 plants per square kilometer. The only truly diverse riparian community in the reach grows in the 1920s–1930s abandoned channel, where tamarisk shares the available space with cottonwood, russian olive, and willow (3A, 3B, and 3C in Figure 10.14). The density of tamarisk in other areas apparently keeps out competitors. Community structure is a useful indicator of the distribution of some geomorphic and sedimentologic features. The west branch of the 1941–1953 channel, for example, is sparsely populated by trees and is easily found in aerial photography and in the field.

The 1941–1953 west channel (2A in Figure 10.14) also has distinct sedimentary characteristics, containing less than 3 percent silt and clay. The flood plain that was associated with the channel (2D in Figure 10.14) contains sediments with an average of 35 percent fines. As in reaches farther upstream, the age of the deposit appears to be inversely related to its percentage of fine material. The flood plain of the 1920s and 1930s contains about 26 percent silt and clay.

The geomorphic and sedimentolgic map of the reach based on aerial-photographic, field, vegetation, and sediment evidence shows that the present channel arrangement bears little connection to the older channels and their deposits. The reach preserves both channel and flood-plain deposits from three separate periods: 1920s–1930s, 1941–1953, and post-1953. The last series, the most likely to contain plutonium if it were available from upstream sources, is the most limited in a geographic and volumetric sense. Given this fact, along with the dilution of sediments by tributary additions and the great distance from upstream sources (313 km from Española), it is unlikely that there are major concentrations of plutonium in the reach.

Summary

The review in Chapters 9 and 10 of detailed channel changes and the resulting fluvial deposits and forms in 11 representative reaches of the Northern Rio Grande produces the following general conclusions:

1. The channel has become consistently narrower over the past several decades, a process that began before the closure of Cochiti Dam and that also appears in places not affected by the dam.
2. The channel has metamorphosed from a braided system to a single-thread system.
3. Riparian vegetation communities have composition and/or structures that are directly linked to underlying forms and sediments. Landforms and materials control the geographical patterns of the vegetation in each reach, although not necessarily in the same way throughout the entire Northern Rio Grande.
4. The characteristics of the sediment in each reach are also directly linked to their fluvial history. The variations are not consistent throughout the study area because of tributary infusions. For example, a flood plain in a northern reach may have particle-size characteristics different from those of a flood plain in a southern reach, but in both cases they contain finer sediments than nearby abandoned channels. The post-1941 sediments are less than 1.5 m deep and usually less than 1.0 m deep.

5. Plutonium from mountain fallout zones and from Los Alamos National Laboratory is most likely to be found in those reaches of the river upstream from Cochiti Dam and possibly in the reaches between Cochiti Dam and the Jemez River. As the downstream distance increases from the confluence with the Jemez River, the contribution of plutonium to the system from Los Alamos is likely to become so small that it is unrecognizable, although plutonium from fallout is likely to be present.

11

Simulation of Sediment and Plutonium Dynamics

Objectives of the Model

The empirical data reviewed in the previous chapters indicate that low levels of plutonium can be found in sediments of the Rio Grande system. It is not readily apparent why the concentrations are low, given that the concentrations in sediments of the upper Los Alamos Canyon are one to three times greater than those in the main river. The explanation of observed concentrations probably lies in the complexities of the water and sediment system. Flash floods on the tributary occasionally evacuate some of the relatively plutonium-rich sediments into the Rio Grande, but when they enter the main river they are subject to two river processes that produce low plutonium concentrations in sedimentary deposits: mixing and dispersal. The concentrations are diluted when the sediments from Los Alamos Canyon combine with the sediments from the Upper Rio Grande and from tributaries (Figure 11.1), which contain fewer contaminants. The dispersal of plutonium on a scale of tens of kilometers along the river also generally lowers the concentrations, although the river processes deposit the contaminated materials in specific places rather than diffusing them completely throughout the river system.

The consequences of the river's complex contaminated sediment processes can be illustrated by means of direct measurement, laboratory experiments, or numerical simulation, but only the last alternative is feasible. Because the mixing, diffusion, and deposition in the Rio Grande cannot be directly observed, no detailed empirical data about them are available. And in order to be feasible, laboratory experiments must duplicate the significant components of the real system using flumes, and physical models of the system require changes in scale that may result in inaccurate representations of the actual system. The sediment in physical models must be smaller than that in their real counterparts, for example, but because the water cannot be "scaled down" in the model, the fine sediment in

Figure 11.1. Dynamic models of river channels may account for channel changes through sedimentation that reduces width and depth, but not location changes such as in the Jemez River below Jemez Dam. (*a*) The view looking east in 1972, showing a braided channel without flow. (H. E. Malde, photo 771, U.S. Geological Survey Photography and Field Records Library, Denver) (*b*) The same view in 1990, showing the channel with water. In this later view, the channel has become narrower, and it has straightened in the middle-right portion of the view. (W. L. Graf, photo 78-X)

the laboratory behaves differently from the coarser sediment it represents in the real system. Therefore, the only possible detailed analysis of the system of contaminant transport and storage in the Rio Grande is through a numerical simulation model.

The term *model* has a variety of meanings, but in this context it refers to an intellectual structure that simplifies reality.[1] The river system is too complex for complete description, even using all the known relationships among river-related variables. The basic adjustable variables of the system are channel width, depth, gradient, velocity of flow, amounts of water and sediment, sediment size, and hydraulic roughness.[2] Although hydraulic and geomorphic theories indicate how these variables are related to one another, the number of equations based on fundamental physical connections is less than the number of variables, so the problem has no finite solution. Simplifications and empirically derived functions substitute for the more desirable finite solutions, and the result is a representation that is an imperfect reflection of the real system. A general model is useful, however, because its operation obeys our best understanding of natural laws and it can clarify the complexities of the real system. The model also permits the user to predict future processes, assuming that the model is adequately calibrated to previously observed conditions and that no new variables intervene in future operations.

A numerical simulation model is a computer program containing formulas that describe the operation of the real-world river system. The numerical components of such models are usually relatively simple mathematical statements that connect parameters describing various attributes of the river, such as its dimensions, amounts of water and sediment in transport, and variables related to energy and momentum in the system. In a dynamic model that accounts for the passage of time, the numerical components also perform simple accounting functions such as keeping track of the total amounts of material that pass through the system or that are internally stored. Simulations use the program to mimic the system's operation, including the changes that result from its own operation. Predictions are based on simulations that extend beyond the currently available data.

The objective of using the numerical model to analyze the concentrations of plutonium in sediments of the Rio Grande is that it be a simple, dynamic, spatially variable sediment transport and storage model based on the distribution of force and resistance in the system.

1. *Simple.* The objective is to construct a program that operates on a personal computer without the need for the support of a larger machine and with an interactive design so that the user does not have to be a specialist. Although some hydraulic models currently in use require detailed survey data, the objective is to make the data demands simple, without the need for expensive data not found in currently available data bases such as stream-gage records and aerial photography.

2. *Dynamic.* The model should be capable of simulating the passage of time by updating itself, taking into account changes in channel geometry resulting from sedimentation or erosion so that the next time unit in the simulation can accurately account for the products of the system's previous operations.
3. *Spatially variable.* The model should take into account the variation from place to place in the physical characteristics of the stream channel and in the nature of the discharge of water and sediment. The model should be more than a simple input–output structure and should depict internal geographic variation along the stream channel.
4. *Transport and storage.* The model should be able to track the amounts of sediment and contaminants that are input, output, and stored at various locations in the system. It is also necessary to account for the remobilization of materials temporarily deposited in the system and near the channel.
5. *Force and resistance.* Because hydraulic force and resistance are the primary physical explanations of river processes, the model should directly represent these aspects of the system and should be able to assess the effects of changes in these aspects on contaminant mobility.

Outline of the Model

The model developed for exploring the dynamics of sediment-borne plutonium in the Northern Rio Grande is the Riverine Accounting and Transport model, or RAT (Figure 11.2). The RAT is a computer program written in Q-BASIC language consisting of about 1,300 statements based on known, simple equations. Each statement performs only one action, so the language it is not as powerful as Pascal, FORTRAN, and some other languages. The advantages of using Q-BASIC are that it is automatically available on most IBM-compatible personal computers without the need for additional compilers and that it can be easily modified by users who wish to enhance its calculations by inserting more sophisticated mathematical statements.

The general structure of the RAT is to define river processes in a series of linked channel segments. In this case, the simulation is limited to the Northern Rio Grande from Otowi to the headwaters of Cochiti Reservoir. Although it is theoretically possible to simulate the river over much greater lengths, the calculations are unwieldy with more than 30 segments. Analyses of current processes and predictions of future processes must focus on the Otowi–Cochiti Reservoir reach, because the reservoir is the termination of the downstream sediment transport. Each channel segment has its own physical characteristics, such as length, width, depth, and gradient. The program mathematically passes water,

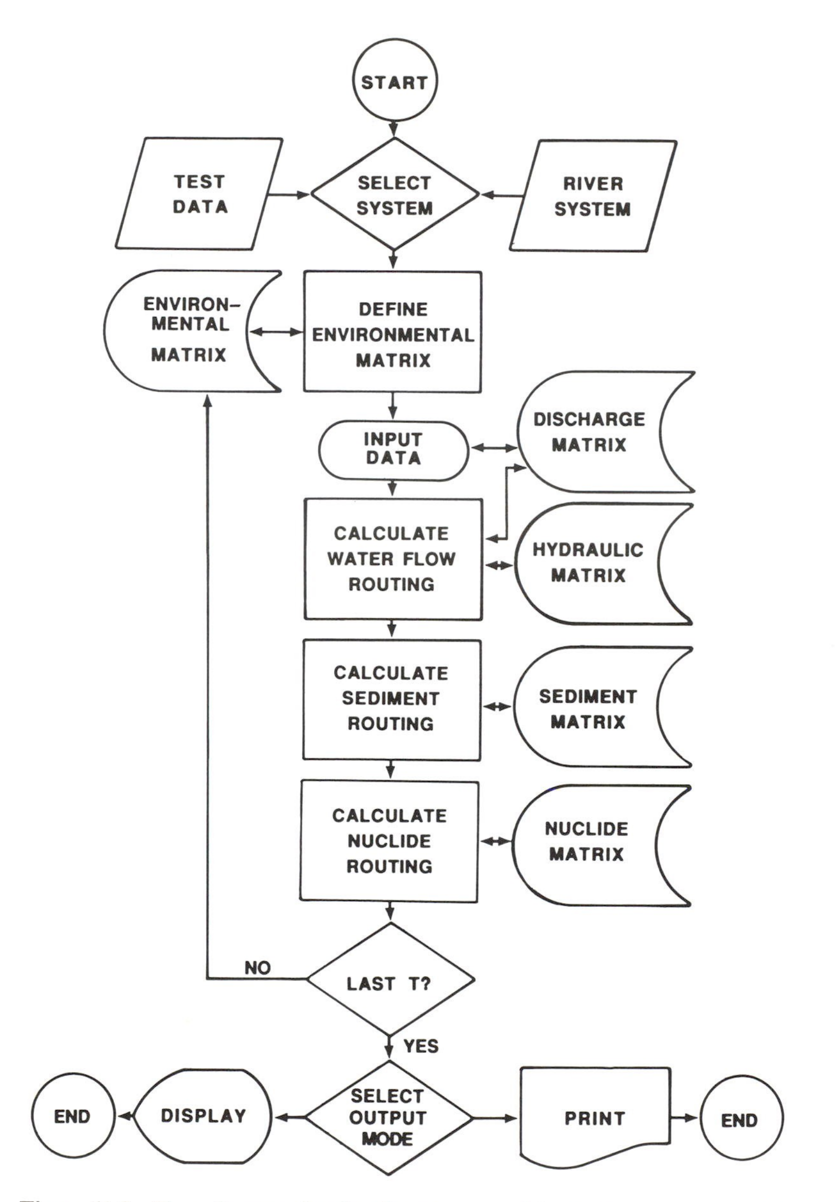

Figure 11.2. Flow diagram showing the structure of the RAT simulation model.

sediment, and contaminants through the sequence of segments, calculating the force available for transport, the resistance of the segment, and the resulting erosion or deposition in each segment.

After calculating the processes for one simulated day, the program changes the physical characteristics of each channel segment to account for erosion or deposition, and then it begins again with calculations for the next day. An internal accounting system tracks the amounts of sediment and contaminants stored in or lost from each segment, as well as the amounts released from the lower end of the last segment (that is, deposited in Cochiti Reservoir under current conditions). The model's calculations are mostly invisible to the user, who sees only a computer screen that asks for various kinds of input and that prompts the user, while the program is running, to choose among a series of options.

The specific structure of the RAT consists of a series of calculations contained in a loop and connected to a series of data matrices or arrays. The program begins by asking the user to select a river system for analysis. Two options are available: a simplistic three-segment synthetic test system for experimental purposes and the Northern Rio Grande. The Northern Rio Grande option includes data describing 22 segments of the stream beginning at Otowi Bridge, immediately upstream from the confluence with Los Alamos Canyon, and ending at the headwaters of Cochiti Reservoir, downstream from the confluence with the Rio de los Frijoles (Figures 11.3 and 11.4). After the user selects the desired option, the program reads the data describing the initial physical characteristics of each section (stored internally in the program) and establishes a geomorphological environmental matrix.

The environmental matrix has three dimensions: stream segment numbers, characteristics of each segment, and time units (Figure 11.5). The segment numbers range from 1 at Otowi Bridge to 22 at Cochiti Reservoir. The characteristics recorded for each segment are the segment number, the downstream distance of the segment's end point from the beginning of the entire set of segments, the segment's length, the elevation of the segment's end point, the mean channel gradient, the mean channel width, the mean channel depth, the hydraulic roughness of the channel, the cumulative channel bed area from the beginning of the entire set to the end point of the segment, a number to connect the segment to the passage of flood waves, and the time unit number. For the sake of simplicity, the calculations assume a rectangular channel (Figure 11.6). The channel segments are 200 to 2,000 m long, with their specific starting and ending points determined by geomorphic conditions along the river. For example, some segments are narrow reaches, whereas others are wider and contain some space for storing sediment. Generally, wide and narrow segments alternate with each other along the channel. The time units range from one day to any maximum number, but practical programming limits the realistic maximum number of simulated days to about 10.

Once the program establishes the initial geomorphological environmental matrix, the user inputs daily mean discharges that enter at the upstream beginning

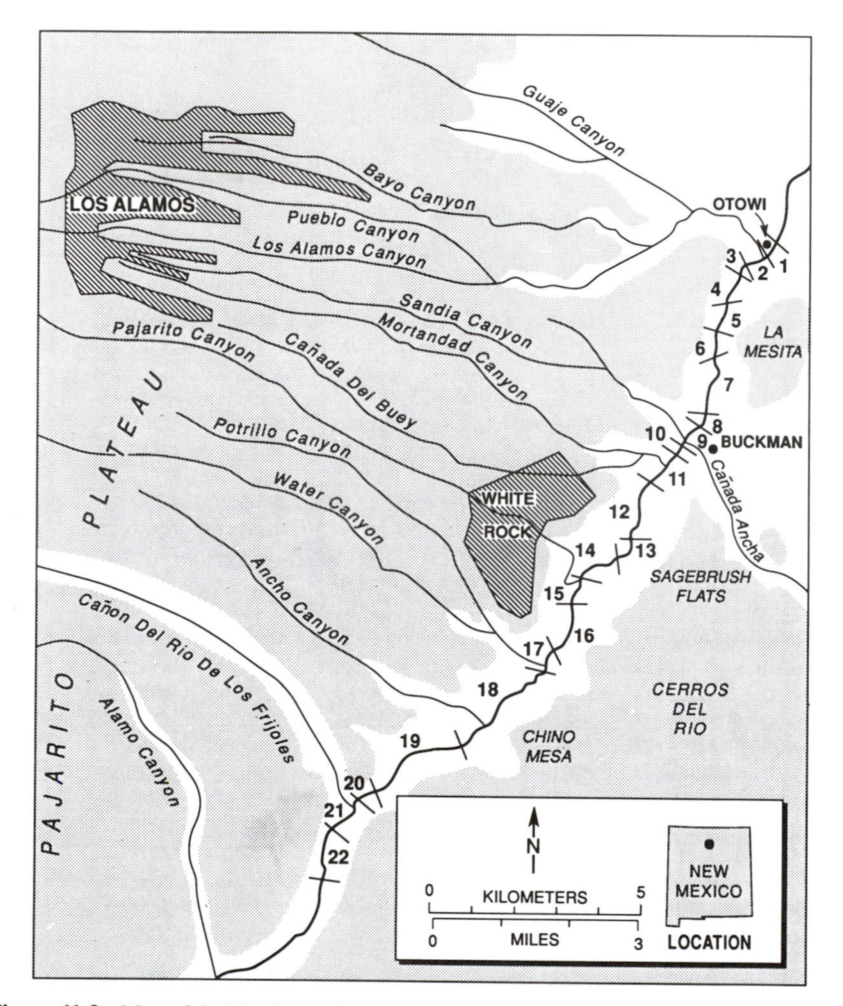

Figure 11.3. Map of the Northern Rio Grande near Los Alamos showing the 22 numbered segments used to define the system for the RAT simulation model.

point of the entire set of segments. The user's input of discharge values causes the program to establish a second data matrix or array for the hydraulic values in each stream segment. The hydraulic matrix, like that for the geomorphic parameters, has three dimensions: segment numbers, segment characteristics, and time units. Again, the segment numbers range from 1 to 22 and the time units from 1 to about 10. The hydraulic characteristics of the channel segments are segment number, an unadjusted discharge value derived from the input data, the total at the end of the segment, an adjusted discharge value that accounts for the transmission losses, depth of flow, velocity of flow, stream power per unit area of the channel bed, and stream power at the downstream end of the cross section. All the hydraulic

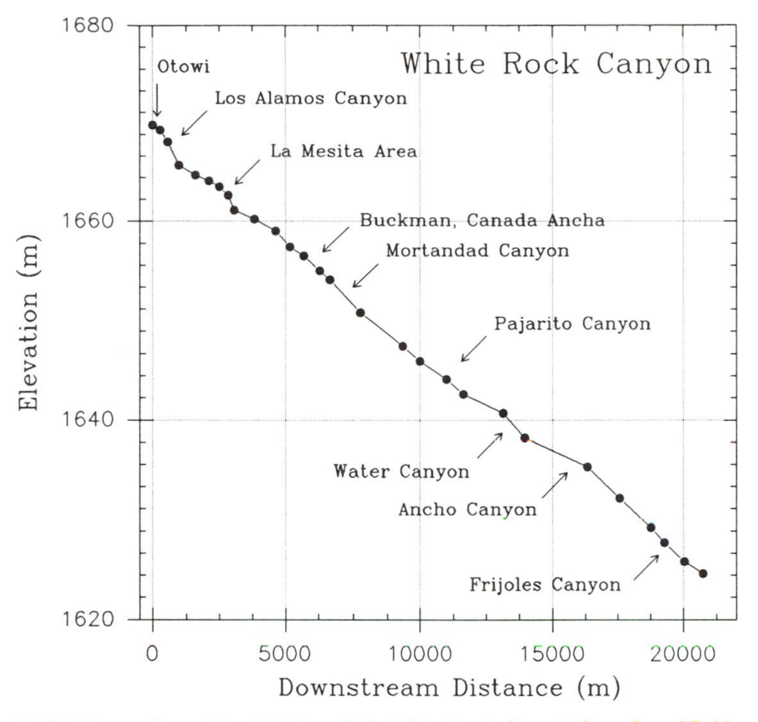

Figure 11.4. The gradient of the Rio Grande in White Rock Canyon from Otowi Bridge to the headwaters of Cochiti Reservoir. The steep reaches generally correspond to narrow sections of the channel, whereas the reaches with more shallow gradients are wider and have some space to store sediment.

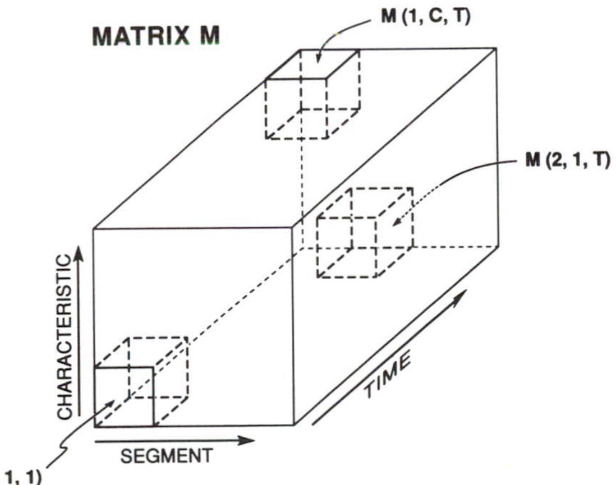

Figure 11.5. Conceptualization of an example data matrix from the RAT simulation model. The matrix is designated M, and each bit of data in the matrix is identified by three values: the segment number, the time unit, and the type of data or characteristic.

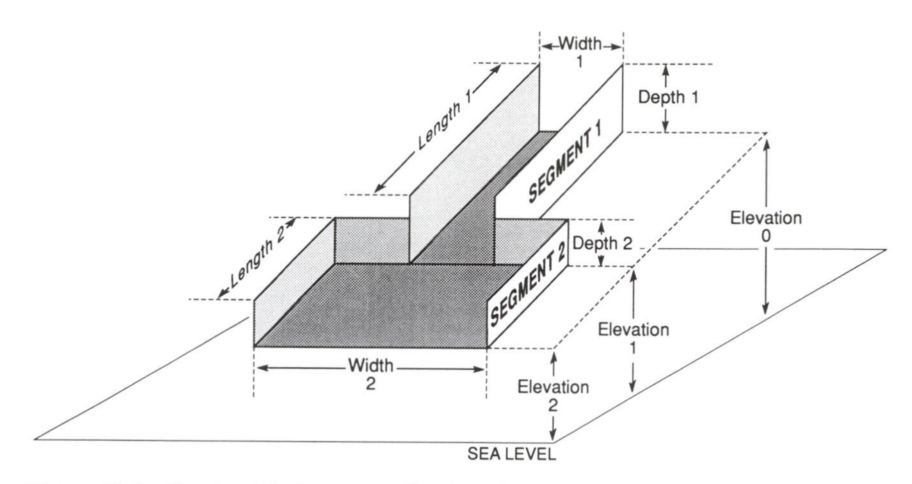

Figure 11.6. The simplified conceptualization of two sample segments of stream channel as used in the RAT simulation model.

calculations and subsequent sediment and plutonium transport calculations use the adjusted discharge. Calculations for the components of the hydraulic matrix use standard, widely accepted formulas employed by geomorphologists and hydraulic engineers (Appendix J).[3]

Using the hydraulic matrix, the RAT program calculates values for a third matrix or array that describes the sediment system segment by segment through the entire series. Like the other matrices, the sediment matrix has three dimensions: segment numbers, characteristics, and time units. As before, the segment numbers range from 1 to 22 and the time units from 1 to about 10. The characteristics recorded in the sediment matrix are segment number, sediment transport capacity, sediment input to the segment, sediment output from the segment, the change in sediment storage during each time unit, a running total to account for the sediment stored in or eroded from the segment from the beginning of the simulation, and the time unit number.

The calculations in the sediment matrix address only bedload because the suspended load passes through the entire system without appreciable losses through storage or additions through erosion. The stream segments are in White Rock Canyon for the most part, and there are no extensive flood plains that might otherwise interact with the suspended load. The program uses standard formulas for bedload transport that are simple and are derived from first principles rather than from empirical data (Appendix J, parts F, G, H, and I). The contaminated sediment from Los Alamos Canyon probably also travels as bedload because of the relatively course materials involved. From a regulatory standpoint, bedload calculations are of greatest interest because the infusions derived from Los Alamos National Laboratory are most likely to be found in the coarse particles carried by the main river as bedload.

The user informs the program about the input of plutonium into the system, by specifying both the location and timing of the input (by segment number and time unit number) and the total mass of sediments and the plutonium concentration of the input. The system therefore simulates an instantaneous release of contaminated sediment into the stream from an accidental spill, or the influx of contaminated sediment introduced by a flash flood on a tributary stream. In the case of the Northern Rio Grande, the simulations depict the introduction of contaminated sediments from Los Alamos Canyon which joins the stream in segment 2. The program assumes that the fallout-derived plutonium in the system is in sediment entering the segment series from upstream in bedload concentrations of about 0.0025 pCi/g.

The RAT program completes its basic calculations by using these input data to track the plutonium in each segment and to construct a fourth matrix or array. The plutonium matrix has the usual three dimensions of segments, characteristics, and time units with the same size ranges as before. The plutonium matrix characteristics are segment number, mass of plutonium input, plutonium concentration of the input sediment, mass of plutonium output, plutonium concentration in the sediment output, mass of plutonium storage in each time unit, total mass of plutonium stored from the beginning of the simulation, concentration of plutonium in the stored sediment, and time unit. These calculations amount to simple accounting (Appendix J, parts J and K).

After completing the calculations and inserting the resulting values in the respective matrices, the RAT determines whether the simulation is complete. If the user has specified the simulation of only one day, the program prepares the data for display. If the user has specified more than one simulated day, the program returns to the environmental matrix and adjusts the initial conditions and segment characteristics to reflect the results of the processes during the first day. The program changes those segments containing stored sediment by making them narrower and more shallow and changes those elevations that ultimately cause changes in the gradient. The program mathematically distributes the stored sediments throughout the segment evenly in the downstream direction and reduces the cross-sectional area so that 80 percent of the reduction is in the depth and only 20 percent is in the width. In other words, when the channel aggrades, it is mostly by vertical accretion. If erosion occurs, the program mathematically enlarges the channel by removing sediments evenly in the long dimension. Cross-sectional changes during erosion cause 80 percent of the sediment to be removed by increasing the channel width and 20 percent to be removed by increasing the channel depth. Thus during erosion, bank erosion is the most important. These deposition and erosion scenarios reflect the changes that have been observed historically in the channel.

During simulations of multiple days, the newly updated environmental matrix is the starting point of a new round of calculations for the hydraulic, sediment, and plutonium matrices, as described earlier. The environmental matrix and the calculations are updated until the program has simulated the requested number of days. At that point, the program prepares the resulting data for output. The RAT either displays these data on the screen or prints them, at which point the program ends.

Model Input

The input supplied by the user for the RAT includes the primary data concerning the water discharge in the main channel and the infusion of sediment and plutonium from Los Alamos Canyon. The daily discharge values for the Rio Grande may be actual values taken from the gaging record at Otowi, so that the resulting simulation represents an interpretation of a known event. The discharge values might also come from hydrologic simulations of past or anticipated future events, or they could be experimental values used to test the system's response to various inputs. Realistic values for the discharge range from tens to hundreds of cubic meters per second (Table 11.1 and Figure 11.7). Stream-gage records show that over periods of 10 to 20 days, the discharge in the Rio Grande varies by 15 percent or less under the conditions prevailing during the major infusions from Los Alamos Canyon. At other times, changes of an order of magnitude are possible.

The program assumes for the purposes of simulation that the Rio Grande transports an amount of bedload sediment that is equal to its bedload transport capacity. Therefore, once the user assigns an input discharge, the program calculates the amount of bedload that the discharge can transport (given the considerations of the channel's geomorphology). The calculations use that value as the input and output of bedload sediment for the first segment. For subsequent segments downstream, however, the amounts of input and output may be different for each segment, because the input depends on the conditions upstream and the

Table 11.1. Discharge of the Rio Grande at Otowi

Event	Discharge ($m^3 s^{-1}$)
Daily minimum flow	2
Maximum flow of record	623
Daily mean flow	43
Flood flows:	
1-year	48
2-year	217
5-year	342
10-year	429
25-year	606
50-year	635
100-year	700

Source: Flood flows estimated from annual flood series; other values from U.S. Geological Survey data.

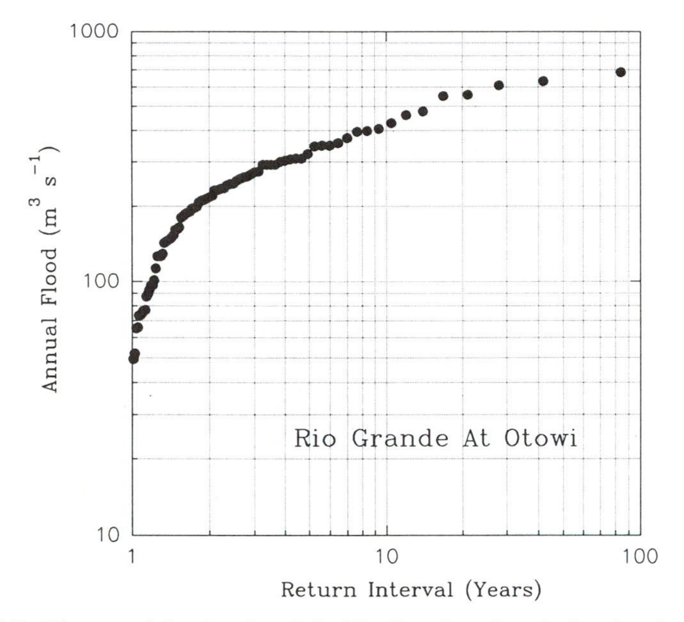

Figure 11.7. The annual flood series of the Rio Grande at Otowi, showing the probable return interval of annual floods of various magnitudes. For example, an annual flood of about 400 m^3 s^{-1} is likely to occur about once every 10 years.

output depends on the conditions in the segment. The differences between input and output lead to erosion or sedimentation within each segment.

The user specifies the nature of the infusion of sediments into the main channel from Los Alamos Canyon in four ways: location, time, amount of sediment, and plutonium concentration. Because of the spatial structure of the model, Los Alamos Canyon enters the main stream in segment 2, and the time unit refers to the simulated day of the infusion, generally the first day. The quantities of sediment emptied into the main channel by flash floods on the tributary, all assumed to enter on a single day, would be reasonably 5,000 to 15,000 Mg. The concentrations of plutonium during the four years of primary concern (1951, 1952, 1957, and 1968) were probably 1.700 to 3.000 pCi/g (Table 11.2), with lower quantities of sediment and lower concentrations in other events. The amount of water contributed to the main channel by flash floods in Los Alamos Canyon is small compared with the amount in the main channel, so the program does not take into account the tributary contribution.

Model Output

The program presents data resulting from the calculations as tables of numbers extracted from the various matrices or arrays in the program. Four types of

Table 11.2. Tributary inputs for Riverine Accounting and Transport program simulation of critical years

	Sediment		Total Plutonium	
Year	**Tons**	**Mg**	**mCi**	**fCi/g**
1951	9,814	8,903	16.90	1,898
1952	6,316	5,730	17.61	3,073
1957	16,470	14,942	43.95	2,941
1968	14,120	12,810	21.82	1,708

Source: Data from unpublished calculations by L. J. Lane.

data are available, and in each case they are organized by stream segment and time unit.

1. *Geomorphologic data* describe the channel characteristics such as width and depth and are derived from the environmental matrix. These data can reveal the responses of the channel to various inflows of water and sediment.
2. *Hydraulic data* describe the behavior of the water flow in the channel segments and are useful mostly for program calculations. They describe the amount of power available to transport sediment through the system.
3. *Sediment data* describe the movement and storage of bedload sediment in the entire channel system, segment by segment. Changes in storage in a particular segment through time can help explain the dynamics of plutonium in the system.
4. *Plutonium data,* including inventory amounts and the concentrations in bedload in each segment, are the most important output of the model. The data can trace mixing and diffusion processes from the segment in which the infusion occurred.

In presenting these data to the user, the RAT constructs data tables to be displayed on the computer screen or printed on paper. The program organizes the data according to spatial or temporal variation. In the spatial variation, the table displays the type of data chosen by the user for one time period through all the segments (Table 11.3). In this form, the plutonium data can reveal the geographic locations (identified by segment number) of plutonium storage in the system. In the temporal variation, the table shows the values describing conditions in a single segment (chosen by the user) over all the time periods simulated by the program (Table 11.4). In the temporal form, the program shows time-based changes in the inventory and concentration of plutonium in a particular stream segment.

Regardless of the method of organizing the data for output, the program indicates the amount of plutonium that passes completely through the series of

Table 11.3. Example of simulation results showing temporal variation of total plutonium in a single stream segment

Day	Input Amount (mCi)	Output Amount (mCi)	Storage Change (mCi)	Storage Balance (mCi)	Storage Concentration (fCi/g)
1	17.61	0.46	+ 17.14	17.14	1,159.26
2	0.01	0.37	−0.36	16.78	925.49
3	0.01	0.24	−0.23	16.55	608.37
4	0.01	0.18	−0.17	16.37	451.54
5	0.01	0.14	−0.13	16.24	358.28
6	0.01	0.12	−0.11	16.13	296.56
7	0.01	0.10	−0.09	16.04	252.75
8	0.01	0.09	−0.08	15.96	220.08
9	0.01	0.08	−0.07	15.89	194.78
10	0.01	0.07	−0.06	15.84	174.65

Note: Calculations for segment 2 with contaminated sediments introduced to segment 2, the confluence of Los Alamos Canyon with the Rio Grande (Figure 11.3); continuous discharge of Rio Grande during simulation was 68 $m^3 s^{-1}$, a typical early summer, nonflood discharge.

stream segments. This plutonium was probably deposited in and along channels downstream from White Rock Canyon before late 1973, but after that year, it was deposited in Cochiti Reservoir. Simulations of future system behavior must assume that the reservoir will contain this "pass-through" amount. The program also calculates the probable concentration of plutonium in these downstream deposits.

At present there is not enough empirical data to assess completely the accuracy of the model predictions, although some comparisons between model predictions

Table 11.4. Example of simulation results showing geographical variation of total plutonium in a single time period

Stream Segment	Input Amount (mCi)	Output Amount (mCi)	Storage Change (mCi)	Storage Balance (mCi)	Storage Concentration (fCi/g)
1	0.01	0.01	0.00	0.00	0.00
2	0.01	0.37	-0.36	16.78	925.49
3	0.37	0.52	0.00	0.00	0.00
4	0.52	0.16	+0.36	0.83	1,023.15
5	0.16	0.12	+0.03	0.08	1,049.89
6	0.12	0.15	0.00	0.00	0.00
7	0.15	0.29	0.00	0.00	0.00
8	0.29	0.65	0.00	0.00	0.00
9	0.65	0.13	+0.53	1.11	1,098.74
10	0.13	0.16	0.00	0.00	0.00

Note: Calculations for day 2 after contaminated sediments introduced to segment 2, the confluence of Los Alamos Canyon with the Rio Grande (Figure 11.3); continuous discharge of Rio Grande during simulation was 68 $m^3 s^{-1}$, a typical early summer, nonflood discharge; no storage in some segments.

and observed conditions are possible. For example, the model's predictions of the concentration of plutonium in stored bedload sediments at Buckman (segment 8) are the same as those actually found in the sediments by recent sampling. Given that the amounts of plutonium are extremely small in relation to the amounts of sediment involved, the prediction of concentration values within an order of magnitude of actual values is probably the best that can be expected, even with more sophisticated calculations. The inventories of stored plutonium are reasonably accurate because the problem of prediction is actually one of distributing a reasonably well known quantity of input material.

In summary, the RAT is a simple computer program that simulates river processes that occur over a period of days in 22 river segments along the Northern Rio Grande. The program allows the nonexpert user to simulate past events or to experiment with hypothetical future situations. The RAT produces information about plutonium concentrations and inventories that is accurate to at least an order of magnitude. The main value of the simulations is to determine geographic and temporal variations in concentrations or inventories.

Simulation Results

These simulations using the RAT program supply the magnitude and speed of mixing, dispersal, and concentration of plutonium in the bedload sediments of the Northern Rio Grande. The model also can account for simple channel adjustments (Figure 11.8). The remainder of this chapter discusses using the model to describe how plutonium concentrations in stored sediments along the Rio Grande respond to four major controls: (1) variation in the mass of input from Los Alamos Canyon, (2) variation in the concentration of input, (3) the passage of time and the continuing river processes in the main channel, and (4) increasing distance downstream away from Los Alamos Canyon.

Although the results of the simulations could be presented as tables of numbers, the data are easier to interpret if they are shown in graphic form. The numerical results of simulation runs provide data suitable for use with standard, commercially available graphics programs for personal computers. The graphs presented in this chapter are the products of combining the data generated by simulations with SIGMAPLOT Version 4.0, a graph program created by Jandel Scientific, Inc. The results in each graph appear as lines representing interpolations between the data points provided by the simulations. The lines are cubic spline curves that closely approximate the general trends of the data through mathematical smoothing and interpolation between given points. Each graph presents the results of 40 to 50 simulations. Because of the magnitude of potential error in the calculations, simplifications in the simulation process, and the smoothing of data, the graphs do not offer precise predictions. The values for a

Figure 11.8. The RAT simulation model describes the flow of water and sediment through relatively simple river reaches such as the Rio Grande at Coronado State Monument north of Albuquerque, and it can take into account changes in the channel configuration. (*a*) A tourism promotion photo from the late 1940s looking southeast across the river to the Sandia Mountains, showing a broad channel with a developing riparian forest on the opposite bank. (Unknown photographer, photo 52349, Museum of New Mexico) (*b*) The same view in 1991, showing a narrower channel, a mature riparian forest on the opposite bank, and dense grass cover on the near slope. (W. L. Graf, photo 104-12)

particular place at a particular time cannot be visually extracted from the diagrams. But the graphs do represent informative general trends, and they are useful sources of generalizations about the operation of the river and plutonium system.

Varying the Input Mass

Figure 11.9 shows the results of the RAT program for 10 simulated days of river processes in segment 2 of the Rio Grande (located on Figure 11.3). The concentration of plutonium in the injected sediments was 3,000 pCi/g, a representative value (Table 11.2). Throughout this chapter, for precision and simplicity, measurements are in fCi/g (1 fCi/g = 0.001 pCi/g). The graph in Figure 11.9 indicates that if the input mass from Los Alamos Canyon were 2,500 Mg (an amount smaller than the injections during the four most important years) and the discharge of the Rio Grande was 50 m^3/s, after 10 days the concentration of plutonium in sediment stored along the Rio Grande in segment 2 would be about 130 fCi/g. After 10 days of mixing and dispersion, the concentration of plutonium in the bedload would have fallen more than 95 percent from the value for the input from Los Alamos.

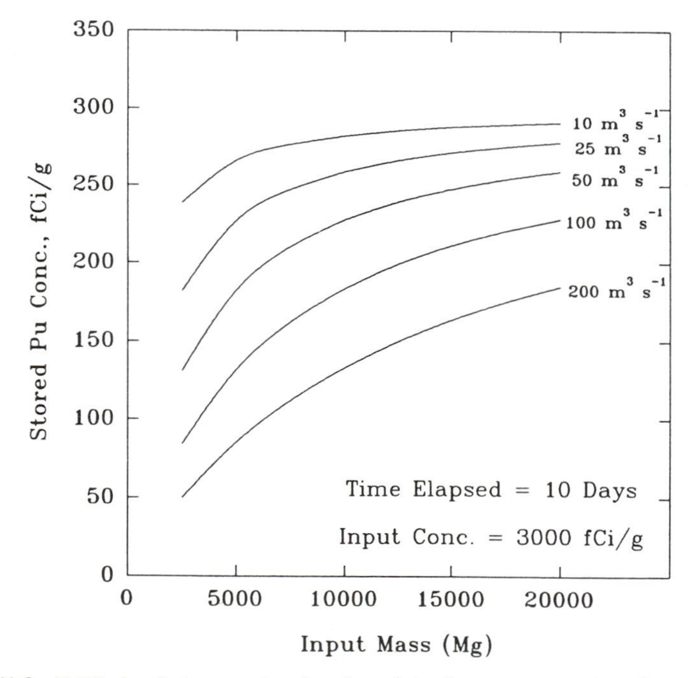

Figure 11.9. RAT simulation results showing plutonium concentrations in stored sediments along the Rio Grande as a result of varying input masses. The individual lines in the diagram represent conditions with different discharge conditions in the Rio Grande, ranging from low flow (10 m^3 s^{-1}) to major flood (200 m^3 s^{-1}).

A simulation for about 10 days is a reasonable approximation of what probably happens in the Northern Rio Grande. In that period of time, the sediment would probably have entered the river, mixed with other materials, and been deposited at some point downstream. Remobilization, remixing, and further dispersal should reduce the concentrations of plutonium even more. Therefore, the simulation results produce maximum probable values for the concentrations of plutonium. In actual measurements, however, lesser concentrations than predicted by the RAT program are probable.

This decline in concentration is the result of mixing with the sediment moving downstream through the main channel in quantities that are huge in comparison with the amount of material introduced from Los Alamos Canyon. Because the bedload sediment descending the river from upstream contains only 2.5 fCi/g of plutonium from fallout, the mixing rapidly decreases the concentrations from the tributary. This mixing process is powerful, because even if we consider input masses as large as 20,000 Mg (a mass more than 50 percent greater than any actual input), the resulting concentration of plutonium in stored bedload in segment 2 increases to only about 260 fCi/g. In other words, although the amount of sediment from Los Alamos Canyon is increased by a factor of 4, the concentrations in stored sediments in the main channel increase by only a factor of 2. This arrangement is common in river processes that are dominated by nonlinear mathematical relationships.[4]

These operations of the model show why plutonium concentrations are relatively low in bedload deposits along the main channel. The simulation indicates that when using realistic values for the 1952 infusions from Los Alamos Canyon (original concentrations of about 3,000 fCi/g in about 6,000 Mg of input to the main river, flowing at about 68 m^3/s), the concentrations of plutonium in stored sediment along the Rio Grande would be about 190 fCi/g after 10 days. Because the system probably operated for a longer period before the final deposition of these sediments, further mixing is likely, and so the value of 190 fCi/g is an expected maximum.

The amount of discharge in the Rio Grande also strongly influences the concentration of stored plutonium because the low discharges lead to relatively slow mixing and the high discharges accelerate the mixing and produce decreasing concentrations. For example, continuing the example of the 1952 conditions but assuming a discharge of 200 m^3/s, the resulting stored plutonium concentrations would not be 190 fCi/g but would be depressed to about 80 fCi/g. At high discharges, immense quantities of sediment (with low plutonium concentrations) enter the system from upstream.

In another example, using realistic values for the 1957 infusions (instead of those of 1952) from Los Alamos Canyon (a concentration of about 3,000 fCi/g, about 15,000 Mg of input, and a main channel discharge of about 120 m^3/s), the simulations represented in Figure 11.9 suggest storage concentrations of about 200 fCi/g. These results explain why the deposits dating from 1952 (with concentrations of about 190 fCi/g) and 1957 may reasonably be expected to contain similar concentrations of plutonium, even though the 1952 sediments derived from an infusion mass that was only one-third as large as the 1957 infusion. In the latter year, the river discharge was twice as great as that in the earlier example, nearly compensating for the larger input mass by more vigorous mixing in the main river.

Varying the Input Concentration

The variation in plutonium concentration in releases from Los Alamos Canyon also affects the ultimate concentrations in the sedimentary deposits along the main channel. Simulations with the RAT program show that concentrations of plutonium in bedload decline by about 50 percent within a single day of their introduction into the main channel (Figure 11.10). The dispersion of the sediments probably explains this precipitous decline, with dilution by main-channel sediments with low plutonium concentrations also playing a role. The increasingly high discharges in the main channel cause increasingly large declines in plutonium concentrations through mixing, even on a restricted time scale of one day. This variation in discharge has a more depressive effect on the higher concentrations, as shown in Figure 11.10.

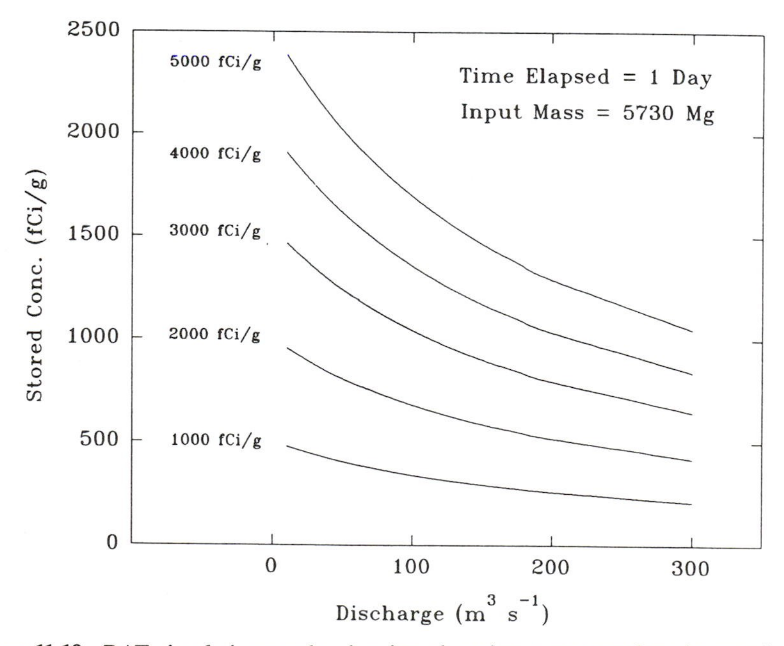

Figure 11.10. RAT simulation results showing plutonium concentrations in stored sediments along the Rio Grande as a result of varying discharges in the Rio Grande. The individual lines in the diagram represent different concentrations in the input masses discharged from Los Alamos Canyon, ranging from 1,000 fCi/g to 5,000 fCi/g.

The range of probable discharge values during the actual infusions of plutonium from Los Alamos Canyon in the years 1951, 1952, 1957, and 1968 was about 15 to 120 m^3/s. The mass used in the simulations shown in Figure 11.10 was the amount from 1952 when the concentration in the tributary sediments was close to 3,000 fCi/g. After a single simulated day, the concentrations in main-channel sediments in segment 2 were about 1,200 fCi/g. Subsequent operations of the system showed further decreases in plutonium concentrations, and the materials were not likely to have been deposited until at least several days later when the discharges have decreased.

The simulation results shown in Figure 11.10 indicate that the discharges in the Rio Grande dampen the range of variation in input concentration. For example, consider the input concentrations that are 1,000 to 5,000 fCi/g. At discharges of less than about 20 m^3/s, plutonium concentrations in main-channel deposits have a range of nearly 2,000 fCi/g. At discharges of more than 200 m^3/s, however, channel-deposit concentrations have a range of only about 1,000 fCi/g.

The Effect of Time

Once materials from Los Alamos Canyon enter the Rio Grande, they are mixed and dispersed downstream quickly as a result of surprisingly powerful mainstream processes, even at low discharges. Simulations using the RAT program to mimic the system's operation over simulated 10-day periods show that during this period, plutonium concentrations in bedload sediment fall by about 80 to 90 percent, depending on the discharge in the main channel (Figure 11.11). They decline because the amounts of sediment from the tributary are small with respect to the amounts in transit in the main channel and because the plutonium concentrations in the bedload contributed by the main channel from upstream fallout are relatively low.

The 1952 case is a useful example of the changes over time in plutonium concentration in segment 2 (where Los Alamos Canyon joins the Rio Grande). In that year, Los Alamos Canyon contributed, by means of flash flood, 5,730 Mg of bedload sediment containing 3,073 fCi/g of plutonium to the Rio Grande when the discharge in the main stream was about 68 m^3/s. After 10 simulated days, the plutonium concentration in bedload sediments in segment 2 was about 200 fCi/g. Although this scenario is probably a simplification of actual events, it indicates why the resulting deposits along the main channel were likely to contain relatively low concentrations of plutonium, even though they began their journey in Los Alamos Canyon with relatively high concentrations. All samples of sediment collected in the field from near-channel deposits along the Rio Grande show plutonium concentrations below this probable maximum value.[5]

As in previous simulations, increasingly high discharges result in decreasing plutonium concentrations in the main channel's bedload sediments. Irrespective of

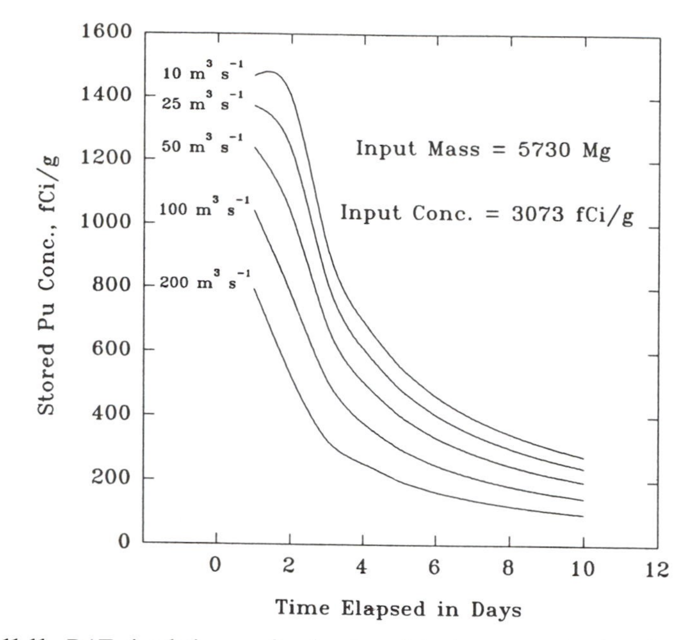

Figure 11.11. RAT simulation results showing plutonium concentrations in stored sediments along the Rio Grande as a result of varying lengths of time for mixing and dispersion. The individual lines in the diagram represent results with different discharge conditions in the Rio Grande, ranging from low flow (10 m^3 s^{-1}) to major flood (200 m^3 s^{-1}).

the magnitude of discharge, however, the greatest amount of mixing and dispersion (with attendant declines in plutonium concentrations) came during the first four to six days (Figure 11.11). Thereafter, the changes were more gradual in all the discharge examples. The slight increase in concentrations between days 1 and 2 for the discharge of 10 m^3/s in Figure 11.11 is an artifact of the cubic spline curve used to represent the data, rather than an actual increase.

Over several days, the operations of the main river cause infusions of different concentrations to drop toward a common, relatively low value. Figure 11.12 shows the results of simulations for the 1952 case by plotting the plutonium concentrations in segment 2 for a variety of initial concentrations. The line in Figure 11.12 representing 3,000 fCi/g is close to the probable actual value for 1952. Although the range in concentrations on simulated day 1 was 4,000 fCi/g (that is, the range from 1,000 to 5,000 fCi/g), after 10 simulated days this range had fallen to only 230 fCi/g. The deposits along the main channel from a variety of infusions from Los Alamos may therefore not exhibit radically different plutonium concentrations even if they had such differences at the times of injection. The principal differences, up to an order of magnitude, that are likely to be observable in deposits are those between materials derived from Los Alamos and those from the upper basins of the Rio Grande. Such

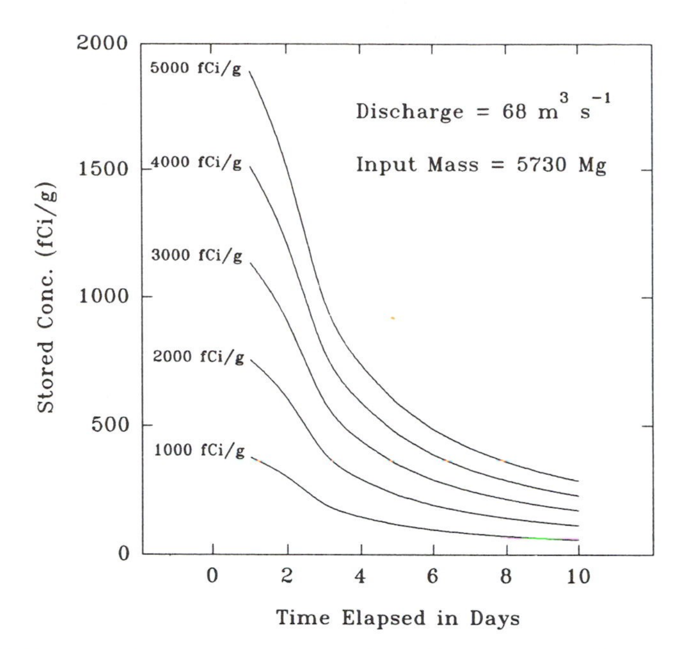

Figure 11.12. RAT simulation results showing plutonium concentrations in stored sediments along the Rio Grande as a result of varying lengths of time for mixing and dispersion. The individual lines in the diagram represent different concentrations of plutonium in the input masses from Los Alamos Canyon.

differences were in fact observed in the deposits described in Chapter 9 for an area at Buckman.

Geographic Variation

The calculations just described referenced processes in a single segment of the Rio Grande, the confluence area with Los Alamos Canyon. As sediment and plutonium move downstream through a series of successive segments in the Rio Grande, some segments store materials, and others merely transport the materials through to the next segment downstream. The result is a substantial geographic variation in plutonium concentrations and inventory in sediments along the Rio Grande.

As an example, calculations using the RAT program provided data for three simulated days and the 1952 case values (input mass = 5,730 Mg, input plutonium concentration = 3,073 fCi/g, input plutonium amount = 17.6 mCi, and Rio Grande discharge = 68 m^3/s). The calculations indicated the plutonium inventory for each of the 22 segments of the Rio Grande between Otowi and the headwaters of Cochiti Reservoir, as well as the inventory of plutonium exiting the canyon and depositing in the reservoir sediments (Figure 11.13).

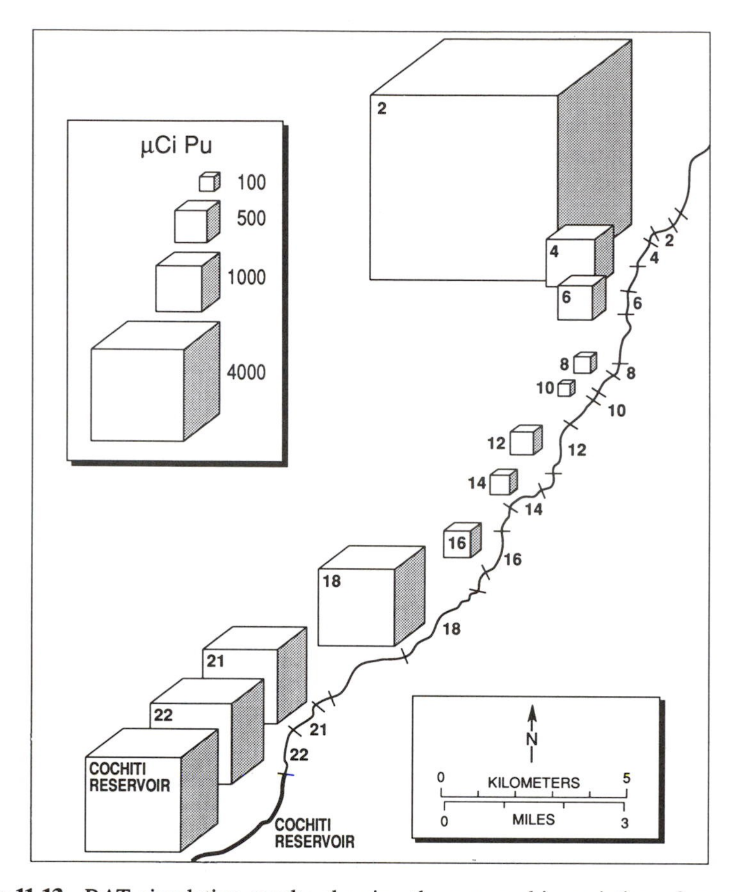

Figure 11.13. RAT simulation results showing the geographic variation of amounts of stored plutonium resulting from three simulated days of transport and storage after an injection of contaminated sediment from a flash flood in Los Alamos Canyon, a tributary in segment 2. The simulation parameters were similar to the conditions likely in 1952. The sizes of the cubes are proportional to the amount of plutonium stored in each segment, with the number in the cube identifying its corresponding segment. Segments without stored sediment or plutonium are not represented by cubes. For a more detailed location map, see Figure 11.3.

On the first level of analysis, there is an alternating pattern of those segments that store materials with those that do not. This gross pattern is the result of the geomorphology of White Rock Canyon described earlier in this chapter and incorporated in the model. In the field, these alternating segments are obvious because those with storage have shallow gradients, sediments along the channel edges and, sometimes, mid-channel bars at low-water periods. The segments without storage have steep gradients, no mid-channel bars, and canyon walls that descend to the channel's margins.

Several variables control the amounts of sediment and associated plutonium in those segments with storage. Although the amounts of sediment moving downstream have some influence on inventories, the geomorphic and hydraulic characteristics of the storage segments also are important. One segment may store large quantities simply because space is available for the material on the canyon floor as is the case in segment 2 at the mouth of Los Alamos Canyon or segment 8 near Buckman (Figure 11.3). Other segments, such as segment 18 in the depths of White Rock Canyon, have relatively little storage (even though they are long) because the steep slopes near the channel restrict the space available for deposition.

The distribution of stored plutonium shown in Figure 11.13 that resulted from three simulated days of river processes shows that the greatest inventory of stored materials is in segment 2 at the mouth of Los Alamos Canyon, but there is no consistent decrease in the inventories downstream from the entry segment. The inventories generally decrease to segment 10 but are larger thereafter. The amount stored in Cochiti Reservoir segment is larger than that in segments immediately upstream because the reservoir segment collects all the material exiting the canyon system. This happens under the present conditions with the reservoir in place, but the dispersed deposition of most of the bedload from White Rock Canyon was probably in Cochiti Pueblo–Santo Domingo Pueblo reaches of the Rio Grande (typified by the Peña Blanca representative reach described in Chapter 10) in the prereservoir period.

Longer simulated periods show that without large floods, there is a slow movement of plutonium from the large inventory at the mouth of Los Alamos Canyon through White Rock Canyon to the reservoir segment at the downstream end. Large floods accelerate the movement. The proportion of the total plutonium inventory stored in each segment remains consistent, with the largest amounts (in addition to the mass at Los Alamos Canyon and immediately downstream to Buckman) located between the confluences of the Rio Grande with Pajarito and Water canyons and below the confluence with Frijoles Canyon (Figures 11.3 and 11.13). The ultimate fate of the plutonium stored along the main channel downstream from Otowi is transport into Cochiti Reservoir, a process likely to require decades or even centuries, depending on the discharge regime of the Rio Grande. Higher discharges produce more rapid movement. The trends established by simulations show that if dispersion and mixing continue over relatively long time periods (approaching 100 days), the plutonium concentrations will decline to values that approach the levels resulting from fallout contributions alone (Figure 11.14).

The concentrations of plutonium in the stored sediments depend on the concentrations introduced from Los Alamos Canyon and the amount of stored sediment. Simulations show that the concentrations are generally low, in many cases approaching the range of less than 100 fCi/g after 20 simulated days of river processes. The plutonium concentrations in sediments affected by a single event

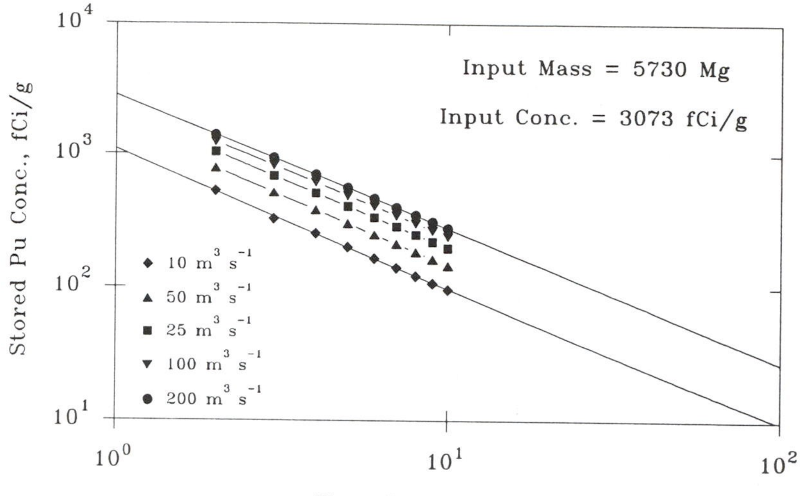

Figure 11.14. Long-term results of the RAT simulation showing the decline in plutonium concentrations in sediments stored along the Rio Grande. Each line of symbols represents different discharge conditions in the Rio Grande, ranging from low flow (10 m^3 s^{-1}) to the unlikely case of a sustained major flood (200 m^3 s^{-1}). The solid lines define the envelope of extrapolated concentrations ranging from one simulated day of mixing and dispersion to 100 simulated days.

vary by about 50 percent as a result of differential mixing, deposition, and then subsequent remobilization. Therefore, not all the deposits resulting from one release from Los Alamos Canyon and stored in several different segments of the Rio Grande may have the same concentrations of plutonium.

Conclusions from Simulations

Simulations of river processes using the RAT program produced the following generalizations about the fate of bedload plutonium released into the main stream by flash floods in Los Alamos Canyon:

1. After being mixed with large quantities of sediment from the Northern Rio Grande Basin and dispersed downstream, the concentration of plutonium is lower in those sediments affected by contributions from Los Alamos Canyon. River processes and the geomorphology of White Rock Canyon concentrate the stored plutonium at particular locations along the main channel, and so the downstream concentrations may not be lower.
2. The concentrations of plutonium in sediments released from Los Alamos Canyon decrease by about 50 percent upon their entry into the Rio Grande.

3. If discharges in the Rio Grande do not decline precipitously, the concentrations of plutonium will decrease rapidly within the first 4 to 6 days after the release, with more gradual decreases thereafter.
4. By the time that plutonium-bearing sediments are deposited, the concentrations of plutonium have decreased to a range of < 100 fCi/g.
5. Because of the complexity of differing input masses, input concentrations, time to deposition in the main channel, and discharge variation in the Rio Grande, different releases from Los Alamos Canyon may have similar or radically different concentrations of plutonium after deposition, irrespective of their initial concentrations.
6. Of the plutonium released from Los Alamos Canyon in bedload, most is likely to remain in the White Canyon segments of the Rio Grande. Before Cochiti Dam was closed in 1973, some sediments probably moved to the Cochiti Pueblo–Santo Domingo Pueblo reaches, and after the dam was closed, the materials moved into the reservoir.
7. Over a period of decades or centuries, almost all the plutonium now stored in White Rock Canyon and at the mouth of Los Alamos Canyon will probably have moved to deposition sites in the reservoir.

12

General Lessons and Conclusions

General Guidelines

The initial investigations reported here offer some broad lessons from the analysis of a specific example. Application of the techniques and results in the Los Alamos case to other areas or cases would require modifications, and even the conclusions about plutonium in the Northern Rio Grande are more first approximations than final answers. After reviewing the lessons of the Los Alamos work, this chapter summarizes some of the natural scientific lessons, with particular reference to the changing Northern Rio Grande and some observations about the interactions between natural science for plutonium and the associated public policy and politics surrounding the issue. Because this work is a beginning rather than an ending, this chapter concludes with some speculations on the future of plutonium in the Northern Rio Grande.

The lessons from the particular case of Los Alamos and the Rio Grande extend far beyond northern New Mexico. As a "test bed," the laboratory, its plutonium releases, and the data-rich Rio Grande provide generalizations useful to researchers, monitors, regulators, decision makers, and managers of other systems and locations. This chapter reviews these lessons as two distinct groups: general guidelines and specific sediment-sampling procedures. First are several general guidelines that should direct any effort at assessing the plutonium system of a river affected by industrial-waste disposal, an accidental release during transportation, or distribution from a nuclear detonation involving surface materials or from atmospheric fallout. Second, because of the overriding importance of sediment in the transport and storage of plutonium, several specific procedures should be followed when sampling soils and sediment, to ensure accurate interpretations of the results. A refined sampling and monitoring program for plutonium in sediment should be driven by a philosophy that has the following general principles.

1. *Obtain an accurate inventory of sediment-bound plutonium at the source location.* If the source of contaminated sediments in a river system is a known mass, such as a well-defined waste-disposal site, a tailings accumulation, or another readily measured mass of material, a reasonable estimate of the total inventory of plutonium in the source is possible. A precise assessment of the original mass of polluted sediment is critical to accurate predictions of potential concentrations downstream in the natural river system. In the case of Los Alamos Canyon, a truly accurate assessment is not yet possible, and so there is a fairly broad envelope of possible error in predicting downstream contamination. Estimates of the course of future events in the contamination of the Rio Grande also are imprecise. In other applications, the total amount of plutonium involved may be better known, but if it is not, efforts should be made to specify the amount injected into the natural environment. Transportation accidents, for example, might involve clearly defined amounts of plutonium, making the applications of predictive models easier than in the Los Alamos case. Fallout-derived plutonium or industrial releases as sources of river pollution may be more difficult to specify, but the expenditure of time and financial resources in evaluating the beginning amounts in a pollution episode are wise investments likely to produce more accurate predictions of downstream impacts.

2. *Obtain an accurate understanding of the temporal trends in the erosion of plutonium from the source area.* Plutonium moves into river sediment systems from the source by means of erosion processes that are reasonably well understood. Agricultural erosion models and direct-monitoring measurements can estimate the rate at which plutonium bound to sediment is entering the regional river system. Without knowing this rate of introduction, subsequent research on the main river and on receiving areas is not likely to be accurate, especially when projections are required for several hundred or several thousand years. Because the remedial measures employed to reduce the loss of plutonium from the pollution source also must be assessed for their effectiveness, measures of the loss rate are important management tools.

3. *Become familiar with the geography of plutonium in sediments in the vicinity of the source of pollution, and be able to make detailed maps of sediments and their plutonium content.* The results of my work indicate that the plutonium in sediments is likely to be spatially variable. Most of the monitoring efforts by regulatory agencies and by the originators of the plutonium in the environment have heretofore focused on temporal variability in a few monitoring sites. In regard to the plutonium in sediments, the geographic location of the deposit (combined with its age) seems to be the most important predictive factor in determining its potential to contain plutonium. The sediments in a river system should therefore be sampled and mapped on a scale that reflects this variation, rather than merely sampling "a channel area" and "a flood-plain area." I have tried to demonstrate that detailed mapping with periodic

updates is necessary for the depositional areas downstream from plutonium sources.

4. *Investigate the distribution of sediments and plutonium in the nearest downstream reservoir.* The first reservoir downstream from the source of plutonium is likely to be a major storage location for polluted sediments. Reservoir deposition produces a complex distribution of deltaic and deep-water sediments that are likely to contain varying concentrations of plutonium. A sampling scheme for reservoir sediments must therefore take into account the process history of the materials (such as delta, turbidity current deposit, and deep-water fine deposit) and specify the location on the reservoir floor. The locations of deposits in reservoirs with large vertical fluctuations are likely to differ from one year or flood to another, further complicating the sampling strategy. Long-term monitoring after a release of plutonium into a river system should include sampling the nearest downstream reservoir as a major indicator of pollution. Other reservoirs in the region not affected by the spill or release in question should also be monitored as control sites for "background" levels of plutonium contamination by surface transport of atmospheric fallout.

5. *In explaining and predicting river transport of sediment-bound plutonium, use a two-part approach that includes engineering-modeling activities and empirical-scientific verification of the conclusions.* The environmental processes with which a monitoring and surveillance program must deal can be explained through engineering models or empirical-scientific methods. Both approaches are necessary. Engineering models such as those used by the Hydraulic Engineering Center to estimate water-surface profiles, sedimentation, and scour (known by their acronyms HEC-2 and HEC-6) provide useful and detailed predictions. Their major weaknesses are that they include many assumptions not borne out in an unstable channel environment, and they require large amounts of high-resolution input. Scientific models based on empirical data (such as this book) offer more convincing conclusions because the data are "real" rather than generated, but the empirical approach is severely limited by the lack of data from critical locations.

The weak point of both approaches is sometimes having to make weakening assumptions such as stationarity for time or spatial series. That is, they must assume that climatologic and hydrologic conditions do not change significantly throughout the period of interest or along the river system. As a result, long-term averages substitute for variable data, and potentially important details are lost. For these reasons, a combination of the two approaches is the best way of explaining the movement and storage of sediment-borne plutonium.

6. *Develop simple, computer-based models that accept ordinary surveillance data and that produce annual assessments of plutonium flux and storage in the stream systems.* For a monitoring and sampling program, the explanation of sediment and sediment-borne plutonium mobility and storage should lead to general model that nonspecialists can use. The model should be simple, because

detailed specifications for the environment of its application are likely to be limited. For example, although maps with contour intervals of less than 1 m would be most useful in the hydrologic and geomorphic analysis of rivers downstream from a plutonium source, they are not likely to exist for most applications. In any case, maintaining such maps would be prohibitively expensive, because on that scale of analysis, the configuration of the channel of most rivers changes with each major flood. Therefore, a model that is based on simplifying assumptions and that makes calculations for mean conditions rather than specific events is probably the most effective. Such a model would accept as input the information from stream-gaging sites for water and sediment discharges as well as the products of the ongoing plutonium sampling.

The model should be adaptable to an interactive program designed for standard, low-level personal computers so that it can be used by nonspecialists and can be explained in nontechnical language to regulators and the public. Although it is tempting to create sophisticated computer models in pursuit of more accurate results, any engineering or scientific gains would be negated by the increasing isolation of the so-called experts from the decision makers. A more productive strategy is to have both groups speaking the same language and sharing experiences in understanding the environment in question.

7. *Establish a regional context for plutonium movement and storage that provides a clear picture of the environment without the inputs from the source in question.* There are no established national standards for chemical and radiological quality in sediment, and so any reported contamination and the resulting concentrations and amounts of flux and storage of plutonium are difficult to determine. Without sediment-quality standards for plutonium, the solution is to provide a context for the reported data. Reporting not only the storage and flux of plutonium from the local source of pollution but also the likely values without the pollution source can be informative. Because fallout plutonium is ubiquitous and because natural processes tend to concentrate it in certain river deposits, plutonium exists in all rivers in variable concentrations. By discovering the change in the preexisting condition brought about by pollution from a particular source, environmental managers can quantify two concepts: first, the degree of environmental degradation resulting from the particular source and, second, a logical target for concentrations after cleanup operations, which might reasonably be expected to restore the system to its condition before the releases from the particular pollution source. In the long term, the development of national sediment quality standards for plutonium is a desirable but distant goal.

8. *Develop a dynamic budget, including movement and storage in geographically correct compartments.* Putting concentration data into static, aspatial compartments (for example, in vegetation, water, and sediments, without ascribing exact geographic locations with residence times) creates an incomplete picture of regional processes. Annual (or some other meaningful time increment) mass

fluxes and mass storage paint a more informative picture. The geographic locations of storage sites should be included because of the radical variability and discontinuous distribution of contaminants in sediments. Mapping the concentrations and annual fluxes is essential to the responsible monitoring and cleanup of any plutonium release involving river sediments.

9. *Evaluate isotopic ratios in order to distinguish the plutonium contributions of the particular source from general atmospheric fallout.* The current budget analysis and computer model show that the contributions of plutonium from fallout and from Los Alamos National Laboratory are within an order of magnitude of each other in the general system. For particular years, the contribution of Los Alamos to specific bedload deposits immediately below Otowi may be larger, and in other deposits the contribution of fallout may dominate. The contributions to Cochiti Reservoir, whatever their origin, are higher than in other reservoirs in the immediate area. There are many depositional environments around the Rio Grande and elsewhere where it is impossible to distinguish, by concentration alone, between the contributions from a particular spill or release and those from atmospheric fallout. In many countries, including the United States, if a pollutant can be identified as coming from a particular source, the operators of the source are legally responsible for the contaminant.

Plutonium in the natural environment can be tagged or fingerprinted, and the source of plutonium in a particular deposit can be defined by isotopic ratios. For worldwide fallout, the ratio between plutonium 238 and plutonium 239 and 240 is about 21. The plutonium released from Los Alamos National Laboratory has a ratio close to 100, and so the plutonium discovered in a sedimentary deposit can be evaluated and the total amount partitioned according to source. Although the approach is not yet common in plutonium studies, the experience at Los Alamos reported here indicates the practicality of the method. A similar approach using lead-zinc-copper ratios in a dryland stream in Arizona,[1] lead–cadmium ratios in a Sierra Nevada stream,[2] and a variety of ratios in humid-region British rivers[3] all have proved successful in differentiating heavy-metal sources.

Specific Sediment-Sampling Procedures

The Los Alamos experience demonstrates that specific procedures for collecting sediment samples in the field for plutonium analysis can ensure the utility of the samples for researchers and managers. Elsewhere, these provisions enable the results to be compared with established data sets and improve the likelihood of accurate evaluations of the plutonium-sediment system.

1. *Select sample sites to enhance dynamic interpretations.* Wherever possible, when collecting samples in regard to a particular spill or release, in order to analyze the plutonium content of river sediments, the samples should be collected

at stream-gage sites. The samples should include suspended sediments and bedload sediments at gaging sites on minor as well as major streams in the vicinity of the particular plutonium source. These collection sites provide the framework for determining the movement and storage of plutonium in the sediments near the site of interest. Although stream-gage data will not be available for every desired location, when they are available they should be used because they provide irreplaceable empirical information regarding the system's water and sediment fluxes. These fluxes must be estimated for ungaged sites, of course, but where gages exist, their physical data should be supplemented with radiochemical data. In most cases, additional supplemental data for ungaged sites will almost certainly be required.

2. *Sample both suspended and bedload materials.* Previous analyses by the Environmental Surveillance Group at Los Alamos have shown that the plutonium content of suspended sediment is greater than that of bedload sediment. My work shows that the difference is critical to understanding the regional dynamics of plutonium as it moves through the river system. Because the inputs from particular pollution sources in other cases may preferentially affect one type of sediment, the concentrations in both types must be known for an accurate overall assessment. Low concentrations in one type of sediment may be significant for the total transport budget of plutonium if there are large quantities of sediment involved. Therefore, both types of sediment should be sampled from each collection site on the river.

In the past, the majority of "river sediment" samples measured for plutonium content were bedload samples drawn from the margin of the active channel. The plutonium content of these samples is useful information, but it should be supplemented with data concerning the sediments in suspension that have a different size distribution and different plutonium concentrations. Having data for both types of sediment will make the interpretation of reservoir data more informative. For example, we would expect that the sediments on the floor of Cochiti Reservoir (derived mostly from suspended sediments) have a plutonium content that differs from that of what have been referred to as "river sediment" samples (heretofore mostly bedload) collected in White Rock Canyon upstream.

3. *Evaluate particle-size data for sediment and soil samples.* Without knowing the particle sizes of the sediment, the interpretation of plutonium concentration data regarding the suspended sediment, bedload sediment, and soil samples is severely limited. Because the samples are often collected by biological or chemical specialists, particle-size data are frequently ignored. The reporting process should include for each sample some indication of the particle-size distribution in order to determine the potential physical mobility of the particles and their associated plutonium. As I have pointed out, the physical mobility of the plutonium associated with sediment is more geographically significant than is its chemical mobility. Changes in physical location may result in changes in chemical mobility as the plutonium enters different temperature, pH, or oxidation regimes.

The simplest (and cheapest) way to analyze particle size is to calculate the percentage of fine material in the sample: the percentage by weight of the sample composed of particles smaller than 63 microns in diameter. Particle-size data are critical, because most of the fallout plutonium is associated with the smallest particles, whereas the plutonium from a particular plutonium pollution source may be associated with larger particles. In any case, particles smaller than 63 microns in diameter have behavior patterns in flowing water that are strikingly different from those of larger particles.

4. *Document the geography of the sample sites.* Investigations and reports of plutonium assays of sediment samples should be accompanied by documentation that includes accurate topographic or photographic maps showing the exact geographic location of each sample site. Locational data are important because future analysts may find it necessary to relocate the same sites and sample them again to establish a time series of evaluations. Such an approach is common in monitoring and regulatory programs for plutonium near possible sources of contamination. If subsequent samples are not taken from exactly the same location, any variations in the samples may not be due to the influence of time but to variation in location. My work on the Rio Grande shows that on flood plains, for example, two samples taken at the same time but separated by only a few meters may have radically different plutonium concentrations if that few meters includes an important sedimentologic boundary.

Topographic maps provide useful records of sample sites. The ideal scale of the maps for sample site locations should be on the order of about 1:10,000, but maps of this detail often are not available. Maps on a scale of 1:24,000 in the United States and 1:25,000 in European countries are commonly available and are reasonable substitutes. Geographic information systems and computer programs combining map and statistical data improve the transmission and exchange of data. Marked aerial photographs are also useful in relocating sample sites in the field. Unfortunately, many of the previously collected plutonium data for sediment often lack these precise locational aides, and thus geographic interpretation of the data is difficult. No matter how they are stored, these accurate geographic data would also improve interpretation of the reported plutonium concentrations by future researchers, regulators, and the public.

5. *Establish a precise protocol for sample site selection.* When attempting to assess the pollution of river sediments by plutonium from accidental spills, industrial releases, or atmospheric fallout, consistency in site selection requires decision rules. The exact protocol is less important than that it be detailed and be followed without exception when collecting samples. For example, plans for sampling of "river sediment" without further definition are unacceptable for a clear interpretation of the results. A more effective protocol is consistently to take samples above major tributaries, from both suspended and bedload sediments, from flood-plain deposits above the annual high-water mark, from the finest

material available, and always from the same depth below the surface unless vertical variation is being analyzed. Samples of bedload sediments should always be taken from pools and similar repositories of fine and heavy particles.

6. *Specify the original sediment samples from outside the active-river environment.* Each sample drawn to analyze plutonium in the "soil" should be reported along with a statement on the geomorphic origin of the soil material. Hillslope materials usually contain less plutonium than do materials found at the foot of the same slope, because of erosion and the deposition of shallow materials carrying fallout. Soil samples from inactive flood plains, as shown in this book, may contain varying amounts of plutonium, depending on the date of emplacement. Soils on flood-plain surfaces established during the 1950s or 1960s, for example, might be expected to contain more plutonium than do flood-plain materials deposited in other decades. Subsequent interpretation of plutonium concentrations must take into account the geomorphic history of the materials containing the plutonium. Without this geomorphic history, fallacious interpretations are likely.

7. *Identify control sites for river sediment and soil samples.* The evaluation of unaffected control sites is important to determining the plutonium loading in river sediments from accidental spills, industrial releases, and unusual atmospheric fallout. There is probably some plutonium in the soils and sediments in most areas of the world, and as shown here, active rivers tend to decrease plutonium concentrations in active sediments rapidly and over fairly short distances. It is only in certain places that concentrations are relatively high. Because the same generalizations describe the distribution of "background" concentrations, the control sites should be as nearly similar to the monitoring sites as possible. For example, if the critical monitoring sites are located on an active flood plain of a river with about 10,000 sq km of drainage area upstream, the matching control sites should be located on an active flood plain of a nearby stream of a similar drainage area that has not been affected by the plutonium source in question. If the critical monitoring sites are hillslope soils, the control sites should be located on hillslopes of a similar size, orientation, vegetation, land use, and geology.

Reservoir control sites have to be selected with special care. Ideally, the control reservoir should be the same size, with a watershed similar in size to the reservoir in question. The watershed upstream from the reservoirs should have similar geology, soils, vegetation, and land use. Often, such perfect matches are not possible, but the control site data are better compared with the critical site data if specific data in addition to the plutonium concentrations are included in the analysis. The sediment's particle size, organic content, and carbonate content of sediments all are known to influence plutonium concentrations. These data are easily and cheaply found from sediment samples and can supplement interpretations of plutonium concentrations from the same samples.

The Los Alamos Case as an Example

My research on the plutonium and river sediments near Los Alamos National Laboratory can be a useful guide for workers in other parts of the world dealing with similar issues. The general approach of assessing geomorphic and hydrologic processes, followed by targeted sampling and surveillance, is likely to be effective in determining the impacts of spills, industrial releases, and atmospheric fallout on rivers and their associated riparian environments. The data collected in other locations can be compared with the Rio Grande data in initial assessments of the relative severity of pollution.

There are limits to the utility of the Los Alamos experience, however. The Northern Rio Grande is in a semiarid environment, and streams in humid regions dominated by fine-grained sediments and soils are likely to behave differently than the Rio Grande does. The vertical accretion found on the Rio Grande's flood plains might be replaced by lateral accretion in other regions, for example, and so sampling schemes and their interpretation would have to be adjusted accordingly. The downstream decline in discharge in the Northern Rio Grande is not likely to be found in all other streams, so the internal storage process and regional fluxes may be different elsewhere. The background fallout might be significantly different, although the isotopic ratios for the fallout are likely to be similar to those found in northern New Mexico.

On the whole, the general approaches used in the Los Alamos study can be used for rivers in other environments. The general importance of sediment, as opposed to water or air, as a transporter of large quantities of plutonium in the natural environment is broadly applicable. A general philosophy that includes a geographic appreciation of plutonium and sediments as well as a radioecologic appreciation of the element in various compartments of the ecosystem is the surest path to an effective system of impact assessment and environmental management.

Natural Science for Plutonium

The physics and chemistry of plutonium are better known than those for almost any other element, but the "natural science" of the plutonium is still poorly understood. Life scientists have investigated its interactions with plants and animals, but earth scientists have yet to explore extensively its dynamics in the earth-surface environment. This book has attempted to begin that exploration, and the results have implications for further research.

The Los Alamos experience shows the importance of context when assessing any release of plutonium into the natural environment. Los Alamos National Laboratory, or any industrial site or spill involving plutonium, does not exist in isolation from its surroundings. The magnitude of the hazard posed by the site must be weighed along with the hazards from other sources already in the surrounding environment. In some cases, such as the disaster at the Chernobyl nuclear generating station, the amount of radioactive material injected into the surrounding environment is overwhelming in comparison with the materials already present. In the case of the Northern Rio Grande, about 90 percent of the plutonium in the river is from atmospheric fallout, with only 10 percent from the laboratory. The distribution of that relatively small contribution is of concern, however, because the Los Alamos contribution is concentrated in only a few places. Those places would have significantly less plutonium without the input from the laboratory.

Surface water is the main driving force behind the movement of plutonium through the surface system of northern New Mexico. The energy represented by the water is expended partly by the moving sediments and associated plutonium from one place to another and partly by the mixing and dispersion of contaminants. Therefore, an understanding of past events and the prediction of future events rest largely on knowing the magnitude, frequency, and geographic distribution of this water-related energy. In other environments, more or less water may be available in the surface environment, but its analysis is likely to be a key to predicting the fate of plutonium released from facilities, nuclear detonations, or accidental spills.

Sediment is the primary carrier of plutonium through the natural environment, and depositional areas are the ultimate resting place for most of the plutonium in the environment. A regional sediment budget is a prerequisite to discovering where the plutonium is likely to go, how soon it will arrive, and how much is involved. In the Northern Rio Grande, on average, each year, only 50 percent of the suspended sediment and 20 percent of the bedload that enter the system leave to be deposited in Elephant Butte Reservoir. The rest, 50 percent of the suspended load and 80 percent of the bedload, remains in internal storage. The precise numbers may be different in other systems, but the basic conclusion is likely to be the same. Because a great deal of sediment is stored within the river systems rather than being transmitted to the oceans, a great deal of plutonium (whether from fallout or other sources of greater local importance) is likely also to be stored internally.

The stored sediment, along with its piggybacked plutonium, is not randomly or evenly distributed throughout the river system. Rather, sediment is stored in particular and predictable locations such as flood plains and reservoirs. Because plutonium is a heavy metal, it behaves in rivers as do other heavy metals, such as gold and silver. The location and identification of plutonium placers are, from a hydraulic perspective, no different from the location and identification of gold or silver placers. The near-channel landforms in the Northern Rio Grande are the surface expressions of these deposits, and they are fairly obvious on the landscape. Once the plutonium is geographically located, standard geologic and radiochemical techniques can easily determine the amount of metal in each deposit.

In the Northern Rio Grande, the change in the river channel over the past half-century has created a complex series of deposits and landforms, some of which are more likely to contain plutonium than others. Not all flood plains or all parts of individual flood plains are equally likely to host plutonium deposits. The geomorphic history, painstakingly unraveled a year or two at a time, is required to identify the most likely depositories of the element. In this sense, the Northern Rio Grande is typical of most rivers, with the complexity and dimension of the changes similar to those found in rivers studied by others elsewhere in North America, Europe, Asia, and Australia.[4]

Few river reaches in the developed world are still entirely natural. In the United States, for example, only about 2 percent of all rivers are in their original, natural condition.[5] For this reason, predicting the fate of plutonium bound to river sediment depends not only on knowing the natural changes in channel processes related to climatological and hydrological adjustments, but also on knowing the engineered changes in the channels. Engineering works such as levees may change the size of the currently active flood plain, but areas now outside the levees may also be important if plutonium-bearing sediments were deposited there before the levee was built. A portion of the flood plain near Ranchito, New Mexico, is a case in point: it now lies outside a levee constructed in the early 1960s, and it now seems totally unconnected to the Rio Grande. In the late 1950s when plutonium was in the system, however, it was an integral part of the river.

This book shows why the laboratory's annual sampling program does not generally detect laboratory-derived plutonium in the Rio Grande. The monitoring program searches in the wrong places. The samples are taken from active Rio Grande sediments at times when Los Alamos Canyon is not discharging sediment and plutonium. Also, there has been no systematic sampling of flood-plain deposits, which are the principal storage sites for laboratory plutonium. It is only in the sediments of Cochiti Reservoir that routine sampling by the laboratory is likely to have encountered its own contribution.

The mapping of flood plains and their components is a task made easier by aerial photography and ground-based research relying on surface form and sedimentary characteristics such as particle sizes. Riparian vegetation, although not a major plutonium reservoir, is intimately related to the distribution of water, soils, sediments, and landforms beneath the plants. In the Northern Rio Grande, the biogeography of trees proved to be a useful aid in analyzing the physical landscape. In other environments, the distribution of other plant types such as grasses and sedges, marsh vegetation, or perhaps some combination in defined communities might be more useful, but the basic approach usually is helpful.

The plutonium itself also has information about its past history and origins. Isotopic ratios, such as plutonium 238 and plutonium 239 and 240 or plutonium 240 and plutonium 241, act as identification tags for the source of the plutonium because the ratios are different for each nuclear detonation in the atmosphere and for each industrial source. It therefore is possible to assign an origin and thus legal responsibility to all the plutonium through isotopic ratio "fingerprinting."

Finally, there are two philosophical methods of understanding and predicting the fate of plutonium in river systems: empirical and model. In empirical science, the objective is to define and explain the past events as guides to predicting the future. Much of this book has adopted the empirical approach, in part because the Northern Rio Grande is rich in geophysical and radiochemical data. The advantage of this method is that it rests on specific data, and its disadvantage is that it allows the analysis of only a limited sequence of events under the actual conditions. But there is no assurance that the past is a completely reliable guide to the future. Conversely, a model approach using simulation, as outlined in the preceding two chapters, allows great flexibility in experimenting with a variety of scenarios. The simulation approach is, however, limited by the model that produces it, and verifying it is difficult. In the end, prudence suggests that using both the empirical and modeling approaches produces the most reliable conclusions.

Public Policy for Plutonium

Given the present state of knowledge, none of the data in my project suggest that plutonium exists at hazardous levels outside the laboratory. But even the question of hazardous levels is subject to debate, because of uncertainties about the risks of being exposed to low levels over long periods. Because what we now consider as safe may in the future be regarded as unsafe, we should find out as much as we can about the distribution and concentration of plutonium. Because of the hazards of plutonium, public policy in the United States provides for the intensive monitoring and surveillance of the radiochemical environment in and around nuclear installations such as Los Alamos National Laboratory.

The laboratory operates under legal requirements such as sampling and monitoring the environment for the potential effects of its activities, including the release of plutonium. Other legal requirements are complying with more than 20 federal and state laws related to environmental quality, such as the federal Clean Water Act and Clean Air Act.[6] The laboratory bases its surveillance and monitoring on specific U.S. Department of Energy orders: General Environmental Protection Program,[7] and Environmental Protection, Safety, and Health Protection Information Reporting Requirements.[8] The general purpose of these regulations is to document compliance, identify temporal trends, provide knowledge to the public, and enhance the general understanding of environmental processes. Because of plutonium's hazards to human health, the program emphasizes sampling air, water, soils, and foods.

The laboratory focuses on three geographic areas. First, workers sample a series of "regional" sites in northern New Mexico in order to gather information about environments not affected by laboratory operations. Second, "perimeter" sites within about 4 km of the laboratory boundaries offer data on areas that might be affected by the laboratory and that serve as a "trip wire" to detect any significant movement of contaminants from the laboratory. These sites include areas frequently used by the general public. Third, "on-site" locations within the laboratory boundaries provide samples of areas that might be affected but are not generally accessible to the public.

The results of my work on the Northern Rio Grande indicate that the United States public policy for surveillance and monitoring of plutonium should contain a more refined program in regard to sediment quality. For example, the movement of sediment from Los Alamos Canyon and other potentially contaminated canyons is the most likely vector by which plutonium will leave the laboratory site. This process is relatively slow but inevitable, and so eventually all the plutonium stored in canyon alluvium will move into the Rio Grande. My study shows that these materials are likely to be stored a short distance down the main river as alluvial deposits either directly accessible to the public (as at Buckman) or indirectly accessible to the public as reservoir deposits (in Cochiti Reservoir). The environments surrounding other nuclear installations probably are similar.

Politics of Plutonium

Like most issues related to public health and safety, scientific and policy approaches to plutonium in the natural environment are inextricably bound up with politics. It was, after all, the politics of international warfare that led to the development of Los Alamos and the employment there of plutonium for weapons. The laboratory's operations are highly politicized because of the laboratory's reliance on public funding, and its attention to environmental quality is related to a different sort of political pressure. There is nothing inherently wrong with this connection between science and politics, but sometimes the political aspects of the problem of plutonium in the environment make dealing with the issue more difficult.

During World War II and most of the Cold War, the military secrecy surrounding the laboratory also surrounded its environmental impact. Whether the secrecy pertaining to environmental contamination was driven solely by military considerations or also by attempts to obfuscate irresponsible behavior is not the point; rather, it is that relevant data were not widely available for external review until the late 1960s, thereby precluding the sort of research on which this book is based. The careful cataloging and public release of data beginning in the early 1970s and the further declassification of documents in the early 1980s enabled a more complete understanding of the system than had been possible previously. That the laboratory itself promoted this policy of openness as part of a more general openness in American government is testimony to the value of scientific freedom in this society. That more analyses like mine are not available demonstrates the difficulties of dealing with a sensitive subject.

In making available to the public the data used in this work, the laboratory is responding to a new kind of political pressure. Whereas before, the objective was to remain as inconspicuous as possible so as not to arouse undue attention, the new strategy is to reveal data before they become embarrassing and difficult to handle. If there can be a criticism of the laboratory, it is that even though the data may have been available, the research to interpret them has been slow to develop. The laboratory should seek the resources to resolve the problems without being forced to do so by others who uncover the problems first. This strategy offers control to the laboratory in an uncertain legal and regulatory environment, and it also reflects the ethic of a new generation of scientists in the laboratory who are truly concerned with the larger issues of environmental quality and public health.

Critics of the laboratory correctly point out earlier mistakes in management and gaps in present knowledge. The news media have effectively fulfilled their function as watchdogs representing the public's interest. In some cases, however, sensational reporting, poorly researched stories, and the failure to put the facts in a context so that readers or listeners can evaluate the information have devalued the media's contribution. The information in my research and the data in the appendices of this book provide a backdrop that can be used by the educated layperson when dealing with news accounts or laboratory reports.

The American public plays an important role in dealing with plutonium in the natural environment. Most members of the general public are frightened of plutonium (and rightfully so), but unreasoned fear serves no useful purpose. Plutonium is everywhere in our environment; it is even part of our bodies. The important issues are its locations and concentrations. Hysterical reactions to plutonium contamination will not be helpful and are likely to force the expenditure of resources that could be better employed elsewhere. Well-reasoned and well-informed policies for surveillance, monitoring, and cleanup will provide maximum protection for the most reasonable investment. Such a scenario depends on continued openness by officials and increasing sophistication by the public.

Futures

Environmental plutonium research is like all other types of science—the researchers want more data and more research. Given the hazards of plutonium, these desires seem justified, but only for specific and convincing objectives. Because of the continuing proliferation of peaceful uses of atomic energy as well as plutonium for nuclear weapons, we must learn much more about the probable fate of the plutonium that may be released to the natural environment through accident or by weapons. More research similar to that for this book but conducted in different environments would permit more informed responses to these future releases that seem specifically unpredictable as to time or place but generally certain to happen somewhere, sometime.

The specific case of Los Alamos and the Northern Rio Grande has more limited objectives. A continuing monitoring and surveillance program is an absolute requirement mandated by law and licensing procedures, to say nothing of common sense. The distribution and concentration of plutonium in stored sediments in Los Alamos Canyon (and in other similar canyons nearby) and in Los Alamos National Laboratory should be better understood because these sediments are now sources for plutonium in the Rio Grande. More data are needed about the physical operations of sediment in Cochiti Reservoir to detail the fate of the plutonium in the system, and small deposits in White Rock Canyon should be inventoried as intermediate way stations for the plutonium. More sophisticated modeling techniques can improve the primitive version developed in this book.

Exact predictions of the behavior of plutonium in the Northern Rio Grande are impossible, but some aspects of the river's future operation are certain. I have shown that the fallout plutonium is eroding and being transported through the system at a rate not likely to exhaust the total inventory (including the fallout on the landscape of the upper watershed) for more than 2,000 years. If the rates of erosion and transport in Los Alamos Canyon observed over the past 40 years prevail, the canyon will contribute plutonium from Los Alamos for 100 to 600 years, depending on the magnitude of the original inventory. The rates of transfer of sediment and plutonium to the Rio Grande will continue to be sporadic, with radical variations in the magnitude of the contribution to the main stream. For the next several hundred years, Cochiti Reservoir will continue to store sediments and plutonium in increasing amounts from upstream sources. The continued pollution, albeit by small amounts, is a historical inevitability. Even in the rarified landscape of northern New Mexico, water runs downhill and carries contaminated sediment with it. Over the same time period, the channel of the Rio Grande is likely to continue in a depositional mode with aggradation, but it is unlikely that upstream dams will completely eliminate large floods and channel adjustments.

The form in which the Rio Grande existed a century ago is not likely to be seen again (Figure 12.1). But changes in the channel with the establishment of a braided system followed by a gradual return to the present geometry are possible within a few hundred years. The sediment and plutonium stored in flood plains, abandoned channels, and channel side bars will be remobilized during these changes, resuming their movement downstream. The result will be temporarily higher rates of deposition in Cochiti Reservoir.

Long-term economic development may bring more emphasis on control of the channel of the Rio Grande in the Española and Santa Clara Pueblo areas immediately upstream from Otowi. If such development produces engineering efforts that result in a straight, narrow channel of the river there, the potential erosion impact on Otowi and Buckman would need to be reassessed. Previously stored

Figure 12.1. An early-nineteenth-century woodcut showing the character of the Rio Grande in northern and central New Mexico before the channel changes and engineering works of the past century. (Denver Public Library, Western History Collection, photo 07031)

sediment and plutonium were remobilized along the Rio Grande downstream from channel works elsewhere in the system (as shown by the regional plutonium budget for the 1970s, which responded to engineering works completed in the 1960s). A similar response is likely in the Española-Buckman reach, with more deposition in Cochiti Reservoir if the Española area is channelized.

The natural environment of the river system is unstable over long periods, as is the regulatory environment in which Los Alamos National Laboratory must operate. Over the past 30 years, environmental regulations regarding rivers have become progressively more restrictive,[9] reflecting a national culture that has become increasingly conscious of environmental quality. It seems reasonable that this trend will continue. Part of any regulatory program is the assignment of responsibility, making the producers of potential pollutants accountable for their contributions to the environment. It seems likely that Los Alamos National Laboratory will someday have to determine how much of the total plutonium loading in Cochiti Reservoir has come from the laboratory. Refining the isotopic ratio method of identifying the source of plutonium would be a wise investment in anticipation of the more rigorous legal and licensing requirements inevitable in the near future. My work defining the regional plutonium budget indicates that the outcome of such efforts will be twofold. First, the contribution by Los Alamos National Laboratory to the total regional plutonium system is a small fraction of the contribution by fallout products. Second, we will someday be able to determine the contribution by Los Alamos to the plutonium in some specific deposits, meaning that we also will be able to determine the legal responsibility.

The public demands a more accurate accounting of the plutonium in the earth-surface environment. In the 1940s, General Leslie Groves and others at Los

Alamos National Laboratory concluded that it was not possible to explain the environmental fate of the element. Given the lack of financial and intellectual resources to solve the problem of protecting environmental quality, Groves reasoned that the problem would have to be left for the future. Fifty years later, environmental quality has become part of American culture, politics, and science, and for the issue of environmental plutonium, the future is now.

Appendixes

Appendix A.1. Prefix terms for units of measure

Term	Power of 10	Symbol
Exa-	10^{18}	E
Peta-	10^{15}	P
Tera-	10^{12}	T
Giga-	10^{9}	G
Mega-	10^{6}	M
Kilo-	10^{3}	k
Hecto-	10^{2}	h
Deca-	10^{1}	da
Deci-	10^{-1}	d
Centi-	10^{-2}	c
Milli-	10^{-3}	m
Micro-	10^{-6}	u
Nano-	10^{-9}	n
Pico-	10^{-12}	p
Femto-	10^{-15}	f
Atto-	10^{-18}	a

Source: R. C. Weast, *CRC Handbook of Chemistry and Physics* (Boca Raton, Fla.: CRC Press, 1988), p. F-158.

Appendix A.2. Units of measure for isotopic decay rates

Medium	Picocuries	Common Usage	International System
Air	10^{-12} uCi/ml	1 pCi/m^3	0.037 Bq/m^3
	10^{-15} uCi/ml	0.001 pCi/m^3	0.000037 Bq/m^3
	10^{-18} uCi/ml	10-6 pCi/m^3	3.7 × 10-8 Bq/m^3
Liquids	10^{-9} uCi/ml	1 pCi/l	37 Bq/m^3
	10^{-12} uCi/ml	0.001 pCi/l	0.037 Bq/m^3
Solids	1 pCi/g	1 pCi/g	37 Bq/kg
	1 fCi/g	0.001 pCi/g	0.037 Bq/kg

Appendix A.3. Concentration and river flow conversions

	Metric Units	English Units
Concentrations	1 mg/l, 1 g/m^3	1 ppm
	1ug/l, 1 mg/m^3	1 ppb
River flows	1 m^3	0.00084 ac ft
	1,220 m^3	1 ac ft, 43,560 ft^3
	1 l/s, 1 dm^3/s	15.9 gal/min, 0.0353 ft^3/s
	1 m^3/s	35.3 ft^3/s, 2.28 × 107 gal/d
	0.028 m^3/s	1 ft^3/s

Appendix B.1. Water and sediment data from stream gages

Site	Gage ID
Rio Grande at Embudo	2795

Year	Water (ac ft)	Flood (cfs)	Sediment (tons)	Year	Water (ac ft)	Flood (cfs)	Sediment (tons)
1890	1,030,000	6,070	–	1938	790,400	5,440	–
1891	1,280,000	8,550	–	1939	522,100	2,410	–
1892	1,040,000	6,660	–	1940	316,600	1,990	–
1893	633,000	5,100	–	1941	1,341,000	12,000	–
1894	529,000	–	–	1942	1,503,000	10,800	–
1895	867,000	5,010	–	1943	418,500	2,220	–
1896	495,000	2,980	–	1944	955,000	8,770	–
1897	992,000	8,740	–	1945	67,4700	5,380	–
1898	969,000	4,700	–	1946	308,300	1,950	–
1899	348,000	1,620	–	1947	474,300	4,080	–
1900	540,000	5,410	–	1948	993,700	10,200	–
1901	542,000	7,400	–	1949	859,600	9,990	–
1902	306,000	2,500	–	1950	341,200	1470	–
1903	1,020,000	15,900	–	1951	246,600	710	–
1904	250,000	–	–	1952	777,300	8,720	–
1905	1,500,000	–	–	1953	373,300	2,000	–
1906	1,200,000	–	–	1954	267,600	1,860	–
1907	2,000,000	–	–	1955	283,400	2,200	–
1908	670,000	–	–	1956	242,000	1,020	–
1909	1,300,000	–	–	1957	753,400	5,000	–
1910	890,000	–	–	1958	860,500	6,840	–
1911	1,100,000	–	–	1959	253,900	2,760	–
1912	1,500,000	–	–	1960	427,900	2,320	–
1913	479,000	2,080	–	1961	408,200	2,340	–
1914	953,000	7,190	–	1962	599,300	3,980	–
1915	962,400	7,330	–	1963	280,800	966	–
1916	995,000	8,560	–	1964	226,200	925	–
1917	1,220,000	8,600	–	1965	719,200	5,200	–
1918	485,000	3,580	–	1966	527,600	1,950	–
1919	975,000	7,280	–	1967	366,300	3,550	–
1920	1,430,000	12,700	–	1968	561,800	3,270	–
1921	1,170,000	14,400	–	1969	590,500	3,140	–
1922	911,000	7,500	–	1970	583,900	2,250	–
1923	772,600	4,640	–	1971	391,100	1,860	–
1924	1,227,000	8,780	–	1972	324,800	1,090	–
1925	430,400	1,580	–	1973	879,700	6,620	–
1926	814,100	5,500	–	1974	329,000	1,050	–
1927	877,200	9,500	–	1975	615,900	3,700	–
1928	673,300	5,180	–	1976	503,600	2,290	–
1929	735,600	5,850	–	1977	223,100	2,480	–
1930	592,300	2,240	–	1978	323,200	1,490	–
1931	362,000	1,740	–	1979	1,128,000	9,000	–
1932	907,000	7,180	–	1980	762,300	5,080	–
1933	443,000	4,300	–	1981	252,300	2,930	–
1934	282,100	832	–	1982	615,800	5,010	–
1935	643,300	5,900	–	1983	861,700	5,660	–
1936	522,900	3,630	–	1984	734,500	6,010	–
1937	892,500	6,690	–	1985	1,297,000	8,420	–

Appendix B.1. Water and sediment data from stream gages *(continued)*

			Site		Gage ID		
			Rio Chama at Chamita		2900		
Year	Water (ac ft)	Flood (cfs)	Sediment (tons)	Year	Water (ac ft)	Flood (cfs)	Sediment (tons)
1913	258,000	–	–	1950	304,100	3,100	1,085,000
1914	469,000	–	–	1951	144,500	1,410	580,201
1915	666,000	5,980	–	1952	566,500	5,880	3,390,906
1916	645,000	6,000	–	1953	167,200	1,330	616,075
1917	526,000	4,600	–	1954	176,600	2,000	1,061,698
1918	25,000	2,630	–	1955	142,300	5,970	1,663,344
1919	602,000	5,500	–	1956	145,600	1,300	537,593
1920	710,000	15,000	–	1957	522,900	4,280	3,344,250
1921	384,700	2,850	–	1958	561,000	4,830	5,467,859
1922	329,300	3,310	–	1959	236,800	1,580	1,020,528
1923	409,000	–	–	1960	364,400	2,470	1,466,070
1924	547,700	–	–	1961	244,700	8,000	1,635,055
1925	255,400	–	–	1962	411,600	3,870	1,017,578
1926	488,400	–	–	1963	269,600	1,830	613,467
1927	604,600	6,180	–	1964	147,100	2,390	702,790
1928	337,200	5,300	–	1965	421,900	5,730	2,283,381
1929	420,000	10,400	–	1966	385,700	3,590	2,227,058
1930	381,600	5,170	–	1967	209,100	8,150	3,016,743
1931	195,000	3,810	–	1968	278,900	2,170	2,013,011
1932	820,000	7,700	–	1969	416,700	10,000	984,180
1933	327,000	8,040	–	1970	265,300	2,270	677,598
1934	115,600	1,600	–	1971	188,200	3,880	373,822
1935	354,100	7,100	–	1972	170,000	2,350	319,352
1936	505,700	5,510	–	1973	437,900	3,310	668,771
1937	769,700	6,610	–	1974	337,600	1,180	283,163
1938	446,600	3,360	–	1975	409,100	2,790	–
1939	321,100	3,600	–	1976	385,500	2,210	–
1940	232,800	3,360	–	1977	199,200	3,880	–
1941	875,900	9,910	–	1978	343,800	3,150	–
1942	851,800	8,350	–	1979	471,300	3,410	–
1943	289,600	6,100	–	1980	667,500	4,330	–
1944	327,100	3,190	–	1981	283,600	2,220	–
1945	436,100	5,490	–	1982	481,500	3,730	–
1946	144,200	2,360	–	1983	569,200	4,140	–
1947	260,700	2,670	–	1984	533,700	4,840	–
1948	358,800	2,500	2,165,000	1985	536,200	3,920	–
1949	428,000	2,580	1,650,000				

Site	Gage ID
Rio Grande at Otowi	3130

Year	Water (ac ft)	Flood (cfs)	Sediment (tons)
1895	8,630–	–	
1896	777,000	5,250	–
1897	1,750,000	15,300	–
1898	1,260,000	7,010	–
1899	567,000	6,710	–
1900	739,000	7,500	–
1901	848,000	8,400	–
1902	456,000	6,980	–
1903	1,650,000	19,600	–
1904	360,000	–	–
1905	2,190,000	19,800	–
1906	–	–	–
1907	–	–	–
1908	–	–	–
1909	–	–	–
1910	1,320,000	12,700	–
1911	1,660,000	12,300	–
1912	2,150,000	21,700	–
1913	728,000	9,250	–
1914	1,430,000	9,140	–
1915	1,640,000	–	–
1916	1,700,000	14,100	–
1917	1,770,000	10,400	–
1918	736,000	–	–
1919	1,600,000	13,300	–
1920	2,161,000	24,400	–
1921	1,585,000	17,000	–
1922	1,281,000	10,400	–
1923	1,232,000	7,550	–
1924	1,813,000	14,200	–
1925	731,200	5,340	–
1926	1,330,000	–	–
1927	1,512,000	10,950	–
1928	1,069,000	8,990	–
1929	1,236,000	11,500	–
1930	1,041,000	6,520	–
1931	605,000	11,000	–
1932	1,750,000	14,500	–
1933	800,000	6,400	–
1934	414,000	3,180	–
1935	1,015,000	8,220	–
1936	1,062,000	9,350	–
1937	1,703,000	10,800	–
1938	1,264,000	7,740	–
1939	888,700	5,770	–
1940	554,400	2,330	–

Year	Water (ac ft)	Flood (cfs)	Sediment (tons)
1941	2,311,000	25,000	–
1942	2,405,000	16,400	–
1943	717,300	7,100	–
1944	1,287,000	10,400	–
1945	1,140,000	10,400	–
1946	456,000	2,610	–
1947	730,300	5,740	–
1948	1,362,000	12,400	4,306,000
1949	1,304,000	10,700	3,681,000
1950	663,400	4,590	1,733,000
1951	395,400	3,440	900,743
1952	1,378,000	9,700	4,473,402
1953	548,600	3,300	732,109
1954	450,600	3,100	1,329,497
1955	432,000	5,140	2,430,691
1956	377,100	1,850	714,319
1957	1,297,000	6,650	4,557,348
1958	152,600	11,000	7,562,178
1959	509,800	2,740	1,424,491
1960	821,000	4,490	2,074,261
1961	675,600	8,700	1,971,899
1962	1,040,000	7,400	3,252,975
1963	559,800	2,670	862,093
1964	383,700	2,720	946,606
1965	1,178,000	7,660	3,377,878
1966	944,800	3,600	2,255,645
1967	580,500	9,520	2,650,962
1968	855,700	4,490	2,573,694
1969	1,038,000	6,760	1,823,935
1970	906,500	5,860	1,939,043
1971	589,500	7,820	1,105,621
1972	513,500	3,440	1,464,464
1973	1,394,000	8,350	3,997,338
1974	687,400	1,760	823,348
1975	1,066,000	5,070	1,525,534
1976	936,400	4,480	1,839,982
1977	435,500	2,620	896,924
1978	700,700	4,020	1,170,813
1979	1,706,000	5,200	2,595,964
1980	1,490,000	8,270	1,538,125
1981	560,100	2,340	436,468
1982	1,126,000	5,460	1,578,260
1983	1,482,000	8,760	1,466,723
1984	1,342,000	9,790	1,468,574
1985	1,935,000	12,400	2,727,140

Appendix B.1. Water and sediment data from stream gages *(continued)*

		Site		Gage ID			
		Rio Grande at San Felipe		3190			
Year	Water (ac ft)	Flood (cfs)	Sediment (tons)	Year	Water (ac ft)	Flood (cfs)	Sediment (tons)
1926	1,387,000	–	–	1956	366,300	8,700	–
1927	1,648,000	14,000	–	1957	1,276,000	12,000	–
1928	1,154,000	11,000	–	1958	1,499,000	10,200	–
1929	1,274,000	22,600	–	1959	490,500	4,400	–
1930	1,077,000	5,840	–	1960	803,400	4,450	–
1931	613,000	10,500	–	1961	670,900	7,440	–
1932	1,720,000	14,200	–	1962	1,007,000	7,380	–
1933	847,000	6,200	–	1963	543,400	3,730	–
1934	409,300	2,260	–	1964	364,900	2,350	–
1935	1,051,000	22,000	–	1965	1,143,000	10,900	–
1936	1,108,000	10,900	–	1966	927,400	10,600	–
1937	1,844,000	27,300	–	1967	606,400	13,700	–
1938	1,194,000	12,500	–	1968	899,400	4,940	–
1939	885,700	6,650	–	1969	1,079,000	5,940	–
1940	569,800	3,880	–	1970	949,600	3,650	–
1941	2,463,000	22,600	–	1971	611,600	6,540	–
1942	2,456,000	18,900	–	1972	504,200	6,220	–
1943	785,500	10,600	–	1973	1,432,000	9,270	–
1944	1,317,000	11,600	–	1974	656,500	1,810	–
1945	1,208,000	11,200	–	1975	1,010,000	4,710	–
1946	466,500	4,830	–	1976	862,700	3,130	–
1947	719,900	5,580	–	1977	395,900	2,720	–
1948	1,343,000	12,500	–	1978	651,300	3,640	–
1949	1,285,000	10,500	–	1979	1,666,000	7,550	–
1950	663,000	3,950	–	1980	1,450,000	7,130	–
1951	363,900	3,210	–	1981	528,400	8,850	–
1952	1,350,000	11,500	–	1982	1,098,000	5,180	–
1953	527,400	10,200	–	1983	1,516,000	7,100	–
1954	427,900	4,660	–	1984	1,315,000	9,220	–
1955	447,300	17,400	–	1985	1,755,000	8,290	–

		Site			Gage ID		
		Jemez River below Jemez Dam			3290		
Year	**Water (ac ft)**	**Flood (cfs)**	**Sediment (tons)**	**Year**	**Water (ac ft)**	**Flood (cfs)**	**Sediment (tons)**
1943	–	16,300	–	1965	38,310	618	–
1944	45,910	2,420	–	1966	29,870	1,800	–
1945	75,360	7,360	–	1967	31,310	1,850	–
1946	13,540	7,670	–	1968	51,550	1,290	–
1947	20,180	3,510	–	1969	56,410	1,340	–
1948	42,220	3,570	–	1970	43,370	3,000	–
1949	54,930	1,300	502,800	1971	14,070	670	–
1950	10,210	6,430	255,600	1972	18,670	1,190	–
1951	13,840	7,000	790,239	1973	129,000	2,920	–
1952	33,020	2,700	515,327	1974	16,060	320	–
1953	7,640	316	61,695	1975	83,370	1,750	–
1954	20,180	1,930	690,970	1976	14,640	276	–
1955	19,730	1,020	768,258	1977	14,150	208	–
1956	13,280	1,230	228,489	1978	36,820	626	–
1957	35,050	704	319,488	1979	100,000	4,260	–
1958	111,000	7,870	687,843	1980	68,290	713	–
1959	27,980	1990	–	1981	18,920	473	–
1960	47,830	772	–	1982	44,180	662	–
1961	53,070	1,190	–	1983	101,600	1,380	–
1962	43,840	1,630	–	1984	55,110	473	–
1963	20,910	606	–	1985	102,200	1,180	–
1964	15,310	530	–				

Appendix B.1. Water and sediment data from stream gages *(continued)*

		Site		Gage ID			
		Rio Grande at Albuquerque		3300			
Year	Water (ac ft)	Flood (cfs)	Sediment (tons)	Year	Water (ac ft)	Flood (cfs)	Sediment (tons)
1942	2,455,000	25,000	–	1964	210,800	1,920	–
1943	540,100	4,490	–	1965	1,049,000	8,720	–
1944	1,183,000	11,400	–	1966	812,800	6,650	–
1945	1,065,000	11,700	–	1967	439,900	13,300	–
1946	282,900	4,700	–	1968	761,800	4,360	–
1947	549,900	6,460	–	1969	938,700	6,480	–
1948	1,229,000	13,100	–	1970	833,300	5,840	3,444,903
1949	1,206,000	10,800	–	1971	500,200	6,650	2,544,489
1950	490,400	5,200	–	1972	383,000	4,380	2,428,212
1951	241,400	8,940	–	1973	1,421,000	8,570	8,017,443
1952	1,269,000	9,600	–	1974	524,700	2,080	1,045,303
1953	375,200	7,920	–	1975	979,200	6,160	2,807,587
1954	278,400	5,720	–	1976	744,400	3,340	1,621,401
1955	288,000	7,960	–	1977	257,700	2,190	491,805
1956	242,600	4,880	–	1978	531,800	4,580	774,449
1957	1,199,000	8,780	–	1979	1,638,000	8,650	1,846,126
1958	1,585,000	12,700	–	1980	1,412,000	7,600	759,946
1959	368,600	2,070	–	1981	431,100	2,750	393,773
1960	372,700	4,800	–	1982	1,015,000	5,460	849,813
1961	559,200	6,770	–	1983	1,390,000	7,700	1,325,704
1962	923,600	6,520	–	1984	1,167,000	9,500	1,197,098
1963	431,700	2,480	–	1985	1,661,000	9,370	1,443,353

		Site		Gage ID			
		Rio Puerco near Bernardo		3530			
Year	Water (ac ft)	Flood (cfs)	Sediment (tons)	Year	Water (ac ft)	Flood (cfs)	Sediment (tons)
1940	–	7,200	–	1963	19,890	1,210	3,026,277
1941	123,500	18,800	–	1964	18,590	2,640	2,917,346
1942	47,080	12,900	–	1965	30,410	3,210	3,807,918
1943	30,970	11,100	–	1966	19,260	1,800	3,528,635
1944	31,410	11,000	–	1967	77,780	7,860	12,257,979
1945	21,720	6,260	–	1968	27,630	3,420	4,940,551
1946	30,760	5,800	–	1969	26,210	3,580	4,919,348
1947	65,280	9,020	–	1970	26,710	6,940	2,822,326
1948	10,500	1,570	1,634,000	1971	9,130	1,300	1,888,661
1949	28,260	3,220	5,760,000	1972	61,510	9,220	9,490,327
1950	12,000	4,140	2,753,000	1973	60,310	3,920	7,958,371
1951	23,060	4,450	4,613,496	1974	6,100	2,980	1,141,388
1952	13,350	1,820	2,953,332	1975	39,160	3,520	6,829,451
1953	34,150	5,490	6,953,247	1976	7,950	2,280	1,734,453
1954	78,340	7,920	14,778,779	1977	24,040	3,010	4,355,708
1955	85,270	8,000	18,315,560	1978	3,960	1,330	478,526
1956	12,280	5,200	3,423,769	1979	25,700	1,960	3,496,564
1957	85,960	5,680	18,054,868	1980	18,650	2,450	1,810,746
1958	44,150	5,340	8,072,030	1981	15,880	1,620	2,093,080
1959	21,470	4,020	5,038,884	1982	32,300	3,460	4,408,094
1960	7,560	3,880	4,156,507	1983	17,170	1,580	1,876,008
1961	22,150	2,470	4,548,008	1984	20,550	1,690	2,678,574
1962	10,150	900	1,448,532	1985	34,990	1,400	3,398,587

Appendix B.1. Water and sediment data from stream gages *(continued)*

Site	Gage ID
Rio Salado near San Acacia	3540

Year	Water (ac ft)	Flood (cfs)	Sediment (tons)	Year	Water (ac ft)	Flood (cfs)	Sediment (tons)
1948	1,651	1,830	–	1967	17,469	17,400	673,140
1949	7,063	4,050	–	1968	6,296	10,400	108,200
1950	4,491	8,500	–	1969	6,084	10,100	352,260
1951	12,091	13,200	–	1970	11,863	4,980	–
1952	5,886	7,800	–	1971	4,110	–	1,246,800
1953	8,022	16,600	–	1972	64,905	–	59,468
1954	20,058	11,000	–	1973	18,689	–	13,300
1955	22,262	4,500	–	1974	2,041	–	216,038
1956	4,254	7,100	–	1975	11,753	–	234,870
1957	19,392	626	–	1976	680	–	141,869
1958	837	15,200	–	1977	11,974	–	165,950
1959	7,087	6,000	–	1978	316	–	703
1960	4,733	4,420	–	1979	507	–	562
1961	12,812	10,900	–	1980	9,494	–	220,100
1962	7,831	6,820	–	1981	2,054	–	31,200
1963	16,497	15,300	–	1982	17,319	–	320,000
1964	8,596	10,000	–	1983	895	–	–
1965	16,449	36,200	–	1984	7,730	–	–
1966	8,592	3,880	782,342				

		Site		Gage ID			
		Rio Grande at San Marcial		3584, 3585			
Year	Water (ac ft)	Flood (cfs)	Sediment (tons)	Year	Water (ac ft)	Flood (cfs)	Sediment (tons)
1895	–	17,900	–	1941	2,440,000	24,600	–
1896	588,500	5,000	–	1942	2,322,000	18,400	–
1897	1,548,000	21,800	–	1943	441,600	4,500	–
1898	1,160,000	16,800	–	1944	982,500	9,650	–
1899	241,400	4,660	–	1945	851,500	9,620	–
1900	487,000	8,500	–	1946	224,900	2,010	1,442,100
1901	612,000	9,110	–	1947	419,200	5,680	3,104,449
1902	240,300	14,100	–	1948	1,036,000	11,700	4,295,000
1903	1,301,000	19,000	–	1949	1,031,000	9,560	4,462,000
1904	178,000	7,900	–	1950	364,100	2,570	1,211,000
1905	2,839,000	50,000	–	1951	132,900	1,760	1,045,075
1906	1,413,000	15,300	–	1952	967,000	7,910	5,445,922
1907	2,227,000	20,600	–	1953	286,800	2,200	1,867,315
1908	868,700	4,990	–	1954	198,500	3,980	3,016,557
1909	1,210,000	14,700	–	1955	257,900	4,160	4,857,440
1910	961,100	9,360	–	1956	174,800	1,160	–
1911	1,347,000	12,000	–	1957	972,300	8,590	11,636,118
1912	1,924,000	15,300	–	1958	1,391,000	9,570	18,137,243
1913	493,800	11,400	–	1959	341,900	1,820	2,052,034
1914	1,111,000	8,380	–	1960	563,400	4,080	3,723,688
1915	1,463,000	14,400	–	1961	437,700	4,300	2,298,082
1916	1,421,000	15,800	–	1962	748,100	5,350	2,454,975
1917	1,305,000	11,400	–	1963	405,500	1,760	136,447
1918	379,000	3,500	–	1964	164,200	2,260	69,596
1919	1,527,000	13,200	–	1965	294,600	3,720	5,939,886
1920	1,970,000	18,000	–	1966	4,890	1,510	263,992
1921	1,470,000	19,400	–	1967	63,010	6,160	2,633,789
1922	1,044,000	–	–	1968	201,600	4,240	6,661,067
1923	964,000	9,900	–	1969	163,900	3,180	2,888,864
1924	1,662,000	13,000	–	1970	69,810	5,120	1,843,799
1925	321,000	3,370	–	1971	851	500	34,691
1926	1,120,000	10,900	–	1972	18,900	2,800	1,400,686
1927	1,180,000	13,900	–	1973	486,100	6,300	10,561,149
1928	773,000	7,500	–	1974	14,650	1,180	410,605
1929	1,240,000	47,000	–	1975	637,800	6,680	7,791,192
1930	930,000	5,880	–	1976	419,900	2,920	2,014,487
1931	418,000	8,130	–	1977	243,200	3,230	4,460,201
1932	1,440,000	12,800	–	1978	384,100	3,130	1,595,183
1933	717,000	20,600	–	1979	1,416,000	6,510	8,519,600
1934	298,300	9,910	–	1980	1,269,000	6,210	5,873,066
1935	917,600	15,000	–	1981	327,000	3,180	2,716,584
1936	872,900	9,550	–	1982	709,300	5,120	10,760,893
1937	1,597,000	30,000	–	1983	1,154,000	5,620	2,888,052
1938	1,004,000	7,410	–	1984	457,100	5,620	1,482,351
1939	615,700	4,850	–	1985	1,062,000	8,110	3,293,362
1940	333,100	2,910	–				

Source: U.S. Geological Survey data.

Appendix B.2. Water, sediment, and plutonium data for Los Alamos Canyon

Year	Water (ac ft)	Flood (cfs)	Sediment (tons)	Σ Pu (mCi)	Pu (yr) (mCi)
1943	22	66	466	0.00	0
1944	198	631	8,393	2.80	2.798
1945	0	0	61	2.83	0.03
1946	28	80	611	3.15	0.32
1947	1	2	65	3.20	0.05
1948	0	0	61	3.24	0.04
1949	0	0	61	3.29	0.05
1950	6	20	77	3.35	0.06
1951	236	687	9,814	20.25	16.9
1952	209	386	6,316	37.86	17.61
1953	2	4	12	37.89	0.03
1954	40	129	1,006	40.75	2.86
1955	91	283	2,783	49.58	8.83
1956	0	0	0	49.58	0
1957	433	649	16,470	93.53	43.95
1958	63	203	2,062	100.89	7.36
1959	33	59	532	102.63	1.74
1960	0	0	154	103.38	0.75
1961	18	53	443	105.78	2.4
1962	0	1	138	106.66	0.88
1963	88	283	2,772	116.73	10.07
1964	0	0	0	116.73	0
1965	124	233	3,163	126.61	9.88
1966	10	32	165	127.14	0.53
1967	129	361	4,197	137.38	10.24
1968	287	924	14,120	159.20	21.82
1969	124	149	2,899	164.16	4.96
1970	0	0	0	164.16	0
1971	16	42	247	164.58	0.42
1972	0	0	0	164.58	0
1973	109	349	3,955	172.79	8.2099
1974	6	20	129	173.12	0.3301
1975	4	6	99	173.42	0.3
1976	6	20	77	173.65	0.23
1977	1	4	8	173.68	0.0299
1978	108	293	3,198	179.69	6.0101
1979	10	312	426	181.18	1.4899
1980	0	0	183	181.98	0.8001
1981	0	–	0	181.98	0
1982	0	–	0	181.98	0
1983	43	–	24,357	184.76	2.7800
1984	0	–	0	184.76	0
1985	43	–	41,461	186.84	2.0800
1986	3	–	2,460	188.43	1.5900

Sources: 1943–1980 data from calculations by L. J. Lane in support of L. J. Lane, W. D. Purtymun, and N. M. Becker, *New Estimating Procedures for Surface Runoff, Sediment Yield, and Contaminant Transport in Los Alamos County, New Mexico*, Los Alamos National Laboratory Report LA-10335-MS, UC-11 (Los Alamos, N.M.: Los Alamos National Laboratory, 1985); 1981–1986 data from W. D. Purtymun, P. Peters, and M. N. Maes, *Transport of Plutonium in Snowmelt Runoff*, Los Alamos National Laboratory Report LA-11795-MS, UC-902 (Los Alamos, N.M.: Los Alamos National Laboratory, 1990) using different techniques. The compatibility of the two data sets is questionable.

Appendix B.3. Summaries and correlations for gaging stations

Station	Annual Mean (Standard Deviation)			Correlation Coefficient			Years
	Water Yield	Flood	Sediment	Wat/Fl	Wat/Sed	Fl/Sed	
Rio Chama near Chamita	390,623 (184,966)	4,451 (2,597)	1,513,500 (1,201,135)	0.56	0.64	0.43	73
Rio Grande at Embudo	717,154 (367,802)	5,044 (3,312)	–	0.88	–	–	96
Rio Grande at Otowi	1,059,493 (543,678)	8,526 (4930)	2,163,396 (1,431,926)	0.81	0.29	0.68	87
Los Alamos Canyon near Otowi	63 98	165 (233)	2,241 (3,927)	0.90	0.98	0.93	38
Rio Grande at San Felipe	1,011,400 (489,610)	8,997 (5,537)	–	0.61	–	–	60
Jemez River below Jemez Dam	42,689 (30,878)	2,443 (3,034)	482,071 (254,543)	0.20	0.34	0.39	43
Rio Grande at Albuqurque	834,593 (502,424)	7,233 (4,102)	1,936,963 (1,845,187)	0.72	0.26	0.36	44
Rio Puerco near Bernardo	32,739 (25,351)	4,751 (3,637)	5,114,814 (4,318,308)	0.73	0.96	0.76	46
Rio Salado near San Acacia	10,351 (11,163)	9,861 (7,408)	285,425 (342,250)	0.23	*	*	37
Rio Grande at San Marcial	825,399 (609,206)	9,593 (8,446)	3,981,758 (3,786,497)	0.75	0.61	0.63	91

Note: –, no data; *, insufficient data; data in English units (metric in parenthesis).

Appendix B.4. Mass budget data summary near Los Alamos, 1951

Station	Water Yield (ac ft) and (10^6 m^3)	Maximum Flood (ft^3 s^{-1}) and (m^3 s^{-1})	Suspended Sediment (t) and (Mg)	Bedload Sediment (t) and (Mg)	Total Pu (mCi)
Rio Chama near Chamita	144,500 (176.290)	1,400 (39.2)	580,201 (526,358)	92,832 (84,217)	0.14
Rio Grande at Embudo	246,600 (300.852)	710 (19.9)	320,542 (290,796)	51,287 (46,528)	1.28
Rio Grande at Otowi Bridge	395,400 (428.388)	3,440 (96.3)	900,743 (817,154)	144,119 (130,745)	1.42
Los Alamos Canyon	236 (0.288)	687 (19.2)	9,814 (8,903)	9,814 (8,903)	16.90
Rio Grande in White Rock Canyon	395,636 (482.676)	3,440 (96.3)	910,557 (826,057)	153,933 (139,648)	18.32

Appendix B.5. Mass budget data summary near Los Alamos, 1952

Station	Water Yield (ac ft) and (10^6 m^3)	Maximum Flood (ft^3 s^{-1}) and (m^3 s^{-1})	Suspended Sediment (t) and (Mg)	Bedload Sediment (t) and (Mg)	Total Pu (mCi)
Rio Chama near Chamita	566,500 (691.130)	5,880 (164.6)	3,390,906 (3,076,230)	542,545 (492,197)	0.82
Rio Grande at Embudo	777,300 (948.306)	8,720 (244.2)	1,082,496 (982,040)	173,199 (157,126)	6.24
Rio Grande at Otowi Bridge	1,378,000 (1,681.160)	9,700 (271.6)	4,473,402 (4,058,270)	715,744 (649,323)	7.06
Los Alamos Canyon	209 (0.255)	386 (10.8)	6,316 (5,730)	6,316 (5,730)	17.61
Rio Grande in White Rock Canyon	1,378,236 (1,681,448)	9,700 (271.6)	4,483,216 (4,067,174)	725,558 (658,226)	24.67

Appendix B.6. Mass budget data summary near Los Alamos, 1957

Station	Water Yield (ac ft) and (10^6 m^3)	Maximum Flood (ft^3 s^{-1}) and (m^3 s^{-1})	Suspended Sediment (t) and (Mg)	Bedload Sediment (t) and (Mg)	Total Pu (mCi)
Rio Chama near Chamita	522,900 (637.938)	4,280 (119.8)	3,344,250 (3,033,904)	535,080 (485,425)	0.81
Rio Grande at Embudo	753,400 (919.148)	5,000 (140.2)	1,213,090 (1,100,515)	194,096 (176,084)	6.39
Rio Grande at Otowi Bridge	1,297,000 (1,582.340)	6,650 (186.2)	4,557,348 (4,134,426)	729,176 (661,508)	7.19
Los Alamos Canyon	443 (0.528)	649 (18.8)	16,470 (14,941)	16,470 (14,941)	43.95
Rio Grande in White Rock Canyon	1,378,236 (1,582,868)	9,700 (186.2)	4,483,216 (4,149,368)	725,558 (646,450)	51.14

Appendix B.7. Mass budget data summary near Los Alamos, 1968

Station	Water Yield (ac ft) and (10^6 m^3)	Maximum Flood (ft^3 s^{-1}) and (m^3 s^{-1})	Suspended Sediment (t) and (Mg)	Bedload Sediment (t) and (Mg)	Total Pu (mCi)
Rio Chama near Chamita	278,900 (340.258)	2,170 (60.8)	2,013,011 (1,826,203)	322,082 (292,193)	0.49
Rio Grande at Embudo	366,300 (446.886)	3,500 (99.4)	560,683 (508,652)	89,709 (82,384)	3.58
Rio Grande at Otowi Bridge	855,700 (1,043.954)	4,490 (125.7)	2,573,694 (2,334,855)	411,791 (373,577)	4.06
Los Alamos Canyon	287 (0.350)	924 (25.9)	14,120 (12,810)	14,120 (12,810)	21.82
Rio Grande in White Rock Canyon	855,987 (1,044,304)	4,490 (125.9)	2,587,814 (2,347,665)	425,911 (386,386)	25.88

Appendix B.8. Assumptions and data sources for mass budget calculations

Station	Drainage Area (km^2)	Years of Record	Annual Mean: Water (10^6 m^3)	Annual Mean: Flood (m^3s^{-1})	Annual Mean: Suspended (mg)	Annual Mean: Bedload (mg)	Annual Mean: Pu (mCi)
Rio Chama	1	1	1	1	1	4	8
Embudo	1	1	1	1	1	5	5
Otowi	1	1	1	1	1	4	8
Los Alamos	2	3	3	3	3	6	3
San Felipe	1	1	1	1	—	—	—
Jemez River	1	1	1	1	1	6	8
Albuquerque	1	1	1	1	1	4	9
Rio Puerco	1	1	1	1	1	7	10
Rio Salado	1	1	1	1	1	6	10
San Marcial	1	1	1	1	1	4	9

Sources:
1. U.S. Geological Survey data, WATSTORE and EarthInfo, Inc.
2. Measured by digitizing U.S. Geological Survey topographic maps.
3. Data from L. J. Lane, used in preparation of L. J. Lane, W. D. Purtymun, and N. M. Becker, *New Estimating Procedures for Surface Runoff, Sediment Yield, and Contaminant Transport in Los Alamos County, New Mexico*, Los Alamos National Laboratory Report LA-10335-MS, UC-11 (Los Alamos, N.M.: Los Alamos National Laboratory, 1985).
4. Calculated as 14% of total load (16% of suspended load): R. J. Garde and K. G. Raju, *Mechanics of Sediment Transportation and Alluvial Stream Problems* (New York: Wiley, 1977), p. 62.
5. Calculated as the difference between the Rio Chama and Otowi stations.
6. Calculated as bedload equal to suspended load.
7. Calculated as 71% of total load (41% of suspended load): D. B. Simons, R-M Li, L-Y Li, and M. J. Ballantine, *Erosion and Sedimentation Analysis of Rio Puerco and Rio Salado Watersheds*, Technical Report, Simons, Li & Associates for U. S. Army Corps of Engineers, Albuquerque District Office (Fort Collins, Colo.: Simons, Li & Associates, 1981), p. 53.
8. Calculated by multiplying concentrations by mass for suspended sediment and bedload and summing for total; concentrations calculated as mean values published by Los Alamos National Laboratory in various surveillance reports.
9. Calculated as in 8 using data for Rio Grande at Bernalillo.
10. Calculated as in 8 using data for Frijoles Canyon, the most similar environment for which data exist.

Appendix B.9. Annual budget calculations for suspended sediment, 1948–1985 (tons/year)

	Transport						Storage		
Year	Otowi	Alamos	Jemez	Albuquerque	Puerco	Marcial	Oto–Alb	Alb–Marc	Oto–Marc
1948	4,306,000	61	514,537	4,364,260	1,634,000	4,295,000	456,338	1,703,260	2,159,598
1949	3,681,000	61	502,800	3,538,210	5,760,000	4,462,000	645,651	4,836,210	5,481,861
1950	1,733,000	77	255,600	1,961,030	2,753,000	1,211,000	27,647	3,503,030	3,530,677
1951	900,743	9,814	790,239	3,492,734	4,613,496	1,045,075	−1,791,938	7,061,155	5,269,217
1952	4,473,402	6,316	515,327	3,057,870	2,953,332	5,445,922	1,937,175	565,280	2,502,455
1953	732,109	12	61,695	3,031,635	6,953,247	1,867,315	−2,237,819	8,117,567	5,879,748
1954	1,329,497	1,006	690,970	2,290,725	14,778,779	3,016,557	−269,252	14,052,947	13,783,695
1955	2,430,691	2,783	768,258	3,103,649	18,315,560	4,857,440	98,083	16,561,769	16,659,852
1956	714,319	0	228,489	2,007,069	3,423,769	1,113,186	−1,064,261	4,317,652	3,253,391
1957	4,557,348	16,470	319,488	2,804,035	18,054,868	11,636,118	2,089,271	9,222,785	11,312,056
1958	7,562,178	2,062	687,843	3,983,688	8,072,030	18,137,243	4,268,395	−6,081,525	−1,813,130
1959	1,424,491	532	447,031	896,440	5,038,884	2,052,034	975,614	3,883,290	4,858,904
1960	2,074,261	154	450,960	1,892,193	4,156,507	3,723,688	633,182	2,325,012	2,958,194
1961	1,971,899	443	471,793	2,489,954	4,548,008	2,298,082	−45,819	4,739,880	4,694,061
1962	3,252,975	138	466,335	2,158,715	1,448,532	2,454,975	1,560,733	1,152,272	2,713,005
1963	862,093	2,772	397,709	1,004,865	3,026,277	136,447	257,709	3,894,695	4,152,404
1964	946,606	0	385,539	945,426	2,917,346	69,596	386,719	3,793,176	4,179,895
1965	3,377,878	3,163	429,616	2,880,804	3,807,918	5,939,886	929,853	748,836	1,678,689
1966	2,255,645	165	445,454	2,279,176	3,528,635	263,992	422,088	5,543,819	5,965,907

1967	2,650,962	4,197	449,386	4,956,701	12,257,979	2,633,789	−1,852,156	14,580,891	12,728,735
1968	2,573,694	14,120	471,369	1,475,209	4,940,551	6,661,067	1,583,974	−245,307	1,338,667
1969	1,823,935	2,899	481,511	2,134,149	4,919,348	2,888,864	174,196	4,164,633	4,338,829
1970	1,939,043	0	501,599	3,444,903	2,822,326	1,843,799	−1,004,261	4,423,430	3,419,169
1971	1,105,621	247	386,979	2,544,489	1,888,661	34,691	−1,051,642	4,398,459	3,346,817
1972	1,464,464	0	409,308	2,428,212	9,490,327	1,400,686	−554,440	10,517,853	9,963,413
1973	3,997,338	3,955	654,953	8,017,443	7,958,371	10,561,149	−3,361,197	5,414,665	2,053,468
1974	823,348	129	381,365	1,045,303	1,141,388	410,605	159,539	1,776,086	1,935,625
1975	1,525,534	99	541,266	2,807,587	6,829,451	7,791,192	−740,688	1,845,846	1,105,158
1976	1,839,982	77	377,627	1,621,401	1,734,453	2,014,487	596,285	1,341,367	1,937,65
1977	896,924	8	374,944	491,805	4,355,708	4,460,201	780,071	387,312	1,167,383
1978	1,170,813	3,198	427,122	774,449	478,526	1,595,183	826,684	−342,208	484,476
1979	2,595,964	426	637,629	1,846,126	3,496,564	8,519,600	1,387,893	−3,176,910	−1,789,017
1980	1,538,125	183	486,550	759,946	1,810,746	5,873,066	1,264,912	−3,302,374	−2,037,462
1981	436,468	–	390,591	393,773	2,093,080	2,716,584	433,286	−229,731	203,555
1982	1,578,260	–	441,433	849,813	4,408,094	10,760,893	1,169,880	−5,502,986	−4,333,106
1983	1,466,723	–	564,609	1,325,704	1,876,008	2,888,052	705,628	313,660	1,019,288
1984	1,468,574	–	456,294	1,197,098	2,678,574	1,482,351	727,770	2,393,321	3,121,091
1985	2,727,140	–	560,426	1,443,353	3,398,587	3,293,362	1,844,213	1,548,578	3,392,791

Grand totals

Storage, Otowi to Albuquerque = 12,369,316

Storage, Albuquerque to San Marcial = 130,247,695

Storage, Otowi to San Marcial = 142,617,011

Note: Storage for Otowi–Albuquerque reach overestimated because of regression inaccuracies that estimated values for Albuquerque gage during years when data were not collected.

Appendix B.10. Annual budget calculations for bedload sediment, 1948–1985 (tons/year)

	Transport						Storage		
Year	Otowi	Alamos	Jemez	Albuquerque	Puerco	Marcial	Oto–Alb	Alb–Marc	Oto–Marc
1948	688,960	61	514,537	698,281.6	669,940	687,200	505,276.4	681,021.6	1,186,298
1949	588,960	61	502,800	566,113.6	2,361,600	713,920	525,707.4	2,213,793.	2,739,501
1950	277,280	77	255,600	313,764.8	1,128,730	193,760	219,192.2	1,248,734.	1,467,927
1951	144,118.8	9,814	790,239	558,837.4	1,891,533.	167,212	385,334.4	2,283,158.	2,668,493.
1952	715,744.3	6,316	515,327	489,259.2	1,210,866.	871,347.5	748,128.1	828,777.8	1,576,905.
1953	117,137.4	12	61,695	485,061.6	2,850,831.	298,770.4	−306,217.	3,037,122.	2,730,905.
1954	212,719.5	1,006	690,970	366,516	6,059,299.	482,649.1	538,179.5	5,943,166.	6,481,345.
1955	388,910.5	2,783	768,258	496,583.8	7,509,379.	777,190.4	663,367.7	7,228,773.	7,892,140.
1956	114,291.0	0	228,489	321,131.0	1,403,745.	178,109.7	21,649	1,546,766.	1,568,415.
1957	729,175.6	16,470	319,488	448,645.6	7,402,495.	1,861,778.	616,488.0	5,989,362.	6,605,850.
1958	1,209,948.	2,062	687,843	6,373,90.0	3,309,532.	2,901,958.	1,262,463.	1,044,963.	2,307,426.
1959	227,918.5	532	447,031	43,430.4	2,065,942.	328,325.4	532,051.1	1,881,047.	2,413,098.
1960	331,881.7	154	450,960	302,750.8	1,704,167.	595,790.0	480,244.8	1,411,128.	1,891,373.
1961	315,503.8	443	471,793	398,392.6	1,864,683.	367,693.1	389,347.2	1,895,382.	2,284,730
1962	520,476	138	466,335	345,394.4	593,898.1	392,796	641,554.6	546,496.5	1,188,051.
1963	137,934.8	2,772	397,709	160,778.4	1,240,773.	21,831.52	377,637.4	1,379,720.	1,757,357.
1964	151,456.9	0	385,539	151,268.1	1,196,111.	11,135.36	385,727.8	1,336,244.	1,721,972.
1965	540,460.4	3,163	429,616	460,928.6	1,561,246.	950,381.7	512,310.8	1,071,793.	1,584,104.
1966	360,903.2	165	445,454	364,668.1	1,446,740.	42,238.72	441,854.0	1,769,169.	2,211,023.

1967	424,153.9	4,197	449,386	793,072.1	5,025,771.	421,406.2	84,664.76	5,397,437.	5,482,102.
1968	411,791.0	14,120	471,369	236,033.4	2,025,625.	1,065,770.	661,246.6	1,195,888.	1,857,135.
1969	291,829.6	2,899	481,511	341,463.8	2,016,932.	462,218.2	434,775.7	1,896,178.	2,330,954.
1970	310,246.8	0	501,599	551,184.4	1,157,153.	295,007.8	260,661.4	1,413,330.	1,673,991.
1971	176,899.3	247	386,979	407,118.2	774,351.0	5,550.56	157,007.1	1,175,918.	1,332,925.
1972	234,314.2	0	409,308	388,513.9	3,891,034.	224,109.7	255,108.3	4,055,438.	4,310,546.
1973	639,574.0	3,955	654,953	1,282,790.	3,262,932.	1,689,783.	15,691.2	2,855,939.	2,871,630.
1974	131,735.6	129	381,365	167,248.4	467,969.0	65,696.8	345,981.2	569,520.7	915,501.9
1975	244,085.4	99	541,266	449,213.9	2,800,074.	1,246,590.	336,236.5	2,002,698.	2,338,934.
1976	294,397.1	77	377,627	259,424.1	711,125.7	322,317.9	412,676.9	648,231.9	1,060,908.
1977	143,507.8	8	374,944	78,688.8	1,785,840.	713,632.1	439,771.0	1,150,896.	1,590,667.
1978	187,330.0	3,198	427,122	123,911.8	196,195.6	255,229.2	493,738.2	64,878.22	558,616.4
1979	415,354.2	426	637,629	295,380.1	1,433,591.	1,363,136	758,029.0	365,835.4	1,123,864.
1980	246,100	183	486,550	121,591.3	742,405.8	939,690.5	611,241.6	−75,693.3	535,548.3
1981	69,834.88	0	390,591	63,003.68	858,162.8	434,653.4	397,422.2	486,513.0	883,935.2
1982	252,521.6	0	441,433	135,970.0	1,807,318.	1,721,742.	557,984.5	221,545.7	779,530.2
1983	234,675.6	0	564,609	212,112.6	769,163.2	462,088.3	587,172.0	519,187.6	1,106,359.
1984	234,971.8	0	456,294	191,535.6	1,098,215.	237,176.1	499,730.1	1,052,574.	1,552,305.
1985	436,342.4	0	560,426	230,936.4	1,393,420.	526,937.9	765,831.9	1,097,419.	1,863,251.

Grand totals
Storage, Otowi to Albuquerque = 17,015,267
Storage, Albuquerque to San Marcial = 69,430,363
Storage, Otowi to San Marcial = 86,445,631

Appendix B.11. Annual budget calculations for total sediment, 1948–1985 (tons /year)

Year	Storage		
	Oto–Alb	Alb–Marc	Oto–Marc
1948	961,614.4	2,384,281.6	3,345,896
1949	1,171,358.4	7,050,003.6	8,221,362
1950	246,839.2	4,751,764.8	4,998,604
1951	– 1,406,603.56	9,344,313.8	7,937,710.24
1952	2,685,303.12	1,394,057.8	4,079,360.92
1953	– 2,544,036.16	11,154,689.47	8,610,653.31
1954	268,927.52	19,996,113.27	20,265,040.79
1955	761,450.72	23,790,542.04	24,551,992.76
1956	– 1,042,612	5,864,418.57	4,821,806.57
1957	2,705,759.08	15,212,147.6	17,917,906.68
1958	5,530,858.4	– 5,036,561.5	494,296.9
1959	1,507,665.16	5,764,337.4	7,272,002.56
1960	1,113,426.88	3,736,140.67	4,849,567.55
1961	343,528.2	6,635,262.8	6,978,791
1962	2,202,287.6	1,698,768.52	3,901,056.12
1963	635,346.48	5,274,415.45	5,909,761.93
1964	772,446.8	5,129,420.66	5,901,867.46
1965	1,442,163.84	1,820,629.26	3,262,793.1
1966	863,942.04	7,312,988.79	8,176,930.83
1967	– 1,767,491.24	19,978,328.31	18,210,837.07
1968	2,245,220.6	950,581.63	3,195,802.23
1969	608,971.76	6,060,811.28	6,669,783.04
1970	– 743,599.6	5,836,760.3	5,093,160.7
1971	– 894,634.88	5,574,377.69	4,679,742.81
1972	– 299,331.68	14,573,291.23	14,273,959.55
1973	– 3,345,505.8	8,270,604.15	4,925,098.35
1974	505,520.2	2,345,606.76	2,851,126.96
1975	– 404,451.48	3,848,544.11	3,444,092.63
1976	1,008,961.96	1,989,598.97	2,998,560.93
1977	1,219,842.04	1,538,208.92	2,758,050.96
1978	1,320,422.24	– 277,329.78	1,043,092.46
1979	2,145,922.08	– 2,811,074.6	– 665,152.52
1980	1,876,153.64	– 3,378,067.34	– 1,501,913.7
1981	830,708.2	256,782.04	1,087,490.24
1982	1,727,864.52	– 5,281,440.26	– 3,553,575.74
1983	1,292,800.04	832,847.6	2,125,647.64
1984	1,227,500.16	3,445,895.86	4,673,396.02
1985	2,610,044.92	2,645,997.23	5,256,042.15
Sum	29,384,583.8	199,678,058.7	229,062,642.5

Appendix B.12. Suspended, bedload, and total load sediment budget summary for critical years (Mg)

	Station	1951	1952	1957	1968
Suspended sediment	Otowi	817,154	4,058,270	4,134,426	2,334,855
	Los Alamos	8,903	5,730	14,942	12,810
	Jemez	716,905	467,505	289,840	427,626
	Rio Puerco	4,185,364	2,679,263	16,379,376	4,482,068
	San Marcial	948,092	4,940,540	10,556,286	6,042,920
	Suspended storage	4,780,234	2,270,228	10,262,298	1,214,439
Bedload sediment	Otowi	130,745	649,323	661,508	373,577
	Los Alamos	8,903	5,730	14,942	12,810
	Jemez	716,905	467,505	289,840	427,626
	Rio Puerco	1,715,999	1,098,498	6,715,544	1,837,648
	San Marcial	151,695	790,486	1,689,006	966,867
	Bedload storage	2,420,857	1,430,570	5,331,320	1,684,794
Total sediment load	Otowi	947,899	4,707,593	4,795,934	2,708,432
	Los Alamos	17,806	11,460	29,884	25,620
	Jemez	1,433,810	935,010	579,680	855,252
	Rio Puerco	5,901,363	3,777,761	23,094,920	6,319,716
	San Marcial	1,099,787	5,731,026	12,245,292	7,009,787
	Total storage	7,201,091	3,700,798	15,593,618	2,899,233

Appendix C. Sediment particle-size data

Reach	Sample No.	Map Unit	Landform Type	Sample Depth (cm)	Fines (%)
San Marcial representative reach	1 Marcial	1A	Active channel	30	10.68
	6 Marcial	1A	Active channel	5	30.53
	7 Marcial	1A	Active channel	5	49.08
	8 Marcial	1A	Active channel	5	45.37
	9 Marcial	2A	Abandoned channel	30	2.52
	10 Marcial	2A	Abandoned channel	5	2.55
	11 Marcial	2A	Abandoned channel	5	0.84
	12 Marcial	2A	Abandoned channel	5	2.66
	14 Marcial	2A	Abandoned channel	60	0.99
	16 Marcial	2D	Abandoned flood plain	30	22.85
	17 Marcial	2D	Abandoned flood plain	5	33.84
	18 Marcial	2D	Abandoned flood plain	5	30.59
	19 Marcial	2D	Abandoned flood plain	5	40.01
	20 Marcial	2D	Abandoned flood plain	60	28.09
	21 Marcial	2D	Abandoned flood plain	90	33.43
	22 Marcial	3A	Older abandoned channel	30	11.53
	23 Marcial	3A	Older abandoned channel	5	13.45
	24 Marcial	3A	Older abandoned channel	5	9.14
	25 Marcial	3A	Older abandoned channel	5	16.85
	26 Marcial	3C	Abandoned flood plain or bar	30	17.66
	27 Marcial	3C	Abandoned flood plain or bar	5	28.95
	28 Marcial	3C	Abandoned flood plain or bar	5	27.69
	29 Marcial	3C	Abandoned flood plain or bar	5	22.11
	30 Marcial	3C	Abandoned flood plain or bar	60	30.20
	31 Marcial	3C	Abandoned flood plain or bar	90	54.34
Chamizal representative reach	32 Chamizal	1B	Active flood plain	30	31.79
	33 Chamizal	1B	Active flood plain	5	56.43
	34 Chamizal	1B	Active flood plain	5	54.82
	35 Chamizal	1B	Active flood plain	5	46.85
	36 Chamizal	2A	Abandoned braided channel	30	44.54
	37 Chamizal	2A	Abandoned braided channel	5	6.77
	38 Chamizal	2A	Abandoned braided channel	5	7.34
	39 Chamizal	2A	Abandoned braided channel	5	55.02
	40 Chamizal	2B	Abandoned flood plain	30	45.58
	41 Chamizal	2B	Abandoned flood plain	60	6.69
	42 Chamizal	2B	Abandoned flood plain	75	38.37
	43 Chamizal	2B	Abandoned flood plain	5	42.11
	44 Chamizal	2B	Abandoned flood plain	5	43.53
	45 Chamizal	2B	Abandoned flood plain	5	42.57
	46 Chamizal	3B	Older abandoned flood plain	30	43.07
	47 Chamizal	3B	Older abandoned flood plain	5	30.95
	48 Chamizal	3B	Older abandoned flood plain	5	23.09
	49 Chamizal	3B	Older abandoned flood plain	5	33.57
	50 Chamizal	2B	Abandoned braided channel	30	0.63
	51 Chamizal	2B	Abandoned braided channel	5	0.51
	52 Chamizal	2B	Abandoned braided channel	5	0.51
	53 Chamizal	2B	Abandoned braided channel	5	0.42

Reach	Sample No.	Map Unit	Landform Type	Sample Depth (cm)	Fines (%)
San Geronimo representative reach	54 Geronimo	3B	Abandoned older flood plain	30	42.59
	55 Geronimo	3B	Abandoned older flood plain	60	36.16
	56 Geronimo	3B	Abandoned older flood plain	5	37.10
	57 Geronimo	2B	Abandoned flood plain	30	61.00
	58 Geronimo	3B	Abandoned older flood plain	5	40.43
	59 Geronimo	3B	Abandoned older flood plain	5	47.30
	60 Geronimo	2B	Abandoned flood plain	60	28.75
	61 Geronimo	2B	Abandoned flood plain	75	28.15
	62 Geronimo	2B	Abandoned flood plain	5	40.16
	63 Geronimo	2B	Abandoned flood plain	5	42.65
	64 Geronimo	2B	Abandoned flood plain	5	42.75
	65 Geronimo	2A	Abandoned flood plain	30	20.85
	66 Geronimo	2B	Abandoned flood plain	30	45.64
	67 Geronimo	2B	Abandoned flood plain	5	46.93
	68 Geronimo	2B	Abandoned flood plain	5	52.53
	69 Geronimo	2B	Abandoned flood plain	5	43.78
	70 Geronimo	2A	Abandoned braided channel	5	28.71
	71 Geronimo	2A	Abandoned braided channel	5	28.02
	72 Geronimo	2A	Abandoned braided channel	5	26.67
Los Lunas representative reach	73 Lunas	3A	Abandoned braided channel	30	54.00
	74 Lunas	3A	Abandoned braided channel	45	1.74
	75 Lunas	3A	Abandoned braided channel	5	62.60
	76 Lunas	3A	Abandoned braided channel	5	62.91
	77 Lunas	3A	Abandoned braided channel	5	36.16
	78 Lunas	3B	Abandoned flood plain	30	34.60
	79 Lunas	3B	Abandoned flood plain	5	21.58
	80 Lunas	3B	Abandoned flood plain	5	37.74
	81 Lunas	3B	Abandoned flood plain	5	50.20
	82 Lunas	3B	Abandoned flood plain	5	62.74
	83 Lunas	3A	Abandoned braided channel	5	20.69
	84 Lunas	3A	Abandoned braided channel	30	4.20
	85 Lunas	3A	Abandoned braided channel	45	1.46
	86 Lunas	3A	Abandoned braided channel	5	1.25
	87 Lunas	3A	Abandoned braided channel	5	4.98
	88 Lunas	3A	Abandoned braided channel	5	5.10
Los Griegos representative reach	89 Griegos	2A	Abandoned braided channel	30	35.96
	90 Griegos	2A	Abandoned braided channel	5	56.82
	91 Griegos	2A	Abandoned braided channel	5	42.86
	92 Griegos	2A	Abandoned braided channel	5	11.52
	93 Griegos	2B	Abandoned flood plain	30	32.67
	94 Griegos	2B	Abandoned flood plain	60	27.55
	95 Griegos	2B	Abandoned flood plain	5	18.20
	96 Griegos	2B	Abandoned flood plain	5	41.43
	97 Griegos	2B	Abandoned flood plain	5	30.86
	98 Griegos	3B	Older abandoned flood plain	30	4.62
	99 Griegos	3B	Older abandoned flood plain	60	1.66
	100 Griegos	3B	Older abandoned flood plain	5	1.43
	101 Griegos	3B	Older abandoned flood plain	5	2.25
	102 Griegos	3B	Older abandoned flood plain	5	7.92

Appendix C. Sediment particle-size data *(continued)*

Reach	Sample No.		Map Unit	Landform Type	Sample Depth (cm)	Fines (%)
Coronado representative reach	103	Coronado	3B	Abandoned braided channel	30	47.07
	104	Coronado	3B	Abandoned braided channel	60	58.05
	105	Coronado	3B	Abandoned braided channel	5	20.38
	106	Coronado	3B	Abandoned braided channel	5	23.01
	107	Coronado	3B	Abandoned braided channel	5	33.61
	108	Coronado	3A	Abandoned single channel	30	31.09
	109	Coronado	3A	Abandoned single channel	45	24.49
	110	Coronado	3A	Abandoned single channel	75	23.31
	111	Coronado	3A	Abandoned single channel	90	22.97
	112	Coronado	3A	Abandoned single channel	5	21.82
	113	Coronado	3A	Abandoned single channel	5	21.80
	114	Coronado	3A	Abandoned single channel	5	11.30
	115	Coronado	2A	Abandoned flood plain	30	44.04
	116	Coronado	2A	Abandoned flood plain	60	10.17
	117	Coronado	2A	Abandoned flood plain	90	5.22
	118	Coronado	2A	Abandoned flood plain	5	58.61
	119	Coronado	2A	Abandoned flood plain	5	40.44
	120	Coronado	2A	Abandoned flood plain	5	57.59
	121	Coronado	1A	Active channel	5	37.61
	122	Coronado	1A	Active channel	5	37.79
	123	Coronado	1A	Active channel	5	33.12
Rio Puerco 2 km above Rio Grande	124	Puerco	1A	Active channel	5	29.59
	125	Puerco	1A	Active channel	5	29.56
	126	Puerco	1A	Active channel	5	26.53
Rio Salado 2 km above Rio Grande	127	Salado	1A	Active channel	5	6.71
	128	Salado	1A	Active channel	5	6.97
	129	Salado	1A	Active channel	5	7.15
Peña Blanca representative reach	130	Peña	4A	Older braided channel	5	17.08
	131	Peña	4A	Older braided channel	5	43.25
	132	Peña	4A	Older braided channel	5	4.65
	133	Peña	4A	Older braided channel	5	4.91
	134	Peña	4B	Abandoned braided channel	5	11.33
	135	Peña	4B	Abandoned braided channel	5	17.07
	136	Peña	4B	Abandoned braided channel	5	6.83
	137	Peña	4B	Abandoned braided channel	5	12.48
	138	Peña	5B	Older abandoned flood plain	5	39.11
	139	Peña	5B	Older abandoned flood plain	5	38.84
	140	Peña	5B	Older abandoned flood plain	5	58.87
	141	Peña	5B	Older abandoned flood plain	5	41.47
	142	Peña	X	Abandoned single channel	5	22.92
	143	Peña	X	Abandoned single channel	5	34.38
	144	Peña	X	Abandoned single channel	5	23.72
	145	Peña	X	Abandoned single channel	5	34.52
Otowi representative reach	146	Otowi	2B	Flood bar	5	18.69
	147	Otowi	2B	Flood bar	5	5.03
	148	Otowi	2B	Flood bar	5	7.61
	149	Otowi	2B	Flood bar	5	11.28

Reach	Sample No.	Map Unit	Landform Type	Sample Depth (cm)	Fines (%)
Los Alamos Canyon at Otowi	150 Alamos	T1A	Outlet channel	5	0.94
	151 Alamos	T1A	Outlet channel	5	1.95
	152 Alamos	T1A	Outlet channel	5	0.58
	153 Alamos	T1A	Outlet channel	5	1.39
	154 Alamos	T1A	Active channel	5	1.18
	155 Alamos	T1A	Active channel	5	1.49
	156 Alamos	T1A	Active channel	5	1.35
	157 Alamos	T1A	Active channel	5	1.35
	158 Alamos	T1B	Active flood plain	5	48.18
	159 Alamos	T1B	Active flood plain	5	64.68
	160 Alamos	T1B	Active flood plain	5	52.54
	161 Alamos	T1B	Active flood plain	5	55.39
	164 Alamos	X	Older outlet channel	5	28.05
	165 Alamos	X	Older outlet channel	5	24.69
	166 Alamos	X	Older outlet channel	5	23.86
	167 Alamos	X	Older outlet channel	5	24.50
	168 Alamos	T1A	Active channel	5	14.58
	169 Alamos	T1A	Active channel	5	8.54
	170 Alamos	T1A	Active channel	5	4.23
	172 Alamos	T1A	Active channel	5	3.95
	173 Alamos	T1A	Active channel	5	6.21
	174 Alamos	T1A	Active channel	5	2.72
	175 Alamos	T1A	Active channel	5	2.07
	176 Alamos	T1A	Active channel	5	5.02
	202 Alamos	T1A	Active channel	5	0.61
	203 Alamos	T1A	Active channel	5	9.23
	204 Alamos	T1A	Active channel	5	0.17
	205 Alamos	T1A	Active flood plain	5	25.66
Buckman representative reach	177 Buckman	2B	Inactive bar, lower	5	16.89
	178 Buckman	2B	Inactive bar, lower	5	4.87
	179 Buckman	2B	Inactive bar, lower	5	7.16
	180 Buckman	2B	Inactive bar, lower	5	8.01
	181 Buckman	1A	Active channel	5	46.43
	182 Buckman	1A	Active channel	5	49.60
	183 Buckman	1A	Active channel	5	47.43
	184 Buckman	1A	Active channel	5	46.39
	185 Buckman	3B	Abandoned single channel	5	19.33
	186 Buckman	3B	Abandoned single channel	5	18.71
	187 Buckman	3B	Abandoned single channel	5	21.07
	188 Buckman	3B	Abandoned single channel	5	23.37
	189 Buckman	2B	Inactive bar, upper	5	6.24
	190 Buckman	2B	Inactive bar, upper	5	6.86
	191 Buckman	2B	Inactive bar, upper	5	8.77
	192 Buckman	2B	Inactive bar, upper	5	12.43

Appendix C. Sediment particle-size data *(continued)*

Reach	Sample No.	Map Unit	Landform Type	Sample Depth (cm)	Fines (%)
Frijoles representative reach	195 Frijoles	X	Tributary fan	5	3.48
	196 Frijoles	X	Tributary fan	5	4.22
	197 Frijoles	X	Tributary fan	5	4.17
	199 Frijoles	X	Tributary fan	5	7.45
	198 Frijoles	X	Lacustrine	5	49.85
	200 Frijoles	X	Lacustrine	5	41.08
	201 Frijoles	X	Lacustrine	5	43.13

Note: For map units, see maps of various representative reaches in Chapters 9 and 10.

Appendix D.1. Partitions for channel and flood-plain area measurements

Partition	Starting Point	Ending Point
1.	Old Española Bridge	Midpoint no. 1
2.	Midpoint no. 1	Otowi Bridge
3.	Otowi Bridge	Midpoint no. 3
4.	Midpoint no. 3	Rio de Los Frijoles
5.	Rio de Los Frijoles	Reservoir Center
6.	Reservoir Center	Cochiti Dam
7.	Cochiti Dam	Midpoint no. 5
8.	Midpoint no. 5	Galesteo Bridge
9.	Galesteo Bridge	Midpoint no. 7
10.	Midpoint no. 7	San Felipe Bridge
11.	San Felipe Bridge	Midpoint no. 9
12.	Midpoint no. 9	Jemez River
13.	Jemez River	Midpoint no. 11
14.	Midpoint no. 11	Bernalillo Bridge
15.	Bernalillo Bridge	Midpoint no. 13
16.	Midpoint no. 13	Alameda Bridge
17.	Alameda Bridge	Midpoint no. 15
18.	Midpoint no. 15	Old Town Bridge
19.	Old Town Bridge	Midpoint no. 17
20.	Midpoint no. 17	Barcelona Bridge
21.	Barcelona Bridge	Midpoint no. 19
22.	Midpoint no. 19	I-25 Bridge
23.	I-25 Bridge	Midpoint no. 21
24.	Midpoint no. 21	Los Lunas Bridge
25.	Los Lunas Bridge	Midpoint no. 23
26.	Midpoint no. 26	Tome Grant Boundary
27.	Tome Grant Boundary	Midpoint no. 25
28.	Midpoint no. 25	Belen Bridge
29.	Belen Bridge	Midpoint no. 27
30.	Midpoint no. 27	Bosque Bridge
31.	Bosque Bridge	Midpoint no. 29
32.	Midpoint no. 29	Bernardo Bridge
33.	Bernardo Bridge	Midpoint no. 31
34.	Midpoint no. 31	Cañada Ancha
35.	Cañada Ancha	Midpoint no. 33
36.	Midpoint no. 34	San Acacia Dam
37.	San Acacia Dam	Midpoint no. 35
38.	Midpoint no. 35	Pueblito Bridge
39.	Pueblito Bridge	Midpoint no. 37
40.	Midpoint no. 37	South Socorro Ford
41.	South Socorro Ford	Midpoint no. 39
42.	Midpoint no. 39	San Antonio Bridge
43.	San Antonio Bridge	Midpoint no. 41
44.	Midpoint no. 44	Bosque Curve
45.	Bosque Curve	Midpoint no. 43
46.	Midpoint no. 43	San Marcial Bridge

Note: The areas of flood plain and channel within each partition are given in Appendix D2.

Appendix D.2. Channel and flood-plain area measurements (km and km^2)

Partition	Distance	Cum Dist	Tot Area	Chn Area	Dep Area	Channel (sq km/km)	Flood Plain (sq km/km)
1	7.03975	7.03975	4.69411	0.59045	4.10366	0.083873	0.582926
2	9.91318	16.95293	5.90572	1.41684	4.48888	0.142924	0.452819
3	13.67914	30.63207	1.5048	0.98522	0.51958	0.072023	0.037983
4	13.14933	43.7814	1.17924	1.02673	0.15251	0.078082	0.011598
5	11	54.7814	0	0	0	0	0
6	11.53	66.3114	0	0	0	0	0
7	7.09787	73.40927	6.10936	1.07485	5.03451	0.151432	0.709298
8	3.17856	76.58783	4.68561	0.48142	4.20419	0.151458	1.322671
9	5.95558	82.54341	4.08531	0.86268	3.22263	0.144852	0.541111
10	4.85008	87.39349	4.45269	0.9019	3.55079	0.185955	0.732109
11	5.07425	92.46774	3.04981	0.43824	2.61157	0.086365	0.514671
12	4.14835	96.61609	4.32898	0.72578	3.6032	0.174956	0.868586
13	1.48807	98.10416	2.29926	0.37613	1.92313	0.252763	1.292365
14	5.61298	103.7171	2.93642	1.05513	1.88129	0.187980	0.335167
15	11.08483	114.8019	6.06395	2.43155	3.6324	0.219358	0.327691
16	6.81665	121.6186	4.07511	1.4842	2.59091	0.217731	0.380085
17	4.20661	125.8252	1.97712	0.9031	1.07402	0.214685	0.255317
18	9.49885	135.3240	4.89788	1.74168	3.1562	0.183356	0.332271
19	5.08169	140.4057	2.54363	1.03545	1.50818	0.203760	0.296787
20	2.89497	143.3007	1.37318	0.49593	0.87725	0.171307	0.303025
21	5.55539	148.8561	3.23404	1.02174	2.2123	0.183918	0.398225
22	3.52804	152.3841	2.13323	0.70181	1.43142	0.198923	0.405726
23	5.10473	157.4889	5.00975	0.86754	4.14221	0.169948	0.811445
24	12.61346	170.1023	10.33094	2.22161	8.10933	0.176130	0.642910
25	2.93605	173.0384	2.12035	0.52042	1.59993	0.177251	0.544926
26	2.13768	175.1760	1.74448	0.48079	1.26369	0.224912	0.591150
27	5.83177	181.0078	3.89308	1.17794	2.71514	0.201986	0.465577
28	6.59455	187.6024	3.87315	1.05826	2.81489	0.160474	0.426850
29	7.59559	195.198	4.86347	1.0694	3.79407	0.140792	0.499509
30	5.88289	201.0808	2.99945	0.69388	2.30557	0.117948	0.391911
31	7.99193	209.0728	6.07346	1.13638	4.93708	0.142190	0.617758
32	7.54082	216.6136	5.81143	1.28867	4.52276	0.170892	0.599770
33	6.62606	223.2397	12.49168	1.06536	11.42632	0.160783	1.724451
34	9.38056	232.6202	17.58926	2.07333	15.51593	0.221024	1.654051
35	1.6958	234.3160	0.44114	0.12819	0.31295	0.075592	0.184544
36	5.57141	239.8874	5.83071	1.33317	4.49754	0.239287	0.807253
37	7.62321	247.5106	5.90674	1.19798	4.70876	0.157149	0.617687
38	8.45222	255.9629	7.73216	2.53789	5.19427	0.300263	0.614545
39	6.72717	262.6900	6.81063	1.57685	5.23378	0.234400	0.778006
40	7.21609	269.9061	5.08632	1.46822	3.6181	0.203464	0.501393
41	5.60135	275.5075	5.80941	1.30032	4.50909	0.232144	0.805000
42	3.87442	279.3819	5.68239	0.71605	4.96634	0.184814	1.281827
43	8.97986	288.3617	12.31601	2.03623	10.27978	0.226755	1.144759
44	6.03472	294.3965	7.04696	1.80976	5.2372	0.299891	0.867844
45	7.86891	302.2654	18.45622	1.15001	17.30621	0.146146	2.199314
46	11.47826	313.7436	25.15176	2.2153	22.93646	0.192999	1.998252
Sum	313.7436	–	254.6004	50.87438	203.7260	–	–
Mean	–	–	–	–	–	0.170933	0.671112

Note: For definitions of the downstream partitions, see Appendix D.1.

Appendix E.1. Plutonium concentrations in river water (pCi/l), Northern Rio Grande

		Plutonium 238			Plutonium 239 and 240		
Year	**Sample Site**	**Mean**	**St. Dev.**	**N**	**Mean**	**St. Dev.**	**N**
1977–1988	Rio Chama at Chamita	0.0041	0.0115	17	0.0128	0.0224	17
	Rio Grande at Embudo	0.0103	0.0215	18	0.0106	0.0344	18
	Rio Grande at Otowi	0.0016	0.0125	17	0.0040	0.0133	17
	Rio Grande at Cochiti	0.0028	0.0180	18	−0.0024	0.0194	18
	Rio Grande at Bernalillo	−0.0017	0.0123	18	0.0048	0.0315	18
	Jemez River at Jemez	−0.0629	0.2825	18	0.0004	0.0257	18
	All sites	−0.0078	0.0521	106	0.0050	0.0246	106
1976	All sites	−0.0082		18	−0.0016		18
1975	All sites	0.0060		18	0.0090		18
1974	All sites	0.0030		24	0.0010		24
1974–1988	All data	−0.0048		166	0.0041		166

Source: Data from Los Alamos National Laboratory, published Environmental Surveillance Reports.

Appendix E.2. Plutonium concentrations in river water (pCi/l), Rio Grande tributaries in the vicinity of Los Alamos, 1977–1988

	Plutonium 238			Plutonium 239 and 240		
Sample Site	**Mean**	**St. Dev.**	**N**	**Mean**	**St. Dev.**	**N**
Los Alamos Reservoir	0.0047	0.0215	16	0.0116	0.0228	16
Guaje Canyon in upper area	0.0005	0.0208	15	0.0101	0.0171	15
Frijoles Canyon at park headquarters	0.0126	0.0240	17	0.0142	0.0249	17
Frijoles Canyon at outlet	0.0013	0.0137	8	0.0021	0.0097	8
Pajarito Canyon at outlet	0.0035	0.0126	8	−0.0016	0.0105	8
Ancho Canyon at outlet	0.0057	0.0130	9	0.0079	0.0107	9
All sites	0.0053	0.0191	73	0.0090	0.0178	73

Source: Data from Los Alamos National Laboratory, published Environmental Surveillance Reports.

Appendix E.3. Plutonium concentrations in bedload sediments (pCi/g), Northern Rio Grande, 1974–1986

	Plutonium 238			Plutonium 239 and 240		
Sample Site	**Mean**	**St. Dev.**	**N**	**Mean**	**St. Dev.**	**N**
Rio Chama at Chamita	0.0001	0.0008	15	0.0015	0.0029	15
Rio Grande at Embudo	0.0003	0.0011	16	0.0033	0.0025	16
Rio Grande at Otowi	0.0003	0.0014	16	0.0106	0.0171	16
Rio Grande at Sandia Canyon	−0.0011	0.0027	8	0.0021	0.0072	8
Rio Gande at Pajarito Canyon	−0.0007	0.0024	7	0.0023	0.0031	7
Rio Grande at Ancho Canyon	−0.0014	0.0026	8	0.0069	0.0065	8
Rio Grande at Frijoles Canyon	0.0010	0.0056	6	0.0025	0.0040	6
Rio Grande at Cochiti	−0.0024	0.0079	7	0.0092	0.0149	7
Rio Grande at Bernalillo	−0.0004	0.0014	14	0.0050	0.0041	14
Jemez River at Jemez	0.0005	0.0016	15	0.0040	0.0044	15
All sites	−0.0002	0.0003	113	0.0048	0.0086	113

Source: Data from Los Alamos National Laboratory, published by W. D. Purtymun, R. J. Peters, T. E. Buhl, W. N. Maes, and F. H. Brown, *Background Concentrations of Radionuclides in Soils and River Sediments in Northern New Mexico, 1974–1986*, Los Alamos National Laboratory Report LA-11134-MS, UC-11 (Los Alamos, N.M.: Los Alamos National Laboratory, 1987).

Appendix E.4. Plutonium concentrations in flood-plain sediments (pCi/g), Northern Rio Grande, 1974–1988

	Plutonium 238			Plutonium 239 and 240		
Sample Site	**Mean**	**St. Dev.**	**N**	**Mean**	**St. Dev.**	**N**
Active-bed sediments	0.0018	2.3759	2	0.0027	0.0003	2
Levee, bar, Santa Clara flood plain	0.0006	0.0010	5	0.0049	0.0041	5
Slough, Santa Clara Abandoned channel	0.0013	0.0017	3	0.0097	0.0118	3
Tributary channel across flood plain	0.0002	–	1	0.0006	–	1
Mean "flood plain" (levee and slough above)	0.0009	0.0012	8	0.0067	0.0075	8

Appendix E.5. Plutonium concentrations in reservoir sediments (pCi/g), Northern Rio Grande

		Plutonium 238			Plutonium 239 and 240		
Reservoir	**Year**	**Mean**	**St. Dev.**	**N**	**Mean**	**St. Dev.**	**N**
Rio Grande	1986	0.0009	0.0011	3	0.0177	0.0184	3
Heron	1982	0.0006	0.0007	3	0.0134	0.0122	3
	1984	0.0005	0.0005	3	0.0093	0.0155	3
	1985	0.0005	0.0003	3	0.0112	0.0064	3
	1982–1985	0.0005	0.0005	9	0.0114	0.0114	9
El Vado	1982	0.0003	0.0006	3	0.0095	0.0077	3
	1984	0.0004	0.0001	3	0.0047	0.0072	3
	1985	0.0003	0.0001	3	0.0078	0.0005	3
	1982–1985	0.0003	0.0003	9	0.0073	0.0051	9
Abiquiu	1982	0.0005	0.0003	2	0.0097	0.0048	2
	1984	0.0007	0.0004	3	0.0127	0.0063	3
	1985	0.0007	0.0005	3	0.0088	0.0009	3
	1986	0.0003	0.0001	3	0.0075	0.0017	3
	1987	0.0002	0.0001	3	0.0038	0.0031	3
	1988	0.0003	0.0002	3	0.0074	0.0026	3
	1982–1988	0.0004	0.0003	17	0.0081	0.0031	17
Cochiti	1982	0.0009	0.0004	7	0.0178	0.0072	7
	1984	0.0007	0.0011	3	0.0197	0.0140	3
	1985	0.0016	0.0006	2	0.0241	0.0073	2
	1986	0.0012	0.0005	4	0.0212	0.0061	4
	1987	0.0008	0.0007	3	0.0175	0.0138	3
	1988	0.0017	0.0021	3	0.0121	0.0029	3
	1982–1988	0.0011	0.0013	22	0.0189	0.0082	22
All	1982–1988	0.0007	0.0007	60	0.0127	0.0073	60

Source: Data from Los Alamos National Laboratory, published Environmental Surveillance Reports.

Appendix E.6. Plutonium concentrations in fluvial sediments (pCi/g), Los Alamos Canyon

Canyon	Sample Site (distance upstream from Rio Grande, m)	Plutonium 238			Plutonium 239 and 240		
		Mean	St. Dev.	N	Mean	St. Dev.	N
Acid Canyon*	Acid Wier (17,520)	0.1293	0.1788	12	13.3491	7.1115	12
DP Canyon*	DPS-1 site (14,330)	1.0883	1.4854	23	2.1574	2.4081	23
	DPS-4 site (13,680)	0.1218	0.0576	17	0.4797	0.2701	17
Pueblo Canyon†	Hamilton Spring Bend (12,850)	0.0035	0.0034	14	0.4994	0.3947	14
	Pueblo site 1 (17,440)	0.0066	0.0158	11	0.0639	0.1665	11
	Pueblo site 2 (13,900)	0.0127	0.0094	11	1.8849	1.6705	11
	Pueblo site 3 (9,780)	0.0086	0.0171	12	2.0836	4.4321	12
	Near SR 4 (7,650)	0.0037	0.0046	16	0.6581	0.7022	16
Los Alamos Canyon	Reservoir site (21,780)	0.0007	0.0009	3	0.0060	0.0037	3
	Bridge site (16,600)	0.0010	0.0014	12	0.0021	0.0024	12
	LAO-1 site (15,410)	0.0102	0.0167	17	0.6763	0.9696	17
	GS-1 site (11,000)	0.0622	0.8515	16	0.3372	0.2767	16
	LAO-3 site (12,610)	0.0265	0.0273	16	0.2581	0.2298	16
	LAO-4.5 site (10,800)	0.1033	0.0921	10	0.4068	0.4186	10
	LAO-4 site (11,730)	0.0790	0.0429	6	0.3313	0.1497	6
	Near SR 4 (7,510)	0.0469	0.0355	15	0.2483	0.2032	15
	Totavi (3,800)	0.0137	0.0174	15	0.2218	0.2186	15
	Otowi (400)	0.0063	0.0098	15	0.1271	0.1318	15
Bayo Canyon‡	Near SR 4 (3,660)	0.0015	0.0017	13	0.0015	0.0011	13
Guaje Canyon‡	Near SR 4 (2,490)	0.0008	0.0014	12	0.0023	0.0010	12

Notes: "SR 4" refers to old New Mexico State Route 4.
*Small canyons with waste-disposal sites.
†Connects Acid Canyon to Los Alamos Canyon.
‡Canyons not directly affected by waste sites that are tributary to Los Alamos Canyon.

Appendix F. Topographic maps, U.S. Geological Survey quadrangles, showing the Rio Grande between Española and Elephant Butte

1:24,000 SCALE	1:100,000 SCALE
Española	Los Alamos
Puye	Albuquerque
White Rock	Belen
Frijoles	Socorro
Cochiti Dam	Oscura Mountains
Santo Domingo Pueblo	San Mateo Mountains
Santo Domingo Pueblo SW	Truth or Consequences
San Felipe Pueblo	
Santa Ana Pueblo	
Bernalillo	
Alameda	
Los Griegos	
Albuquerque West	
Isleta	
Los Lunas	
Tome	
Belen	
Veguita	
Abeytas	
Lemetar	
Socorro	
Loma de las Canas	
San Antonio	
Val Verde	
Fort Craig	
Paraje Well	
Romero Canyon	
Lava	
Black Bluffs	
Elephant Butte	

Note: All maps are of New Mexico.

Appendix G. Sources of aerial photographs for the Northern Rio Grande

---

For U.S. Geological Survey, U.S. Bureau of Reclamation, U.S. Army, and U.S. Air Force photographs, write to

U.S. Geological Survey
EROS Data Center
Sioux Falls, S.D. 57198

For U.S. Department of Agriculture, Soil Conservation Service, and Agricultural Stabilization and Commodity Service photographs, write to

U.S. Department of Agriculture
Aerial Photography Field Office
User Services Branch
2222 West, 2300 South
P.O. Box 30010
Salt Lake City, Utah 84125

For pre-1941 U.S. Department of Agriculture and Soil Conservation Service photographs, write to

General Services Administration
National Archives and Records Administration
Cartographic and Architectural Branch
Washington, D.C. 20408

For pre-1950 Fairchild Aerial Photography, write to

Fairchild Aerial Photography Collection
c/o Dallas D. Rhoads
Department of Geology
Whittier College
Whittier, Calif. 90608

Appendix H. Sources of historical ground photographs for the Northern Rio Grande

Sources in New Mexico

Los Alamos Historical Society
Attn: Hedy Dunn
P.O. Box 43
Los Alamos, N.M. 87544

Albuquerque Museum
Albuquerque, N.M. 87114

Albuquerque Public Library
501 Copper Avenue, N.W.
Albuquerque, N.M. 87102

Maxwell Museum of Anthropology
University of New Mexico
Albuquerque, N.M. 87131

Special Collections
Main Library
University of New Mexico
Albuquerque, N.M. 87131

Middle Rio Grande Conservancy District
1931 Second Street, S.W.
Albuquerque, N.M. 87102

New Mexico Museum of Natural History
1801 Mountain Road, N.W.
Albuquerque, N.M. 87104

Photo Archives of New Mexico
110 Washington Avenue
Santa Fe, N.M. 87501

State Archives
404 Montezuma Avenue
Santa Fe, N.M. 87503

U.S. Bureau of Reclamation
505 Marquette Avenue, N.W.
Albuquerque, N.M. 87102

Sources Outside New Mexico

Colorado Historical Society
Historical Photograph Collection
South Broadway and Colfax Avenue
Denver, Colo. 80201

Denver Public Library
Western History Collection
Denver, Colorado 80201

U.S. Geological Survey
Photographic and Field Records File
Box 25046
Denver Federal Center
Denver, Colorado 80225

El Paso Public Library
Southwest Room
El Paso, Texas

Library of Congress
U.S. National Archives
Washington, D.C. 20408

National Anthropological Archives
BAE Holdings
Smithsonian Institution
Washington, D.C. 20560

Pomona Public Library
Special Collections
625 South Garey Avenue
Pomona, Calif. 91766

Southwest Museum
234 Museum Drive
Los Angeles, Calif. 90065

Appendix I. Contact persons for Northern Rio Grande data

For U.S. Army Corps of Engineers, Albuquerque District:

Frank Jaramillo
Chief Engineer
(505) 766-2635

Justice Edge
Sedimentation Specialist
(505) 766-1034

Mark Sifuentes
Environmental Specialist
(505) 766-3577

Frank Collins
Emergency Operations
(505) 766-3829

Kim Zahm
Assistant Chief of Design
(Previously, Sedimentation)
(505) 766-2626

For U.S. Bureau of Reclamation:

Drew Baird
Chief of River Assessment
(505) 766-1758

Middle Rio Grande Conservancy District:

Subhas Shah
District Engineer
(505) 247-0234

Appendix J. Basic formulas used in the Riverine Accounting and Transport program

Equation of Continuity

$$Q = W\,D\,V$$

where Q = original unadjusted discharge ($m^3\,s^{-1}$), W = channel width (m), D = channel depth (m), and v = velocity ($m\,s^{-1}$).

Transmission Losses

$$Q_a = Q - L_r\,(\sum_{1}^{n} A_b)$$

where Q_a = adjusted discharge ($m^3\,s^{-1}$), Q = original unadjusted discharge ($m^3\,s^{-1}$), L^r = transmission loss rate ($m^3\,s^{-1}\,m^{-2}$), $1,n$ = channel-segment number, and A_b = channel-bed area in segment n (m^2).

Velocity of Flow

$$V = n^{-1}\,R^{2/3}\,S^{1/2}$$

where V = velocity ($m\,s^{-1}$), n = Manning's hydraulic roughness coefficient (0.015-0.045), R = hydraulic radius defined below (m), and S = channel gradient (dimensionless).

Hydraulic Radius

$$R = \frac{W\,D}{(2D + W)}$$

where R = hydraulic radius (m), W = channel width (m), and D = channel depth (m).

Depth of Flow

$$D = \frac{Q}{WV}$$

where D = depth of flow (m), Q = discharge ($m^3\,s^{-1}$), W = width of flow (m), and V = velocity of flow ($m\,s^{-1}$). Alternatively, in some cases, the following also might be used:

$$D = \left(\frac{Qn}{(kWS^{1/2})}\right)^{0.6}$$

where k = shape parameter (1.0 for broad, shallow channels, $<$ 1.0 for others), and other symbols as before.

Unit Stream Power

$$\omega = \varrho\,g\,D\,V\,S$$

where ω = unit stream power ($N\,m^{-1}\,s^{-1}$ or $W\,m^{-2}$), ϱ = density of the fluid ($1\,g\,cm^{-3}$), g = acceleration of gravity ($9.81\,m\,s^2$), D = depth of flow (m), V = velocity of flow ($m\,s^{-1}$), and S = flow gradient (dimensionless).

(continued)

Stream Power

$$\Omega = \varrho \, g \, D \, W \, V = \varrho \, Q = \omega \, W$$

where Ω = stream power (N s^{-1} or W), ϱ = density of the fluid (1 g cm^{-3}), g = acceleration of gravity (9.81 m s^2), D = depth of flow (m), W = width of flow (m), V = velocity of flow (m s^{-1}), S = flow gradient (dimensionless), Q = discharge (m^3 s^{-1}), and ω = unit stream power (defined above; N m^{-1} s^{-1} or W m^{-2}).

Sediment Transport Capacity

$$i = \omega \left(\frac{e_b}{\tan\alpha} + 0.01 \frac{\overline{u}}{v_{ss}} \right)$$

where i = total load transport (kg s^{-1}), ω = unit stream power (defined above; N m^{-1} s^{-1} or W m^{-2}), e_b = a bedload efficiency factor (usually $0.11 < e_b < 0.15$), $\tan \alpha$ = coefficient of friction (about 1.0 for bedload particles), u = velocity (m s^{-1}), and v_{ss} = settling velocity of particles for a given size (m s^{-1}). The first term inside the parentheses on the right side of the equation is for bedload, and the second term inside the parentheses is for suspended load.

Sediment Continuity Equation

$$\frac{\partial M_t}{\partial t} = \frac{\partial M_i}{\partial t} - \frac{\partial M_o}{\partial t}$$

where M_t = sediment mass total in storage in a channel segment (Mg), M_i = input mass from channel segment upstream (Mg), M_o = ouput mass to channel segment downstream (Mg), and t = time unit (day).

Plutonium Continuity Equation

$$\frac{\partial P_t}{\partial t} = \frac{\partial P_i}{\partial t} - \frac{\partial P_o}{\partial t}$$

where P_t = plutonium mass in storage in a channel segment (mCi), P_i = input plutonium mass from channel segment upstream (mCi), P_o = output plutonium mass to channel segment downstream (mCi), and t = time unit (day).

Plutonium Concentration in Sediment

$$P_{conc_{t,i,o}} = \frac{P_{t,i,o}}{M_{t,i,o}}$$

where P_{conc} = plutonium concentration in stored, input, or output sediment (fCi/g), $P_{t,i,o}$ = plutonium mass in stored, input, or output sediment mass (fCi/g)
$P_{t,i,o}$ = plutonium mass in stored, input, or output sediment (fCi/g).

Sources: V. T. Chow, *Open Channel Hydraulics* (New York: McGraw-Hill, 1959); W. L. Graf, *Fluvial Processes in Dryland Rivers* (Berlin: Springer-Verlay, 1988).

Notes

1. Introduction and Related Research

1. F. W. Whicker and V. Schultz, *Radioecology: Nuclear Energy and the Environment*, 2 vols. (Boca Raton, Fla.: CRC Press, 1982).

2. T. E. Hakonson, R. L. Watters, and W. C. Hanson, "The Transport of Plutonium in Terrestrial Ecosystems," *Health Physics* 40 (1981): 63–69.

3. A. Stoker, A. J. Ahlquist, D. L. Mayfield, W. R. Hansen, A. D. Talley, and W. D. Purtymun, *Radiological Survey of the Site of a Former Radioactive Liquid Waste Treatment Plant (TA-45) and the Effluent Receiving Areas of Acid, Pueblo, and Los Alamos Canyons, Los Alamos, New Mexico*, Los Alamos National Laboratory and U.S. Department of Energy Report LA-8890-ENV, UC-70 (Los Alamos, N.M.: Los Alamos National Laboratory, 1981).

4. J. W. Nyhan, B. J. Drennon, W. V. Abeele, M. L. Wheeler, W. D. Purtymun, G. Trujillo, W. J. Herrera, and J. W. Booth, *Distribution of Radionuclides and Water in Bandelier Tuff Beneath a Former Los Alamos Liquid Waste Disposal Site After 33 Years*, Los Alamos National Laboratory Report LA-10159-LLWM (Los Alamos, N.M.: Los Alamos National Laboratory, 1984).

5. W. D. Purtymun, *Storm Runoff and Transport of Radionuclides in DP Canyon, Los Alamos County, New Mexico*, Los Alamos National Laboratory Report LA-5744 (Los Alamos, N.M.: Los Alamos National Laboratory, 1974).

6. L. J. Lane, W. D. Purtymun, and N. M. Becker, *New Estimating Procedures for Surface Runoff, Sediment Yield, and Contaminant Transport in Los Alamos County, New Mexico*, Los Alamos National Laboratory Report LA-10335-MS, UC-11 (Los Alamos, N.M.: Los Alamos National Laboratory, 1985).

7. Environmental Surveillance Group, ed., *Environmental Surveillance at Los Alamos* (Los Alamos, N.M.: Los Alamos National Laboratory, 1971–1991); W. D. Purtymun, R. J. Peters, T. E. Buhl, W. N. Maes, and F. H. Brown, *Background Concentrations of Radionuclides in Soils and River Sediments in Northern New Mexico, 1974–1986*, Los Alamos National Laboratory Report LA-11134-MS, UC-11 (Los Alamos, N.M.: Los Alamos National Laboratory, 1987).

8. W. Solomons and U. Förstner, *Metals in the Hydrocycle* (Berlin: Springer-Verlag, 1984).

9. W. C. Pfeiffer, M. Fiszman, and N. Carbonell, "Fate of Chromium in a Tributary of the Irajo River, Rio de Janeiro," *Environmental Pollution* 1 (1980): 117–26.

10. K. H. Larson, J. L. Leitch, W. F. Dunn, J. W. Neel, J. H. Olafson, E. E. Held, J. Taylor, W. J. Cross, and A. W. Bellamy. *University of California Report, Los Alamos*, University of California Report UCLA-108 (Berkeley: University of California, 1951); V. E. Noshkin, "Ecological Aspects of Plutonium Dissemination in Aquatic Environments," *Health Physics* 22 (1972): 537; F. W. Whicker, C. A. Little, and T. F. Winsor, *Symposium on Environmental Surveillance Around Nuclear Installations, 5–9 November 1973* (Warsaw: International Atomic Energy Commission, 1973); T. E. Hakonson, L. J. Johnson, and W. D. Purtymun, "The Distribution of Plutonium in Liquid Waste Disposal Areas at Los Alamos," in *Proceedings of the 3rd International Congress of the Radiation Protection Association, September 9–14, 1973* (Washington, D.C.: Radiation Protection Association, 1973), pp. 248–53; T. E. Hakonson and L. J. Johnson, "Distribution of Environmental Plutonium in the Trinity Site Ecosystem After 27 Years," in *Proceedings of the 3rd International Congress of the Radiation Protection Association*, pp. 242–47.

11. R. C. Dahlman, C. T. Garten, Jr., and T. E. Hakonson, "Comparative Distribution of Plutonium in Contaminated Ecosystems at Oak Ridge, Tennessee, and Los Alamos, New Mexico," in *Transuranic Elements in the Environment*, U.S. Department of Energy Report DOE/TIC-22800, ed. W. C. Hanson (Washington, D.C.: U.S. Department of Energy and National Technical Information Service, 1980), pp. 371–80.

12. J. M. Martin and M. Maybeck, "Elemental Mass Balance of Material Carried by Major World Rivers," *Marine Chemistry* 7 (1979): 173–206.

13. G. C. White and T. E. Hakonson, "Statistical Considerations and Survey of Plutonium Concentration Variability in Some Terrestrial Ecosystem Components," *Journal of Environmental Quality* 8 (1979): 176–82.

14. J. W. Nyhan, T. G. Schofield, G. C. White, and G. Trujillo, *Sampling Soils for Cs-137 Using Various Field Sampling Volumes*, Los Alamos National Laboratory Report LA-8951-MS, UC-11 (Los Alamos, N.M.: Los Alamos National Laboratory, 1981).

15. T. E. Hakonson, "Environmental Pathways of Plutonium into Terrestrial Plants and Animals," *Health Physics* 29 (1975): 583–88.

16. Hakonson et al., "Transport of Plutonium in Terrestrial Ecosystems"; E. M. Romney and J. J. Davis, "Ecological Dissemination of Plutonium in the Environment," *Health Physics* 21 (1972): 551; R. E. Wilding and T. R. Garland, "The Relationship of Microbial Processes to the Fate and Behavior of Transuranic Elements in Soils, Plants, and Animals," in *Transuranic Elements in the Environment*, ed. Hanson, pp. 31–37; R. L. Watters, T. E. Hakonson, and L. J. Lane, "The Behavior of Actinides in the Environments," *Radiochimica Acta* 32 (1983): 89–103.

17. H. Nishita and R. M. Haug, "The Effect of Fulvic and Humic Acids and Inorganic Phase of Soil on the Sorption and Extractability of Pu(IV) 239 from Several Clay Minerals," *Soil Science* 128 (1979): 291–96.

18. H. Nishita and M. Hamilton, "The Influence of Several Soil Components and Their Interaction on Plutonium Extractability from a Calcareous Soil," *Soil Science* 131 (1981): 56–59.

19. J. W. Nyhan, F. R. Mera, and R. E. Neher, "Distribution of Plutonium in Trinity Soils After 28 Years," *Journal of Environmental Quality* 5 (1976): 431–37; Nyhan et al., *Sampling Soils for Cs-137*.

20. L. Jacobson and P. Oversteret, "The Uptake by Plants of Plutonium and Some Productions of Nuclear Fission Adsorbed on Soil Colloids," *Soil Science* 65 (1948): 129–34.

21. W. E. Prout, "Adsorption of Radioactive Wastes on Savannah River Plant Soil," *Soil Science* 86 (1958): 13–17.

22. A. T. Jakubick, "Migration of Plutonium in Natural Soils," in *Transuranic Nuclides in the Environment*, International Atomic Energy Agency Report IAEA-SM-199/3 (Vienna: International Atomic Energy Agency, 1976), pp. 216–27; A. T. Jakubick, "In Situ Plutonium Transport in Geomedia," in Organization for Economic Cooperation and Development, *Migration of Long-lived Radionuclides in the Geosphere* (Paris: OECD, 1979), pp. 201–23.

23. J. L. Means, D. A. Crerar, M. P. Borcsik, and J. O. Duquid. "Adsorption of Co and Selected Actinides by Mn and Fe Oxides in Soils and Sediments," *Geochimica and Cosmochimica Acta* 42 (1978): 1763–73.

24. S. R. Ashton and D. A. Stanners, "Observations on the Deposition, Mobility and Chemical Associations of Plutonium in Intertidal Sediments," in *Techniques for Identifying Transuranic Speciation in Aquatic Environments*, International Atomic Energy Agency Report STI/PUB/613 (Vienna: International Atomic Energy Agency, 1981), pp. 209–17; T. M. Beasley, R. Carpenter, and C. D. Jennings, "Plutonium, Am-241 and Cs-137 Ratios, Inventories and Vertical Profiles in Washington and Oregon Continental Shelf Sediments." *Geochimica and Cosmochimica Acta* 46 (1982): 1831–1946.

25. P. J. Coughtrey, D. Jackson, C. H. Jones, P. Kane, and M. C. Thorne, *Radionuclide Distribution and Transport in Terrestrial and Aquatic Ecosystems*, vol. 4: *Plutonium and Neptunium* (Rotterdam: A. A. Balkema, 1984), p. 6.

26. Prout, "Adsorption of Radioactive Wastes on Savannah River Plant Soil."

27. M. Yamamoto, T. Tanii, and M. Sakanoue, "Characteristics of Fallout Plutonium in Soil," *Journal of Radiation Research* 22 (1981): 134–42.

28. M. J. Frissel, A. T. Jakubick, A. van der Klugt, R. Pennders, P. Poelstra, and E. Zwemmer, *Modelling of the Transport of Radionuclides of Strontium, Caesium, and Plutonium in Soil. Experimental Verification* (Brussels: Commission of the European Communities, 1980); Jakubick, "Migration of Plutonium in Natural Soils."

29. T. E. Hakonson and J. W. Nyhan, "Ecological Relationships of Plutonium in Southwest Ecosystems," in *Transuranic Elements in the Environment*, ed. Hanson, pp. 82–91.

30. T. E. Hakonson, J. W. Nyhan, and W. D. Purtymun, "Accumulation and Transport of Soil Plutonium in Liquid Waste Discharge Areas at Los Alamos," in *Transuranic Nuclides in the Environment*, pp. 175–89.

31. S. M. Wise, "Cesium-137 and Lead-210: A Review of the Techniques and Some Applications in Geomorphology," in *Timescales in Geomorphology*, ed. R. A. Cullingford, D. A. Davidson, and J. Lewin (Chichester: Wiley, 1980), pp. 110–27.

32. R. K. Schultz, R. Overstreet, and I. Barshad, "On the Soil Chemistry of Cesium-137," *Soil Science* 89 (1960): 16–27.

33. A. S. Rogowski and T. Tamura, "Movement of Cesium-137 by Runoff, Erosion, and Infiltration on the Alluvial Captina Silt Loam," *Health Physics* 11 (1965): 1333–40.

34. J. C. Lance, S. C. McIntyre, J. W. Naney, and S. S. Rousseva, "Measuring Sediment Movement at Low Erosion Rates Using Cesium-137," *Soil Science Society of America Proceedings* 50 (1986): 1303–9.

35. J. C. Ritchie, J. R. McHenry, and A. C. Gill, "Dating Recent Reservoir Sediments," *Limnology and Oceanography* 18 (1973): 254–63; B. L. Campbell, R. J. Loughran, G. L. Elliott, and D. J. Shelly, "Mapping Drainage Basin Sediment Sources Using Caesium-137," in *International Symposium on Drainage Basin Sediment Delivery*, Albuquerque, New Mexico, International Association of Hydrological Sciences Publication 159

(Paris: International Association of Hydrological Sciences, 1986), pp. 437–46; R. G. Kachanoski, "Comparison of Measured Soil Cesium-137 Losses and Erosion Rates," *Canadian Journal of Soil Science* 67 (1987): 199–203.

36. R. B. Brown, G. F. Kling, and N. H. Cutshall, "Agricultural Erosion Indicated by Cs-137 Redistribution," *Soil Science Society of America Proceedings* 45 (1981): 1191–97; S. C. McIntyre, J. C. Lance, B. L. Campbell, and R. L. Miller, "Using Cesium-137 to Estimate Soil Erosion on a Clearcut Hillside," *Journal of Soil and Water Conservation* 42 (1987): 117–20; L. W. Martz and E. de Jong, "Using Cesium-137 to Assess the Variability of Net Soil Erosion and Its Association with Topography in a Canadian Prairie Landscape," *Catena* 14 (1987): 439–51.

37. R. J. Loughran, B. L. Campbell, and D. E. Walling, "Soil Erosion and Sedimentation Indicated by Caesium 137, Jackmoor Brook Catchment, Devon, England," *Catena* 14 (1987): 201–12.

38. W. W. Sayre and F. W. Chang, *A Laboratory Investigation of Open-Channel Dispersion Processes for Dissolved, Suspended, and Floating Dispersants*, U.S. Geological Survey Professional Paper 433-E (Washington, D.C.: U.S. Geological Survey, 1968).

39. R. G. Godfrey and B. J. Frederick, *Stream Dispersion at Selected Sites*, U.S. Geological Survey Professional Paper 433-K (Washington, D.C.: U.S. Geological Survey, 1970).

40. P. H. Carrigan, Jr., *Radioactive Waste Distribution in the Clinch River, Eastern Tennessee*, U.S. Geological Survey Professional Paper 433-G (Washington, D.C.: U.S. Geological Survey, 1968); P. H. Carrigan, Jr., *Inventory of Radionuclides in Bottom Sediment of the Clinch River, Eastern Tennessee*, U.S. Geological Survey Professional Paper 433-I (Washington, D.C.: U.S. Geological Survey, 1969).

41. R. E. Glover, *Dispersion of Dissolved or Suspended Materials in Flowing Streams*, U.S. Geological Survey Professional Paper 433-B (Washington, D.C.: U.S. Geological Survey, 1964).

42. W. W. Sayer, H. P. Guy, and A. R. Chamberlain, *Uptake and Transport of Radionuclides by Stream Sediments*, U.S. Geological Survey Professional Paper 433-A (Washington, D.C.: U.S. Geological Survey, 1963); E. A. Jenne and J. S. Wahlberg, *Role of Certain Stream-Sediment Components in Radioion Sorption*, U.S. Geological Survey Professional Paper 433-E (Washington, D.C.: U.S. Geological Survey, 1968).

43. R. J. Pickering, *Distribution of Radionuclides in Bottom Sediment of the Clinch River, Eastern Tennessee*, U.S. Geological Survey Professional Paper 433-H (Washington, D.C.: U.S. Geological Survey, 1969); Carrigan, *Inventory of Radionuclides in Bottom Sediment of the Clinch River*.

44. W. W. Sayre and D. W. Hubbell, *Transport and Dispersion of Labeled Bed Material, North Loup River, Nebraska*, U.S. Geological Survey Professional Paper 433-C (Washington, D.C.: U.S. Geological Survey, 1965).

45. L. B. Leopold (chief, Water Resources Division, U.S. Geological Survey), interview, Santa Fe, N.M., May 24, 1990; and David Dawdy (hydrologist, Water Resources Division, U.S. Geological Survey), interview, Washington, D.C., May 28, 1989.

46. Hakonson et al., "Accumulation and Transport of Soil Plutonium."

47. Hakonson et al., "Transport of Plutonium in Terrestrial Ecosystems."

48. T. E. Hakonson and K. B. Bostick, "Cesium-137 and Plutonium in Liquid Waste Discharge Areas at Los Alamos," in *Radioecology and Energy Resources*, Special Publication no. 1, Ecological Society of America (Stroudsburg, Pa: Dowden, Hutchenson & Ross, 1976), pp. 40–48.

49. Watters et al., "Behavior of Actinides in the Environments," p. 91.

50. M. R. Scott, "Thorium and Uranium Concentrations and Isotopic Ratios in River Sediments," *Earth and Planetary Science Letters* 4 (1968): 245–52; R. J. Gibbs, "Mechanisms of Trace Metal Transport in Rivers," *Science* 180 (1973): 71–73.

51. R. G. Menzel, "Enrichment Ratios for Water Quality Modeling," in *CREAMS: A Field Scale Model for Chemicals, Runoff, and Erosion from Agricultural Management Systems*, U.S. Department of Agriculture Conservation Research Report 26 (Washington, D.C.: U.S. Department of Agriculture, 1980), pp. 486–92.

52. L. J. Lane and T. E. Hakonson, "Influence of Particle Sorting in Transport of Sediment Associated Contaminants," in *Proceedings of the Waste Management Symposium, Tucson, Arizona, 1982* (Tucson: University of Arizona Press, 1982), pp. 543–57.

53. W. C. Weimer, R. R. Kinnison, and J. H. Reeves, *Survey of Radionuclide Distributions Resulting from the Church Rock, New Mexico, Uranium Mill Tailings Pond Dam Failure*, U.S. Nuclear Regulatory Commission Report NUREG/DR-2449, PNL-4122 (Richland, Wash.: Battelle Pacific Northwest Laboratory, 1981).

54. W. L. Graf, *Fluvial Processes in Dryland Rivers* (New York: Springer-Verlag, 1988).

55. W. L. Graf, "Fluvial Dynamics of Thorium-230 in the Church Rock Event, Puerco River, New Mexico," *Annals of the Association of American Geographers* 80 (1990): 327–42.

56. T. J. East, "Geomorphological Assessment of Sites and Impoundments for the Long Term Containment of Uranium Mill Tailings in the Alligator River Region," *Australian Geographer* 16 (1986): 16–22.

57. T. J. East, R. F. Cull, A. S. Murray, and K. Duggan, "Fluvial Dispersion of Radioactive Mill Tailings in the Seasonally Wet Tropics, Northern Australia," in *Fluvial Geomorphology of Australia*, ed. R. F. Warner (Sydney: Academic Press, 1988), pp. 303–22.

58. R. F. Cull, K. Duggan, R. Martin, and A. S. Murray, "Tailings Transport and Deposition Downstream of the Northern Hercules (Moline) Mine in the Catchment of the Mary River, N.T.," in *Environmental Planning and Management for Mining and Energy* (Darwin: Northern Australia Mine Rehabilitation Workshop, 1986), pp. 199–216.

59. P. J. Jarvis, *Heavy Metal Pollution: An Annotated Bibliography, 1976–1980* (East Anglia: Geobooks, 1983), p. 3.

60. Solomons and Förstner, *Metals in the Hydrocycle*.

61. R. C. Weast, *Handbook of Chemistry and Physics* (Boca Raton, Fla.: CRC Press).

62. D. B. Simons and F. Senturk, *Sediment Transport Technology* (Fort Collins, Colo.: Water Resources Publications, 1977).

63. K. E. Carpenter, "Fauna of Rivers Polluted by Lead Mining in the Aberystwyth District of Cardiganshire," *Annals of Applied Biology* 11 (1924): 1–23; K. E. Carpenter, "On the Biological Factors Involved in the Destruction of River Fisheries by Pollution Due to Lead Mining," *Annals of Applied Biology* 12 (1925): 1–13.

64. J. R. E. Jones, "A Study of the Zinc-polluted River Ystwyth in Northern Cardiganshire, Wales," *Annals of Applied Biology* 27 (1940): 363–78; J. R. E. Jones, "A Further Study of the Zinc-polluted River Ystwyth," *Journal of Animal Ecology* 27 (1958): 1–14; L. Newton, "Pollution of the Rivers of West Wales by Lead and Zinc Mine Effluent," *Annals of Applied Biology* 31 (1944): 1–11.

65. J. B. Wertz, "Logarithimic Pattern in River Placer Deposits," *Economic Geology* 44 (1949): 193–209; J. Lewin, B. E. Davies, and P. Wolfenden, "Interactions Between Channel Change and Historic Mining Sediments," in *River Channel Change*, ed. K. J. Gregory (Chichester: Wiley, 1977), pp. 353–67; P. Wolfenden and J. Lewin, "Distribution of Metal Pollutants in Active Stream Sediments," *Catena* 5 (1978): 67–78.

66. W. D. Payne, "The Role of Sulfides and Other Heavy Minerals in Copper Anomalous Stream Sediments" (Ph.D. diss., Stanford University, 1971); V. V. Plikarpochkin, "The Quantitative Estimation of Ore-bearing Areas from Sample Data of the Drainage System," in *Geochemical Exploration* (Toronto: Canadian Institute of Mining and Metallurgy, 1971), pp. 585–86; G. Müller and U. Förstner, "Heavy Metals in Sediments of the Rhine and Elbe Estuaries," *Environmental Geology* 1 (1977): 33–39; W. A. Marcus, "Copper Dispersion in Ephemeral Stream Sediments," *Earth Surface Processes and Landforms* 12 (1987): 117–28.

67. R. C. Lindholm, *Practical Approach to Sedimentology* (London: Allen & Unwin, 1987).

68. S. B. Bradley, "Sediment Quality Related to Discharge in a Mineralized Region of Wales," in *Recent Developments in the Explanation and Prediction of Erosion and Sediment Yield*, International Association of Scientific Hydrology Publication no. 137 (Paris: International Association of Scientific Hydrology, 1982), pp. 341–50; S. B. Bradley, "Flood Effects on the Transport of Heavy Metals," *International Journal of Environmental Studies* 22 (1984): 225–30; S. B. Bradley, "Partitioning of Heavy Metals on Suspended Sediments During Floods on the River Ystwyth, Wales," *Trace Substances in Environmental Health* 18 (1984): 548–57; S. B. Bradley and J. J. Cox, "Heavy Metals in the Hamps and Manifold Valleys, North Staffordshire, U.K.: Distribution in Flood Plain Soils," *Science of the Total Environment* 50 (1986): 103–28; M. G. Macklin and R. B. Dowsett, "The Chemical and Physical Speciation of Trace Metals in Fine Grained Overbank Flood Sediments in the Tyne Basin, North-East England," *Catena* 16 (1989): 135–51; M. G. Macklin, "Flood-Plain Sedimentation in the Upper Axe Valley, Mendip, England," *Transactions of the Institute of British Geographers* 10 (1985): 235–44; M. G. Macklin and D. G. Passmore, "Late Quaternary Sedimentation in the Lower Tyne Valley, North East England," Seminar Paper 55 (University of Newcastle upon Tyne, Department of Geography, Mimeo, 1988).

69. B. E. Davies and J. Lewin, "Chronosequences in Alluvial Soils with Special Reference to Historic Lead Pollution in Cardiganshire, Wales," *Environmental Pollution* 6 (1974): 49–57; P. Wolfenden and J. Lewin, "Distribution of Metal Pollutants in Floodplain Sediments," *Catena* 4 (1977): 309–17; J. C. Knox, "Historic Valley Floor Sedimentation in the Upper Mississippi Valley," *Annals of the Association of American Geographers* 77 (1987): 224–44; W. L. Graf, S. L. Clark, M. T. Kammerer, T. Lehman, K. Randall, and R. Schröder, "Geomorphology of Heavy Metals in the Sediments of Queen Creek, Arizona, U.S.A.," *Catena* 18 (1991): 567–82.

70. D. C. Marron, "Physical and Chemical Characteristics of a Metal Contaminated Overbank Deposit, West Central South Dakota, USA," *Earth Surface Processes and Landforms* 14 (1989): 419–32; Graf et al., "Geomorphology of Heavy Metals in the Sediments of Queen Creek."

71. K. R. Imhoff, P. Koppe, and F. Deitz, "Heavy Metals in the Ruhr River and Their Budget in the Catchment Area," *Progress in Water Technology* 12 (1980): 735–49.

72. W. L. Graf, "Mercury Tansport in Stream Sediments of the Colorado Plateau," *Annals of the Association of American Geographers* 75 (1985): 552–65.

73. H. Leenaers, *The Dispersal of Metal Mining Wastes in the Catchment of the River Geul, Belgium–The Netherlands* (Utrecht: Geografisch Instituut, Rijksuniveriteit, 1989).

74. Marron, "Physical and Chemical Characteristics of a Metal Contaminated Overbank Deposit."

75. G. R. Foster and T. E. Hakonson, "Erosional Losses of Fallout Plutonium," in "Proceedings of the Symposium on Environmental Research on Actinides in the Environment" (U.S. Department of Energy, Mimeo, 1983); G. R. Foster and T. E. Hakonson,

"Predicted Erosion and Sediment Delivery of Fallout Plutonium," *Journal of Environmental Quality* 13 (1984): 595–602.

76. M. A. Devaurs, E. P. Springer, L. J. Lane, and G. J. Langhorst, "Contaminant Transport from Rangeland Watersheds," in American Society of Agricultural Engineers, International Symposium on Modeling Agricultural, Forest and Rangeland Hydrology, Los Alamos National Laboratory Report LA-UR-88-2071 (Los Alamos National Laboratory, Mimeo, 1988).

77. Lane et al., *New Estimating Procedures*.

78. Whicker and Schultz, *Radioecology*.

79. J. A. C. Fortescue, *Environmental Geochemistry: A Holistic Approach* (New York: Springer-Verlag, 1980), p. 249.

80. Coughtrey et al., *Plutonium and Neptunium*, p. 12.

81. Solomons and Förstner, *Metals in the Hydrosphere*.

82. A. A. Levenson, *Introduction to Exploration Geochemistry* (Wilmette, Ill.: Applied Publishing, 1980).

2. Plutonium and Los Alamos

1. R. Rhoads, *The Making of the Atomic Bomb* (New York: Simon and Schuster, 1986); L. Bickel, *The Deadly Element* (London: Stein & Day, 1980); G. Holton, *Thematic Origins of Scientific Thought* (Cambridge, Mass.: Harvard University Press, 1973).

2. E. Curie, *Madame Curie* (Garden City, N.Y.: Doubleday, 1938).

3. M. Oliphant, *Rutherford* (New York: Elsevier, 1972).

4. D. Wilson, *Rutherford* (Cambridge, Mass.: MIT Press, 1983).

5. S. Rozental, ed., *Niels Bohr* (London: North-Holland, 1967).

6. A. Romer, ed., *Radiochemistry and the Discovery of Isotopes* (New York: Dover, 1970).

7. R. H. Stuewer, ed., *Nuclear Physics in Retrospect* (Minneapolis: University of Minnesota Press, 1979).

8. E. Segre, *From X-Rays to Quarks: Modern Physicists and Their Discoveries* (San Francisco: Freeman, 1980).

9. E. Segre, Enrico Fermi, *Physicist* (Chicago: University of Chicago Press, 1970).

10. G. T. Seaborg, *The Transuranic Elements* (New Haven, Conn.: Yale University Press, 1958).

11. G. T. Seaborg, *Early History of Heavy Isotope Production at Berkeley*, Lawrence Berkeley Laboratory Report (Berkeley: University of California, 1976).

12. G. T. Seaborg and W. D. Loveleand, *The Elements Beyond Uranium* (New York: Wiley, 1990).

13. G. T. Seaborg, quoted in Rhoads, *Making of the Atomic Bomb*, p. 414.

14. R. G. Hewlett and O. E. Anderson, *The New World*, 1939–1946 (University Park: Pennsylvania State University Press, 1962).

15. S. Groueff, *Manhattan Project* (Boston: Little, Brown, 1967).

16. A. K. Smith, "Manhattan Project: The Atomic Bomb," in *The Nuclear Almanac: Confronting the Atom in War and Peace*, ed. J. Dennis (Reading, Mass.: Addison-Wesley, 1984), pp. 20–52.

17. Los Alamos Historical Society, *Los Alamos: Beginning of an Era, 1943–1945* (Los Alamos, N.M.: Los Alamos Historical Society, 1986).

18. J. W. Kunettka, *City of Fire: Los Alamos and the Atomic Age, 1943–1945* (Albuquerque: University of New Mexico Press, 1979).

19. F. M. Szaz, *The Day the Sun Rose Twice: The Story of the Trinity Site Nuclear Explosion, July 16, 1945* (Albuquerque: University of New Mexico Press, 1984).

20. Smith, "Manhattan Project," p. 27.

21. P. Goodchild, *J. Robert Oppenheimer: Shatterer of Worlds* (New York: Fromm International, 1985).

22. E. E. Morrison, *Turmoil and Tradition: A Biography of Henry L. Stimson* (Boston: Houghton Mifflin, 1960), p. 620.

23. R. Nash, *The Rights of Nature* (Madison: University of Wisconsin Press, 1989).

24. Los Alamos Scientific Laboratory, "Minutes of the Laboratory Advisory Board, August 19, 1944," file notes, Los Alamos National Laboratory, Los Alamos, N.M.

25. R. S. Stone, "Health Protection Activities of the Plutonium Project," *Proceedings of the American Philosophical Society* 90 (1946): 11–19.

26. E. P. Odum, "Consideration of the Total Environment in Power Reactor Waste Disposal," *Proceedings of the International Conference on Peaceful Uses of Atomic Energy* 13 (1956): 350–58; A. M. Kuzin and A. A. Peredelsky, "Nature Conservancy and Some Aspects of Radioactive–Ecological Relations" (in Russian), *Okhr. Prir. Zap. SSSR* 1 (1956): 65–70.

27. Szasz, *Day the Sun Rose Twice*, p. 63.

28. L. M. Groves, *Now It Can Be Told: The Story of the Manhattan Project* (New York: Harper and Brothers, 1962).

29. A. Stoker, A. J. Ahlquist, D. L. Mayfield, W. R. Hansen, A. D. Talley, and W. D. Purtymun, *Radiological Survey of the Site of a Former Radioactive Liquid Waste Treatment Plant (TA-45) and the Effluent Receiving Areas of Acid, Pueblo, and Los Alamos Canyons, Los Alamos, New Mexico*, Los Alamos National Laboratory and U.S. Department of Energy, Report LA-8890-ENV, UC-70 (Los Alamos, N.M.: Los Alamos National Laboratory, 1981).

30. D. Rai, R. J. Serne, and J. L. Swanson, "Solution Species of Plutonium-239 (V) in the Environment," in *Waste Isolation Safety Assessment Program, Task 4, Second Information Meeting*, Report PNL-SA-7352, ed. R. J. Serne (Washington, D.C.: U.S. Department of Energy, 1978); D. M. Nelson and M. B. Lovett, "Measurement of the Oxidation State and Concentration of Plutonium in Interstitial Waters of the Irish Sea," in *Impacts of Radionuclides Released into the Marine Environment*, Report STI/PB/5651 (Vienna: International Commission on Atomic Energy, 1981), pp. 338–42.

31. R. K. Schulz, "Root Uptake of Transuranic Elements," in *Transuranics in Natural Environments*, Report NVO-178, ed. M. G. White and P. B. Dunaway (Washington, D.C.: U.S. Energy Research and Development Administration, 1977), pp. 92–99.

32. International Commission on Radiological Protection, *The Metabolism of Compounds of Plutonium and Other Actinides*, International Commission on Radiological Protection Publication 19 (New York: Pergamon Press, 1972).

33. W. J. Bair, C. R. Richmond, and B. W. Wachholz, *Radiobiological Assessment on the Spatial Distribution of Radiation Dose from Inhaled Plutonium*, Report WASH-1320 (Washington, D.C.: U.S. Atomic Energy Commission, 1974).

34. W. J. Bair and R. C. Thompson, "Plutonium: Biomedical Research," *Science* 183 (1974): 715.

35. International Commission on Radiological Protection, *Metabolism of Compounds of Plutonium and Other Actinides*.

36. J. H. Olafson and K. H. Larson, "Plutonium: Its Biology and Environmental Persistence," in *Radioecology*, ed. V. Schultz and A. W. Klement (New York: Reinhold, 1963), pp. 161–69.

37. P. Beckmann, *The Health Hazards of Not Going Nuclear* (Boulder, Colo.: Golem Press, 1976).

38. F. W. Whicker and V. Schultz, *Radioecology: Nuclear Energy and the Environment*, 2 vols. (Boca Raton, Fla.: CRC Press, 1982).

39. R. E. Lapp and H. L. Andrews, *Nuclear Radiation Physics* (Englewood Cliffs, N.J.: Prentice-Hall, 1972).

40. G. D. Chase and J. L. Rabinowitz, *Principles of Radioisotope Methodology* (Minneapolis: Burgess, 1967).

41. Bureau of Radiological Health and Training Institute, *Radiological Health Handbook*, U.S. Public Health Service Publication 2016 (Washington, D.C.: U.S. Department of Health, Education, and Welfare, 1972).

42. International Commission on Radiological Protection, *Report of Committee IV on Evaluation of Radiation Doses to Body Tissues from Internal Contamination Due to Occupational Exposures*, International Commission on Radiological Protection Publication 10 (Oxford: Pergamon Press, 1959); Stoker et al., Radiological Survey.

43. H. Leenaars, *The Dispersal of Metal Mining Wastes in the Catchment of the River Geul, Belgium–The Netherlands* (Utrecht: Geografisch Instituut, Rijksuniversiteit, 1989).

3. The Northern Rio Grande Basin

1. *Gage* is the technical hydrologic term for *gauge*, referring to an automatic recording device that measures characteristics of stream flow.

2. P. A. Emery, R. J. Snipes, J. M. Dumeyer, and J. M. Klein, *Water in the San Luis Valley, South-Central Colorado*, Colorado Water Conservation Board Water-Resources Circular 18 (Denver: Colorado Water Conservation Board, 1973); D. Huntley, "Ground-Water Recharge to the Aquifers of Northern San Luis Valley, Colorado–Summary," *Geological Society of America Bulletin* 90 (1979): 707–9; G. A. Hearne and J. D. Dewey, *Hydrologic Analysis of the Rio Grande Basin North of Embudo, New Mexico, Colorado and New Mexico*, U.S. Geological Survey Water-Resources Investigations Report 86-4113 (Washington, D.C.: U.S. Geological Survey, 1988).

3. F. W. Whicker and C. M. Loveless, "Relationships of Physiography and Microclimate to Fallout Deposition," *Ecology* 49 (1968): 363–68.

4. C. B. Hunt, *Natural Regions of the United States and Canada* (San Francisco: Freeman, 1974); K. A. Erickson and A. W. Smith, *Atlas of Colorado* (Boulder: Colorado Associated University Press, 1985); J. L. Williams, ed., *New Mexico in Maps*, 2nd ed. (Albuquerque: University of New Mexico Press, 1986).

5. W. W. Atwood and K. F. Mather, *Physiography and Quaternary Geology of the San Juan Mountains, Colorado*, U.S. Geological Survey Professional Paper 166 (Washington, D.C.: U.S. Geological Survey, 1932).

6. Environmental Surveillance Group, ed., *Environmental Surveillance at Los Alamos During 1987* (Los Alamos, N.M.: Los Alamos National Laboratory, 1987), p. 67.

7. J. E. Upson, "Physiographic Subdivisions of the San Luis Valley, Southern Colorado," *Journal of Geology* 47 (1939): 721–36.

8. N. M. Fenneman, *Physiography of Western United States* (New York: McGraw-Hill, 1931), p. 130.

9. C. E. Seibenthal, *Geology and Water Resources of the San Luis Valley, Colorado*, U.S. Geological Survey Water-Supply Paper 240 (Washington, D.C.: U.S. Geological Survey, 1910).

10. C. K. Cooperrider and B. A. Hendricks, *Soil Erosion and Stream Flow on Range and Forest Lands of the Upper Rio Grande Watershed in Relation to Land Resources and Human Welfare*, U.S. Department of Agriculture Technical Bulletin 567 (Washington, D.C.: U.S. Department of Agriculture, 1937).

11. E. S. Larson and W. Cross, *Geology and Petrology of the San Juan Region*, U.S. Geological Survey Professional Paper 258 (Washington, D.C.: U.S. Geological Survey, 1956).

12. Atwood and Mather, *Physiography and Quaternary Geology*; New Mexico Geological Society, ed., *Guidebook of Southwestern San Juan Mountains, Colorado*, New Mexico Geological Society 8th Annual Field Conference Guidebook (Socorro: New Mexico Geological Society, 1957).

13. V. C. Kelly, "General Geology and Tectonics of the Western San Juan Mountains, Colorado," in *Guidebook of Southwestern San Juan Mountains, Colorado*, ed. New Mexico Geological Society, pp. 154–62.

14. O. Tweto, *Geologic Map of Colorado* (Reston, Va: U.S. Geological Survey, 1979).

15. R. H. De Voto, F. A. Peel, and W. H. Pierce, "Pennsylvanian and Permian Stratigraphy, Tectonism, and History, Northern Sangre de Cristo Range, Colorado," in *Guidebook of the San Luis Basin, Colorado*, New Mexico Geological Society 22nd Annual Field Conference Guidebook, ed. H. L. James (Socorro: New Mexico Geological Society, 1971), pp. 141–64.

16. J. R. Gaca and D. E. Karig, "Gravity Survey in the San Luis Area, Colorado," U.S. Geological Survey Open-File Report (unnumbered) (Denver: U.S. Geological Survey, 1965).

17. Hearne and Dewey, *Hydrologic Analysis of the Rio Grande Basin*.

18. J. E. Upson, "Physiographic Subdivisions of the San Luis Valley, Southern Colorado (Reprinted with Additional Notes)," in *Guidebook of the San Luis Basin*, ed. James, pp. 113–22.

19. W. Lambert, "Notes on the Late Cenozoic Geology of the Taos–Questa Area, New Mexico," in *Guidebook of Taos–Raton–Spanish Peaks Country, New Mexico and Colorado*, New Mexico Geological Society 17th Annual Field Conference Guidebook, ed. S. A. Northrop and C. B. Read (Socorro: New Mexico Geological Society, 1966), pp. 43–50.

20. C. H. Dane and G. O. Bachman, *Geologic Map of New Mexico* (Reston, Va: U.S. Geological Survey, 1965).

21. M. A. Dungan, W. R. Muehlberger, L. Leininger, C. Peterson, N. J. Gunn, M. Lindstrom, and L. Haskin, "Volcanic and Sedimentary Stratigraphy of the Rio Grande Gorge and the Late Cenozoic Geologic Evolution of the Southern San Luis Valley," in *Rio Grande Rift: Northern New Mexico*, New Mexico Geological Society 35th Annual Field Conference Guidebook, ed. W. S. Baldridge, P. W. Dickerson, R. E. Riecker, and J. Zidek (Socorro: New Mexico Geological Society, 1984), pp. 157–70.

22. V. T. McLemore and R. M. North, "Occurrences of Precious Metals and Uranium Along the Rio Grande Rift in Northern New Mexico," in *Rio Grande Rift*, ed. Baldridge et al., pp. 205–12.

23. L. Cordell, "Gravimetric Expression of Graben Foulting in Santa Fe Country and the Española Basin, New Mexico," in *Guidebook of Santa Fe Country, New Mexico*, New Mexico Geological Society 30th Annual Field Conference Guidebook, ed. R. V. Ingersall (Socorro: New Mexico Geological Society, 1979), pp. 59–64.

24. Dane and Backman, *Geologic Map of New Mexico*.

25. S. J. May, "Neogene Stratigraphy and Structure of the Ojo Caliente-Rio Chama Area, Española Basin, New Mexico," in *Guidebook of Santa Fe Country*, ed. Ingersall, pp. 83–88.

26. D. P. Dethier and K. A. Demsey, "Erosional History and Soil Development on Quaternary Surfaces, Northwest Española Basin, New Mexico," in *Rio Grande Rift*, ed. Baldridge et al., pp. 227–33.

27. V. C. Kelly, "Geomorphology of Española Basin," in *Guidebook of Santa Fe Country*, ed. Ingersall, pp. 281–88.

28. W. L. Chenoweth, "Uranium in the Santa Fe Area, New Mexico," in *Guidebook of Santa Fe Country*, ed. Ingersall, pp. 261–64.

29. E. C. Beaumount and C. B. Read, eds., *Guidebook of Rio Chama Country, New Mexico*, New Mexico Geological Society 11th Annual Field Conference Guidebook (Socorro: New Mexico Geological Society, 1960); Dane and Bachman, *Geologic Map of New Mexico*.

30. W. R. Muehlberger, G. E. Adams, T. E. Longgood, Jr., and B. E. St. John, "Stratigraphy of the Chama Quadrangle, Northern Arriba County, New Mexico," in *Guidebook of Rio Chama Country*, ed. Beaumont and Read, pp. 103–9; V. C. Kelly, *Geology of the Española Basin, New Mexico*, New Mexico Bureau of Mines and Mineral Resouces Geologic Map 48 (Socorro: New Mexico Bureau of Mines and Mineral Resources, 1978).

31. L. A. Woodward, "Tectonics of Central-Northern New Mexico," in *Silver Anniversary Guidebook: Ghost Ranch, Central-Northern New Mexico*, New Mexico Geological Society 25th Annual Field Conference Guidebook, ed. C. T. Siemers (Socorro: New Mexico Geological Society, 1974), pp. 123–30.

32. Siemers, ed., *Silver Anniversary Guidebook: Ghost Ranch*.

33. W. L. Chenoweth, "Uranium Occurrences of the Nacimiento-Jemez Region, Sandoval and Rio Arriba Counties, New Mexico," in *Silver Anniversary Guidebook: Ghost Ranch*, ed. Siemers, pp. 309–14.

34. R. L. Griggs, *Geology and Ground-Water Resources of the Los Alamos Area, New Mexico*, U.S. Geological Survey Water-Supply Paper 1753 (Washington, D.C.: U.S. Geological Survey, 1964).

35. Ibid.

36. C. D. Harrington and M. J. Aldrich, Jr., "Development and Deformation of Quaternary Surfaces on the Northeastern Flank of the Jemez Mountains," in *Rio Grande Rift*, ed. Baldridge et al., pp. 235–39.

37. S. A. Northrop, ed., *Guidebook of the Albuquerque Country, New Mexico* Geological Society 12th Annual Field Conference Guidebook (Socorro: New Mexico Geological Society, 1961).

38. R. Y. Anderson, "Physiography, Climate, and Vegeation of the Albuquerque Region," in *Guidebook of the Albuquerque Country*, ed. Northrop, pp. 63–71.

39. J. C. Moore, "Uranium Deposits in the Galisteo Formation of the Hagan Basin, Sandoval County, New Mexico," in *Guidebook of Santa Fe Country*, ed. Ingersall, pp. 265–67.

40. H. R. Joesting, J. E. Case, and L. E. Cordell, "The Rio Grande Trough Near Albuquerque, New Mexico," in *Guidebook of the Albuquerque Country*, ed. Northrop, pp. 148–52.

41. R. H. Tedford, "Neogene Stratigraphy of the Northwestern Albuquerque Basin," in *Albuquerque Country II*, New Mexico Geological Society 33rd Annual Field Conference Guidebook, ed. J. A. Gramblin and S. G. Wells (Socorro: New Mexico Geological Society, 1982), pp. 273–78.

42. Z. Speigel, "Late Cenozoic Sediments of the Lower Jemez River Region," in *Guidebook of the Albuquerque Country*, ed. Northrop, pp. 132–38.

43. C. F. Nordin, Jr., *A Preliminary Study of Sediment Transport Parameters, Rio Puerco Near Bernardo, New Mexico*, U.S. Geological Survey Professional Paper 462-C (Washington, D.C.: U.S. Geological Survey, 1963).

44. D. W. Love and J. D. Young, "Progress Report on the Late Cenozoic Geologic Evolution of the Lower Rio Puerco," in *Socorro Region II*, New Mexico Geological Society 34th Annual Field Conference Guidebook, ed. C. E. Chapin (Socorro: New Mexico Geological Society, 1983), pp. 277–84.

45. J. W. Hawley, D. W. Love, and S. G. Wells, "Summary of the Hydrology, Sedimentology, and Stratigraphy of the Rio Puerco Valley," in *Chaco Canyon Country*, American Geomorphological Field Group Conference Guidebook, ed. S. G. Wells, D. W. Love, and T. W. Gardner (Albuquerque: American Geomorphological Field Group, 1983), pp. 33–36.

46. C. J. Popp, J. W. Hawley, and D. W. Love, *Radionuclide and Heavy Metal Distribution in Recent Sediments of Major Streams in the Grants Mineral Belt, New Mexico*, Final Project Report, Office of Surface Mining (Washington, D.C.: U.S. Department of the Interior, 1983).

47. Chapin, ed., *Socorro Region II*.

48. Williams, *New Mexico in Maps*; Erickson and Smith, *Atlas of Colorado*.

49. Erickson and Smith, *Atlas of Colorado*.

50. L. B. Leopold, W. W. Emmett, and R. M. Myrick, *Channel and Hillslope Processes in a Semiarid Area, New Mexico*, U.S. Geological Survey Professional Paper 352-G (Washington, D.C.: U.S. Geological Survey, 1966), pp. 193–253.

51. Ibid.

52. R. B. Brown, C. H. Lowe, and C. P. Pase, *A Digitized Systematic Classification for Ecosystems with an Illustrated Summary of the Natural Vegetation of North America*, Forest Service General Technical Report Rm-73 (Washington, D.C.: U.S. Department of Agriculture, 1980).

53. Cooperrider and Hendricks, *Soil Erosion and Stream Flow*, p. 38.

54. C. P. Pase and E. F. Layser, "Classification of Riparian Habitat in the Southwest," in *Importance, Preservation, and Management of Riparian Habitat*, Forest Service General Technical Report RM-43 (Washington, D.C.: U.S. Department of Agriculture, 1977), pp. 5–9.

55. A. H. Frazier and W. Heckler, *Embudo, New Mexico, Birthplace of Systematic Stream Gaging*, U.S. Geological Survey Professional Paper 778 (Washington, D.C.: U.S. Geological Survey, 1972).

56. E. M. Shaw, *Hydrology in Practice* (Berkshire: Van Nostrand Reinhold, 1983), p. 301.

57. P. D. Komar, "Sediment Transport by Floods," in *Flood Geomorphology*, ed. V. R. Baker, R. C. Kochel, and P. C. Patton (New York: Wiley, 1988), pp. 97–112.

58. International Commission on Large Dams, *World Register of Dams* (Paris: International Commission on Large Dams, 1976); T. W. Mermel, *Register of Dams in the United States: Completed, Under Construction and Proposed* (New York: McGraw-Hill, 1958).

59. S. W. Trimble and S. W. Lund, *Soil Conservation and the Reduction of Erosion and Sedimentation in the Cook Creek Basin, Wisconsin*, U.S. Geological Survey Professional Paper 1234 (Washington, D.C.: U.S. Geological Survey, 1982).

60. G. C. Lusby, V. H. Reid, and O. D. Knipe, *Effects of Grazing on the Hydrology and Biology of the Badger Wash Basin in Western Colorado*, 1953–1966, U.S. Geological Survey Water-Supply Paper 1532-D (Washington, D.C.: U.S. Geological Survey, 1971).

61. W. L. Graf, "Fluvial Erosion and Federal Public Policy in the Navajo Nation," *Physical Geography* 7 (1986): 97–115.

62. J. C. Knox, "Historical Valley Floor Sedimentation in the Upper Mississippi Valley," *Annals of the Association of American Geographers* 77 (1987): 224–44.

63. Cooperrider and Hendricks, *Soil Erosion and Stream Flow.*

64. W. DeBuys, *Enchantment and Exploitation: The Life and Hard Times of a New Mexico Mountain Range* (Albuquerque: University of New Mexico Press, 1985), pp. 227–30.

65. P. Stewart, "Watershed Characteristics, Part B: Decreases in Flow in the Rio Grande," U.S. Forest Service Report (Taos, N.M.: Carson National Forest, 1981).

66. K. K. Hirschboeck, "Flood Hydroclimatology," in *Flood Geomorphology*, ed. Baker et al., pp. 27–49.

67. R. S. Bradley, R. G. Barry, and G. Kiladis, *Climatic Fluctuations of the Western of the Western United States During the Period of Instrumental Records*, Department of Geography and Geology Contribution 42 (Amherst: Department of Geography and Geology, University of Massachusetts, 1982), p. 67.

68. L. Wilson, *Future Water Issues, Taos County, New Mexico* (Santa Fe, N.M.: Lee Wilson and Associates, 1980).

69. T. R. Karl and R. W. Knight, *Atlas of Monthly Hydrological Drought Indices for the Contiguous United States* (Asheville, N.C.: National Climatic Data Center, 1985).

70. Data suggesting similar conclusions are in L. Wiard, *Floods in New Mexico, Magnitude and Frequency*, U.S. Geological Survey Circular 464 (Washington, D.C.: U.S. Geological Survey, 1962).

4. Fluvial Sediment, Forms, and Processes

1. A. Stoker, A. J. Ahlquist, D. L. Mayfield, W. R. Hansen, A. D. Talley, and W. D. Purtymun, *Radiological Survey of the Site of a Former Radioactive Liquid Waste Treatment Plant (TA-45) and the Effluent Receiving Areas of Acid, Pueblo, and Los Alamos Canyons, Los Alamos, New Mexico*, Los Alamos National Laboratory and U.S. Department of Energy, Report LA-8890-ENV, UC-70 (Los Alamos, N.M.: Los Alamos National Laboratory, 1981); L. J. Lane, W. D. Purtymun, and N. M. Becker, *New Estimating Procedures for Surface Runoff, Sediment Yield, and Contaminant Transport in Los Alamos County, New Mexico*, Los Alamos National Laboratory Report LA-10335-MS, UC-11 (Los Alamos, N.M.: Los Alamos National Laboratory, 1985).

2. A. Goudie, ed., *Geomorphological Techniques* (London: Allen & Unwin, 1981).

3. T. Dunne and L. B. Leopold, *Water in Environmental Planning* (San Francisco: Freeman, 1978), p. 670.

4. G. E. Petts, *Impounded Rivers: Perspectives for Ecological Management* (New York: Wiley, 1984).

5. H. Y. Hammad, "River Bed Degradation After Closure of Dams," *Proceedings of the American Society of Civil Engineers, Journal of the Hydraulics Division* 98, HY9 (1976): 591–607; J. D. Dewey, F. E. Roybal, and D. E. Funderburg, *Hydrologic Data on Channel Adjustments, 1970 to 1975, on the Rio Grande Downstream from Cochiti Dam, New Mexico, Before and After Closure*, U.S. Geological Survey Water Resources Investigation and National Technical Information Service Report PB-301-134 (Washington, D.C.: U.S. Geological Survey, 1979).

6. P. F. Lagasse, *An Assessment of the Response of the Rio Grande to Dam Construction—Cochiti to Isleta Reach*, 7 vols., Albuquerque District Office Technical Report (Albuquerque: U.S. Army Corps of Engineers, 1980).

7. G. P. Williams and M. G. Wolman, *Downstream Effects of Dams on Alluvial Rivers*, U.S. Geological Survey Professional Paper 1286 (Washington, D.C.: U.S. Geological Survey, 1984).

8. E. L. Pemberton, "Channel Changes in the Colorado River Below Glen Canyon Dam," *Proceedings of the Third Interagency Sedimentation Conference* (Denver: Interagency Sedimentation Conference, 1976), pp. 5/61–5/73.

9. D. B. Simons, R-M. Li, L-Y. Li, and M. J. Ballantine, *Erosion and Sedimentation Analysis of Rio Puerco and Rio Salado Watersheds*, Technical Report, U.S. Army Corps of Engineers, Albuquerque District Office (Fort Collins, Colo.: Simons, Li and Associates, 1981), p. 53.

10. Stoker et al., *Radiological Survey*.

11. Dunne and Leopold, *Water in Environmental Planning*, p. 662.

12. R. J. Garde and K. G. Ranga-Raju, *Mechanics of Sediment Transportation and Alluvial Stream Problems* (New York: Wiley, 1977), p. 62.

13. Simons et al., *Erosion and Sedimentation Analysis*, p. 53.

14. Lagasse, *Assessment of the Response of Rio Grande to Dam Construction*.

15. Lane et al., *New Estimating Procedures*.

16. K. J. Gregory, ed., *River Channel Change* (Chichester: Wiley, 1977).

17. S. A. Schumm and H. R. Kahn, "Experimental and Isotopic Ratios in River Sediments," *Geological Society of America Bulletin* 83 (1972): 1755–70; W. L. Graf, *Fluvial Processes in Dryland Rivers* (Berlin: Springer-Verlag, 1988), pp. 201–4.

18. M. R. Karlinger, T. R. Eschner, R. F. Hadley, and J. E. Kirchner, *Relation of Channel-Width Maintenance to Sediment Transport and River Morphology, Platte River, Nebraska*, U.S. Geological Survey Professional Paper 1277-E (Washington, D.C.: U.S. Geological Survey, 1983).

19. H. D. Miser, *The San Juan Canyon: Southeastern Utah*, U.S. Geological Survey Water-Supply Paper 538 (Washington, D.C.: U.S. Geological Survey, 1924); W. L. Graf, *The Colorado River: Change and Basin Management*, Association of American Geographers Resource Paper 85-2 (Washington, D.C.: Association of American Geographers, 1985).

20. G. P. Williams and M. G. Wolman, *Downstream Effects of Dams on Alluvial Rivers*, U.S. Geological Survey Professional Paper 1286 (Washington, D.C.: U.S. Geological Survey, 1984).

21. C. B. Hunt, P. Averitt, and R. L. Miller, *Geology and Geography of the Henry Mountains Region, Utah*, U.S. Geological Survey Professional Paper 228 (Washington, D.C.: U.S. Geological Survey, 1953); W. L. Graf, "Downstream Changes in Stream Power in the Henry Mountains, Utah," *Annals of the Association of American Geographers* 75 (1983): 97–115.

22. R. Hereford, "Modern Alluvial History of the Paria River Drainage Basin, Southern Utah," *Quaternary Research* 25 (1986): 293–311.

23. R. Hereford, "Climate Change and Ephemeral-Stream Processes: Twentieth Century Geomorphology and Alluvial Stratigraphy of the Little Colorado River, Arizona," *Geological Society of America Bulletin* 95 (1984): 654–68.

24. D. E. Burkham, *Channel Changes of the Gila River in Safford Valley, Arizona*, U.S. Geological Survey Professional Paper 655-G (Washington, D.C.: U.S. Geological Survey, 1972).

25. P. Ackers and F. G. Charlton, "Meander Geometry Arising from Varying Flows," *Journal of Hydrology* 11 (1970): 230–52.

26. Hydraulic Engineering Center, *HEC-6: Scour and Deposition in Rivers and Reservoirs*, U.S. Army Corps of Engineers User's Manual 723-G2-L2470 (Davis, Calif.: Hydraulic Engineering Center, 1976).

27. M. G. Wolman and L. B. Leopold, *River Flood Plains: Some Observations on Their Formation*, U.S. Geological Survey Professional Paper 282-C (Washington, D.C.: U.S. Geological Survey, 1957), pp. 87–107; R. Kellerhals, C. R. Meil, and D. I Bray, "Classification and Analysis of River Processes," *American Society of Civil Engineers, Journal of the Hydraulics Division* 107 (1976): 813–29.

28. J. R. C. Allen, *Physical Processes of Sedimentation* (London: Allen & Unwin, 1970).

29. D. J. Cant and R. G. Walker, "Fluvial Processes and Facies Sequences in the Sandy Braided South Saskatchewan River, Canada," *Sedimentology* 25 (1978): 625–48.

30. A. P. Heward, "Alluvial Fan Sequence and Megasequence Models with Examples from Westphalian D–Stephanian B. Coalfields, Northern Spain," in *Fluvial Sedimentology*, ed. A. D. Miall (Calgary: Canadian Society of Petroleum Geologists, 1978), pp. 669–702.

31. Lagasse, *Assessment of the Response of Rio Grande to Dam Construction*.

5. Engineering Works

1. A. Brookes, *Channelized Rivers: Perspectives for Environmental Management* (Chichester: Wiley, 1988).

2. G. E. Petts, *Impounded Rivers: Perspectives for Ecological Management* (New York: Wiley, 1984).

3. F. Wendorf, "A Reconstruction of Northern Rio Grande Prehistory," *American Anthropologist* 56 (1954): 200–221.

4. A. E. Bandelier, *Documentary History of the Rio Grande Pueblos*, pt. 2 of *Indians of the Rio Grande Valley*, ed. A. E. Bandelier and E. L. Hewitt (Albuquerque: University of New Mexico Press, 1929).

5. M. C. Meyer, *Water in the Hispanic Southwest: A Social and Legal History* (Tucson: University of Arizona Press, 1984).

6. A. G. Harper, A. R. Cordova, and K. Oberg, *Man and Resources in the Middle Rio Grande Valley* (Albuquerque: University of New Mexico Press, 1943).

7. C. K. Cooperrider and B. A. Hendricks, *Soil Erosion and Stream Flow on Range and Forest Lands of the Upper Rio Grande Watershed in Relation to Land Resources and Human Welfare*, U.S. Department of Agriculture Technical Bulletin 567 (Washington, D.C.: U.S. Department of Agriculture, 1937); S. A. Happ, "Sedimentation in the Middle Rio Grande Valley, New Mexico," *Geological Society of America Bulletin* 59 (1948): 1191–1215.

8. I. G. Clark, *Water in New Mexico: A History of Its Management and Use* (Albuquerque: University of New Mexico Press, 1987).

9. Ibid., p. 212; U.S. Congress, House of Representatives, *Rio Grande and Tributaries, New Mexico*, 81st Cong., 1st sess., H. Doc. 243 (Washington, D.C.: U.S. Government Printing Office, 1948).

10. H. F. Blaney, P. A. Ewing, O. W. Israelsen, C. Rohwer, and F. C. Scobey, *Investigations in the Upper Rio Grande Drainage Basin, Colorado–New Mexico–Texas* (Washington, D.C.: U.S. Government Printing Office, 1937).

11. National Resources Council, *Regional Planning, Part V—Upper Rio Grande* (Washington, D.C.: U.S. Government Printing Office, 1938).

12. C. E. Hyvarinen, "Channel Stabilization Practices on Middle Rio Grande," in *Symposium on Channel Stabilization Problems*, Technical Report 1 (Vicksburg, Miss.: U.S. Army Corps of Engineers, 1963), chap. 5.

13. House of Representatives, *Rio Grande and Tributaries*.

14. M. E. Welsh, *U.S. Army Corps of Engineers: Albuquerque District, 1935–1985* (Albuquerque: University of New Mexico Press, 1987).

15. J. C. Thompson, "Channel Stabilization, Middle Rio Grande Project," in *Symposium on Channel Stabilization Problems*, Technical Report 1 (Vicksburg, Miss.: U.S. Army Corps of Engineers, 1965), vol. 3, chap. 1.

16. R. C. Woodson, "Stabilization of the Middle Rio Grande," *American Society of Civil Engineers, Waterways and Harbor Division* 87 (1961): 1–15.

17. F. Collins (engineer, U.S. Army Corps of Engineers, Albuquerque District), interview, June 21, 1990.

18. V. M. Simmons, *The San Luis Valley* (Boulder, Colo.: Pruett, 1979).

19. Clark, *Water in New Mexico*, pp. 91–98.

20. Ibid., p. 388.

21. Welsh, *U.S. Army Corps of Engineers*, p. 189.

22. U.S. Bureau of Reclamation, *Annual Report, 1979* (Washington, D.C.: U.S. Bureau of Reclamation, 1979).

23. Environmental Surveillance Group, ed., *Environmental Surveillance at Los Alamos, 1987* (Los Alamos, N.M.: Los Alamos National Laboratory, 1987).

24. Ibid.

6. Riparian Vegetation

1. J. R. Hastings, "Vegetation Change and Arroyo Cutting in Southeastern Arizona," *Journal of the Arizona Academy of Science* 1 (1959): 60–67.

2. D. E. Brown, "Southwestern Wetlands—Their Classification and Characteristics," in *Strategies for Protection and Management of Floodplain Wetlands and Other Riparian Ecosystems*, ed. R. R. Johnson and J. F. McCormick, U.S. Forest Service General Technical Report WO-12 (Washington, D.C.: U.S. Department of Agriculture, 1978), pp. 269–82.

3. G. P. Davis, "Man and Wildlife in Arizona: The Pre-Settlement Era, 1823–1864" (M.S. thesis, Arizona State University, 1973).

4. P. S. White, "Pattern, Process and Natural Disturbance in Vegetation," *Botanical Review* 45 (1979): 229–99; C. R. Rupp, "Stream-Grade Variation in Riparian Forest Ecology Along Passage Creek, Virginia," *Bulletin Torr. Botanical Club* 109 (1982): 488–99.

5. J. W. Abert, "Notes of Lieutenant J. W. Abert," in *Notes of a Military Reconnaissance from Fort Leavenworth in Missouri to San Diego in California, Including Parts of Arkansas, Del Norte, and Gila Rivers*, Executive Document 41, 30th Cong., 1st sess., ed. W. H. Emory (Washington, D.C.: U.S. Government Printing Office, 1848), pp. 386–414; J. W. Abert, "Report of Lieutenant J. W. Abert of His Examination of New Mexico in the Years 1846–1847," in *Notes of a Military Reconnaissance*, ed. Emory, pp. 417–546; V. L. Burkholder, *Report of the Chief Engineer* (Albuquerque: Middle Rio Grande Conservancy District, 1928); H. Fergusson, *Rio Grande* (New York: McGraw-Hill, 1931).

6. J. R. Watson, "Plant Geography of North Central New Mexico," *Botanical Gazette* 54 (1912): 194–217.

7. Fergusson, *Rio Grande*, pp. 7–9.

8. M. Van Cleave, "Vegetation Changes in the Middle Rio Grande Conservancy District" (M.S. thesis, University of New Mexico, 1935).

9. V. C. Hink and R. D. Ohmart, *Middle Rio Grande Biological Survey: Final Report*, Contract Report DACW47-81-C-0015 (Albuquerque: U.S. Army Corps of Engineers, 1984).

10. T. W. Robinson, *Phreatophytes*, U.S. Geological Survey Water-Supply Paper 1423 (Washington, D.C.: U.S. Geological Survey, 1958).

11. E. M. Christensen, "The Rate of Naturalization of *Tamarix* in Utah," *American Midland Naturalist* 68 (1962): 51–57.

12. C. B. Thompson, "Importance of Phreatophytes in Water Supply," *American Society of Civil Engineers, Proceedings of the Irrigation and Drainage Division* 84 (1958): 1502/1–1502/17; T. W. Robinson, *Introduction, Spread, and Areal Effect of Saltcedar* (Tamarix) *in the Western States*, U.S. Geological Survey Professional Paper 491-A (Washington, D.C.: U.S. Geological Survey, 1965).

13. L. D. Potter, "Shoreline Ecology of Lake Powell," in *Program and Abstracts of the Geological Society of America Annual Meetings, Salt Lake City Meeting* (Boulder, Colo.: Geological Society of America, 1975), p. 125; B. L. Everitt, "Ecology of Saltcedar—A Plan for Research," *Environmental Geology* 3 (1980): 77–84.

14. M. D. Freehling, *Riparian Woodlands of the Middle Rio Grande Valley, New Mexico: A Study of Bird Population and Vegetation with Special Reference to Russian Olive (Elaeagnus angustifolia)*, U.S. Fish and Wildlife Service Report, Region 2 Office (Albuquerque: U.S. Fish and Wildlife Service, 1982).

15. C. J. Campbell and W. A. Dick-Peddie, "Comparison of Phreatophyte Communities on the Rio Grande in New Mexico," *Ecology* 45 (1964): 492–502.

16. Hink and Ohmart, *Middle Rio Grande Biological Survey*.

17. Brown, "Southwestern Wetlands."

18. Robinson, *Phreatophytes*.

19. H. C. Fritts, *Tree Rings and Climate* (London: Academic Press, 1976).

20. R. E. Wilding and T. R. Garland, "The Relationship of Microbial Processes to the Fate and Behavior of Transuranic Elements in Soils, Plants, and Animals," in *Transuranic Elements in the Environment*, U.S. Department of Energy Report DOE/TIC-22800, ed. W. C. Hanson (Washington, D.C.: U.S. Department of Energy and National Technical Information Service, 1980), pp. 31-37.

21. E. M. Romney and J. J. Davis, "Ecological Dissemination of Plutonium in the Environment," *Health Physics* 21 (1972): 551-60.

22. D. C. Adriano, A. Wallace, and E. M. Romney, "Uptake of Transuranic Nuclides from Soil by Plants Grown Under Controlled Environmental Conditions," in *Transuranic Elements in the Environment*, ed. Hanson, pp. 31–46.

7. Plutonium in the Rio Grande System

1. Stockhom International Peace Research Institute, *SIPRI Annual Yearbook, World Armaments and Disarmament* (Cambridge, Mass.: MIT Press, 1987).

2. S. Glasstone and J. W. Hawley, *Soils and Geomorphology in the Basin and Range Area of Southern New Mexico—Guidebook to the Desert Project*, New Mexico Bureau of Mines and Mineral Resources Memoir 39 (Socorro: New Mexico Bureau of Mines and Mineral Resources, 1981).

3. L. D. Hamilton, "Fallout and Countermeasures," *Bulletin of the Atomic Scientists* 19 (1963): 36–40.

4. H. Ball, *Justice Downwind: America's Atomic Testing Program in the 1950's* (Oxford: Oxford University Press, 1986).

5. K. L. Miller, *Under the Cloud: The Decades of Nuclear Testing* (New York: Free Press, 1986).

6. D. Deudney, "What Goes Up Must Come Down," *Bulletin of the Atomic Scientists* 40 (1984): 10–11.

7. R. W. Perkins and C. W. Thomas, "Worldwide Fallout," in *Transuranic Elements in the Environment*, U.S. Department of Energy Document DOE/TIC-22800, ed. W. C. Hanson (Washington, D.C.: U.S. Department of Energy and National Technical Information Service, 1980), pp. 8–14.

8. E. P. Harley, P. W. Krey, and H. L. Volchok, "Global Inventory and Distribution of Fallout Plutonium," *Nature* 241 (1973): 444.

9. Stockholm International Peace Research Institute, 1987, *SIPRI Annual Yearbook*, pp. 54–55.

10. Glasstone and Dolan, *Soils and Geomorphology*, p. 449.

11. F. W. Whicker and V. Schultz, *Radioecology: Nuclear Energy and the Environment*, 2 vols. (Boca Raton, Fla.: CRC Press, 1982).

12. G. R. Foster and T. E. Hakonson, "Estimating Erosional Losses of Fallout Plutonium," in *Environmental Surveillance at Los Alamos During 1982*, ed. Environmental Surveillance Group (Los Alamos, N.M.: Los Alamos National Laboratory, 1982), pp. 83–84; G. R. Foster and T. E. Hakonson, "Predicted Erosion and Sediment Delivery of Fallout Plutonium," *Environmental Quality* 13 (1984): 595–602.

13. Harley et al., "Global Inventory."

14. Foster and Hakonson, "Estimating Erosional Losses."

15. J. C. Ritchie, J. R. McHenry, and A. C. Gill, "Dating Recent Reservoir Sediments," *Limnology and Oceanography* 18 (1973): 254–63; J. C. Ritchie and J. R. McHenry, "Fallout Cs-137: A Tool in Conservation Research," *Journal of Soil and Water Conservation* 30 (1975): 283–85; J. R. McHenry and J. C. Ritchie, "Dating Recent Sediment in Impoundments," in *Proceedings of the Symposium on Surface Water Impoundments, American Society of Civil Engineers, Minneapolis, June 2–5, 1980* (Washington, D.C.: American Society of Civil Engineers), pp. 1279–89.

16. W. H. Kingsley, *Survey of Los Alamos and Pueblo Canyon for Radioactive Contamination and Radioassay Test Run on Sewer-Water Samples and Water and Soils Samples Taken from Los Alamos and Pueblo Canyons*, Los Alamos National Laboratory Report LAMS-516 (Los Alamos, N.M.: Los Alamos National Laboratory, 1947).

17. A. Stoker, A. J. Ahlquist, D. L. Mayfield, W. R. Hansen, A. D. Talley, and W. D. Purtymun, *Radiological Survey of the Site of a Former Radioactive Liquid Waste Treatment Plant (TA-45) and the Effluent Receiving Areas of Acid, Pueblo, and Los Alamos Canyons, Los Alamos, New Mexico*, Los Alamos National Laboratory and U.S. Department of Energy Report LA-8890-ENV, UC-70 (Los Alamos, N.M.: Los Alamos National Laboratory, 1981), pp. 11–12.

18. Ibid.

19. L. J. Lane, W. D. Purtymun, and N. M. Becker, *New Estimating Procedures for Surface Runoff, Sediment Yield, and Contaminant Transport in Los Alamos County, New Mexico*, Los Alamos National Laboratory Report LA-10335-MS, UC-11 (Los Alamos, N.M.: Los Alamos National Laboratory, 1985); Stoker et al., *Radiological Survey*, p. 49.

20. Lane et al., *New Estimating Procedures*, p. 33.

21. R. O. Gilbert, L. L. Eberhardt, E. B. Fowler, E. M. Romney, E. H. Essington, and J. E. Kinnear, "Statistical Analysis of Pu239, 240 and Am241 Contamination of Soil and

Vegetation on NAEG Study Sites," in *The Radioecology of Plutonium and Other Transuranics in Desert Environments*, U.S. Atomic Energy Commission and National Technical Information Service Report NVO-153, ed. M. G. White and P. B. Dunaway (Washington, D.C.: U.S. Atomic Energy Commission, 1975), pp. 242–53.

22. Lane et al., *New Estimating Procedures*.

23. Ibid.

24. Data provided to me through personal communications with L. J. Lane, research hydrologist, Agricultural Research, Tucson.

25. W. D. Purtymun, *Storm Runoff and Transport of Radionuclides in DP Canyon, Los Alamos County, New Mexico*, Los Alamos National Laboratory Report LA-5744 (Los Alamos, N.M.: Los Alamos National Laboratory, 1974).

26. W. D. Purtymun, R. J. Peters, T. E. Buhl, W. N. Maes, and F. H. Brown, *Background Concentrations of Radionuclides in Soils and River Sediments in Northern New Mexico, 1974–1986*, Los Alamos National Laboratory Report LA-11134-MS, UC-11 (Los Alamos, N.M.: Los Alamos National Laboratory, 1987).

27. Ibid.

28. R. C. Dahlman, C. T. Garten Jr., and T. E. Hakonson, "Comparative Distribution of Plutonium in Contaminated Ecosystems at Oak Ridge, Tennessee, and Los Alamos, New Mexico," in *Transuranic Elements in the Environment*, ed. Hanson, pp. 371–80; C. A. Little, "Plutonium in a Grassland Ecosystem," in *Transuranic Elements in the Environment*, ed. Hanson, pp. 188–95.

29. T. E. Hakonson and J. W. Nyhan, "Ecological Relationships of Plutonium in Southwest Ecosystems," in *Transuranic Elements in the Environment*, ed. Hanson, pp. 381–92.

30. W. D. Purtymun (hydrogeologist, Los Alamos National Laboratory), interview, September 25, 1990.

31. Purtymun et al., *Background Concentrations of Radionuclides*.

32. G. E. Petts, *Impounded Rivers: Perspectives for Ecological Management* (New York: Wiley, 1984).

33. J. R. C. Allen, *Physical Processes of Sedimentation* (London: Allen & Unwin, 1970).

34. M. R. Leeder, *Sedimentology: Processes and Product* (Boston: Allen & Unwin, 1982).

8. Annual Plutonium Budget for the Rio Grande

1. L. J. Lane, W. D. Purtymun, and N. M. Becker, *New Estimating Procedures for Surface Runoff, Sediment Yield, and Contaminant Transport in Los Alamos County, New Mexico*, Los Alamos National Laboratory Report LA-10335-MS, UC-11 (Los Alamos, N.M.: Los Alamos National Laboratory, 1985).

2. T. Dunne and L. B. Leopold, *Water in Environmental Planning* (San Francisco: Freeman, 1978).

3. G. R. Foster and T. E. Hakonson, "Estimating Erosional Losses of Fallout Plutonium," in *Environmental Surveillance at Los Alamos During 1982*, ed. Environmental Surveillance Group (Los Alamos, N.M.: Los Alamos National Laboratory, 1982), pp. 83–84; G. R. Foster and T. E. Hakonson, "Predicted Erosion and Sediment Delivery of Fallout Plutonium," *Environmental Quality* 13 (1984): 595–602.

4. E. P. Harley, P. W. Krey, and H. L. Volchok, "Global Inventory and Distribution of Fallout Plutonium," *Nature* 241 (1973): 444.

5. Lane et al., *New Estimating Procedures*.

6. L. J. Lane (hydrologic engineer, Agricultural Research Service), interview, Tucson, 1986.

7. W. H. Kingsley, *Survey of Los Alamos and Pueblo Canyon for Radioactive Contamination and Radioassay Test Run on Sewer-Water Samples and Water and Soils Samples Taken from Los Alamos and Pueblo Canyons*, Los Alamos National Laboratory Report LAMS-516 (Los Alamos, N.M.: Los Alamos National Laboratory, 1947).

8. W. D. Purtymun, G. L. Johnson, and E. C. John, *Distribution of Radioactivity in Alluvium of a Disposal Area at Los Alamos, New Mexico*, U.S. Geological Survey Professional Paper 550-D (Washington, D.C.: U.S. Geological Survey, 1966).

9. Lane et al., *New Estimating Procedures*.

10. W. C. Weimer, R. R. Kinnison, and J. H. Reeves, *Survey of Radionuclide Distributions Resulting from the Church Rock, New Mexico, Uranium Mill Tailings Pond Dam Failure*, U.S. Nuclear Regulatory Commission Report NUREG/DR-2449, PNL-4122 (Richland, Wash.: Battelle Pacific Northwest Laboratory, 1981).

9. Plutonium Storage Upstream from Cochiti

1. V. C. Hink and R. D. Ohmart, *Middle Rio Grande Biological Survey: Final Report*, U.S. Army Corps of Engineers Contract Report DACW47-81-C-0051 (Albuquerque: U.S. Army Corps of Engineers, 1984).

2. Ibid.

3. National Resources Council, *Regional Planning, Part V—Upper Rio Grande* (Washington, D.C.: U.S. Government Printing Office, 1938), plate 12.

4. P. P. Church, *The House at Otowi Bridge* (Albuquerque: University of New Mexico Press, 1959); F. Waters, *The Woman at Otowi Crossing* (Athens, Ohio: Swallow Press, 1966).

5. J. A. Gjevre, *The Chili Line: The Narrow Rail Trail to Santa Fe* (Española, N.M.: Rio Grande Sun Press, 1971).

6. T. M. Pearce, *New Mexico Place Names* (Albuquerque: University of New Mexico Press).

7. Hink and Ohmart, *Middle Rio Grande Biological Survey*.

8. R. L. Griggs, *Geology and Ground-Water Resources of the Los Alamos Area, New Mexico*, U.S. Geological Survey Water-Supply Paper 1753 (Washington, D.C.: U.S. Geological Survey, 1964).

9. Environmental Surveillance Group, ed., *Environmental Surveillance at Los Alamos During 1982* (Los Alamos, N.M.: Los Alamos National Laboratory, 1982), p. 177; Environmental Surveillance Group, ed., *Environmental Surveillance at Los Alamos During 1984* (Los Alamos, N.M.: Los Alamos National Laboratory, 1984), p. 41; Environmental Surveillance Group, ed., *Environmental Surveillance at Los Alamos During 1985* (Los Alamos, N.M.: Los Alamos National Laboratory, 1985), p. 171; Environmental Surveillance Group, ed., *Environmental Surveillance at Los Alamos During 1986* (Los Alamos, N.M.: Los Alamos National Laboratory, 1986), p. 67; Environmental Surveillance Group, ed., *Environmental Surveillance at Los Alamos During 1987* (Los Alamos, N.M.: Los Alamos National Laboratory, 1987), pp. 53–54; Environmental Surveillance Group, ed., *Environmental Surveillance at Los Alamos During 1988* (Los Alamos, N.M.: Los Alamos National Laboratory, 1988), pp. 53–54.

10. Environmental Surveillance Group, ed., *Environmental Surveillance at Los Alamos During 1987,* p. 54.

11. Environmental Surveillance Group, ed., *Environmental Surveillance at Los Alamos During 1985,* p. 49.

10. Plutonium Storage Downstream from Cochiti

1. J. R. C. Allen, *Physical Processes of Sedimentation* (London: Allen & Unwin, 1970).
2. C. Meine, *Aldo Leopold: His Life and Work* (Madison: University of Wisconsin Press, 1988), pp. 111–21.
3. National Resources Council, *Regional Planning, Part V—Upper Rio Grande* (Washington, D.C.: U.S. Government Printing Office, 1938), plate 16.
4. Ibid., plate 17.

11. Simulation of Sediment and Plutonium Dynamics

1. R. J. Chorley and P. Haggett, *Physical and Information Models in Geography* (London: Methuen, 1967); M. G. Anderson, "A Review of the Bases of Geomorphological Modelling," in *Modelling Geomorphologic Systems*, ed. M. G. Anderson (Chichester: Wiley, 1988), pp. 1-2D19.
2. W. B. Langbein, "Geometry of River Channels," *Journal of the Hydraulics Division, American Society of Civil Engineers* 90 (1964): 301–12; R. D. Hey, "Dynamic Process-Response Model of River Channel Development," *Earth Surface Processes* 4 (1979): 59–72.
3. V. T. Chow, *Open-Channel Hydraulics* (New York: McGraw-Hill, 1959); W. L. Graf, *Fluvial Processes in Dryland Rivers* (Berlin: Springer-Verlag, 1988).
4. L. B. Leopold, M. G. Wolman, and J. P. Miller, *Fluvial Processes in Dryland Rivers* (San Francisco: Freeman, 1964).
5. W. D. Purtymun, G. L. Johnson, T. E. Buhl, W. N. Maes, and F. H. Brown, *Background Concentrations of Radionuclides in Soils and River Sediments in Northern New Mexico, 1974–1986*, Los Alamos National Laboratory Report LA-11134-MS, UC-11 (Los Alamos, N.M.: Los Alamos National Laboratory, 1987).

12. General Lessons and Conclusions

1. W. L. Graf, S. L. Clark, M. T. Kammerer, T. Lehman, T. Randall, and R. Schröder, "Geomorphology of Heavy Metals in the Sediments of Queen Creek, Arizona, USA," *Catena* 18 (1991): 567–82.
2. Y. Erel, J. J. Morgan, and C. C. Patterson, "Natural Levels of Lead and Zinc in a Remote Mountain Stream," *Geochimica et Cosmochimica Acta* 55 (1991): 707–19.
3. M. G. Macklin and D. G. Passmore, "Late Quaternary Sedimentation in the Lower Tyne Valley, North East England," Seminar Paper 55 (University of Newcastle upon Tyne, Department of Geography, Mimeo, 1988).
4. K. J. Gregory, *River Channel Change* (Chichester: Wiley, 1977); R. Warner, *Fluvial Geomorphology in Australia* (Sydney: Academic Press, 1988).
5. J. D. Echeverria, P. Barrow, and R. Roos-Collins, *Rivers at Risk: The Concerned Citizen's Guide to Hydropower* (Washington, D.C.: Island Press, 1989).
6. Environmental Surveillance Group, ed., *Environmental Surveillance at Los Alamos During 1988* (Los Alamos, N.M.: Los Alamos National Laboratory, 1988), pp. 3–11.
7. U.S. Department of Energy, *U.S. Department of Energy Order 5400.1, General Environmental Protection Program* (Washington, D.C.: U.S. Department of Energy, 1988).
8. U.S. Department of Energy, *U.S. Department of Energy Order 5484.1* (Washington, D.C.: U.S. Department of Energy, 1981), chap. 3.
9. P. E. Black, *Conservation of Water and Related Land Resources*, 2nd ed. (Totowa, N.J.: Rowman & Littlefield, 1987).

References

Abert, J. W. "Notes of Lieutenant J. W. Abert, and Report of Lieutenant J. W. Abert of His Examination of New Mexico in the Years 1846–1847." In *Notes of a Military Reconnaissance from Fort Leavenworth in Missouri to San Diego in California, Including Parts of Arkansas, Del Norte, and Gila Rivers*, edited by W. H. Emory, pp. 386–546. Executive Document 41. 30th Cong., 1st sess. Washington, D.C.: U.S. Government Printing Office, 1848.

Ackers, P., and F. G. Charlton. "Meander Geometry Arising from Varying Flows." *Journal of Hydrology* 11 (1970): 230–52.

Adriano, D. C., A. Wallace, and E. M. Romney. "Uptake of Transuranic Nuclides from Soil by Plants Grown Under Controlled Environmental Conditions." In *Transuranic Elements in the Environment*, edited by W. C. Hanson, pp. 31–46. U.S. Department of Energy Report DOE/TIC-22800. Washington, D.C.: U.S. Department of Energy and National Technical Information Service, 1980.

Allen, J. R. C. *Physical Processes of Sedimentation*. London: Allen & Unwin, 1970.

Anderson, M. G. "A Review of the Bases of Geomorphological Modelling." In *Modelling Geomorphologic Systems*, edited by M. G. Anderson, pp. 1–19. Chichester: Wiley, 1988.

Anderson, R. Y. "Physiography, Climate, and Vegetation of the Albuquerque Region." In *Guidebook of the Albuquerque Country*, edited by S. A. Northrop, pp. 63–71. New Mexico Geological Society 12th Field Conference Guidebook. Socorro: New Mexico Geological Society, 1961.

Ashton, S. R., and D. A. Stanners. "Observations on the Deposition, Mobility and Chemical Associations of Plutonium in Intertidal Sediments." In *Techniques for Identifying Transuranic Speciation in Aquatic Environments*, pp. 209–17. International Atomic Energy Agency Publication STI/PUB/613. Vienna: International Atomic Energy Agency, 1981.

Atwood, W. W., and K. F. Mather. *Physiography and Quaternary Geology of the San Juan Mountains, Colorado*. U.S. Geological Survey Professional Paper 166. Washington, D.C.: U.S. Geological Survey, 1932.

Bagnold, R. A. *An Approach to the Sediment Transport Problem from General Physics*. U.S. Geological Survey Professional Paper 422-J. Washington, D.C.: U.S. Geological Survey 1966.

——. "Bed Load Transport by Natural Rivers." *Water Resources Research* 13 (1977): 303–12.

Bair, W. J., C. R. Richmand, and B. W. Wachholz. *Radiobiological Assessment on the Spatial Distribution of Radiation Dose from Inhaled Plutonium*. U.S. Atomic Energy Commission Report WASH-1320. Washington, D.C.: U.S. Atomic Energy Commission, 1974.

Bair, W. J., and R. C. Thompson. "Plutonium: Biomedical Research." *Science* 183 (1974): 715.

Ball, H. *Justice Downwind: America's Atomic Testing Program in the 1950's*. Oxford: Oxford University Press, 1986.

Bandelier, A. E. *Documentary History of the Rio Grande Pueblos*. Part 2 of *Indians of the Rio Grande Valley*, edited by A. E. Bandelier and E. L. Hewitt. Albuquerque: University of New Mexico Press, 1929.

Beasley, T. M., R. Carpenter, and C. D. Jennings. "Plutonium, Am-241 and Cs-137 Ratios, Inventories and Vertical Profiles in Washington and Oregon Continental Shelf Sediments." *Geochimica and Cosmochimica Acta* 46 (1982): 1831–1946.

Beaumont, E. C., and C. B. Read, eds. *Guidebook of Rio Chama Country, New Mexico*. New Mexico Geological Society 11th Annual Field Conference Guidebook. Socorro: New Mexico Geological Society, 1960.

Beckmann, P. *The Health Hazards of Not Going Nuclear*. Boulder, Colo.: Golem Press, 1976.

Bickel, L. *The Deadly Element*. London: Stein & Day, 1980.

Black, P. E. *Conservation of Water and Related Land Resources*. 2nd ed. Totowa, N.J.: Rowman & Littlefield, 1987.

Blaney, H. F., P. A. Ewing, O. W. Israelsen, C. Rohwer, and F. C. Scobey. *Investigations in the Upper Rio Grande Drainage Basin, Colorado–New Mexico–Texas*. Washington, D.C.: U.S. Government Printing Office, 1937.

Bradley, R. S., R. G. Barry, and G. Kiladis, G. *Climatic Fluctuations of the Western United States During the Period of Instrumental Records*. Department of Geology and Geography, University of Massachusetts, Contribution 42. Amherst: University of Massachusetts, 1982.

Bradley, S. B. "Flood Effects on the Transport of Heavy Metals." *International Journal of Environmental Studies* 22 (1984): 225–30.

——. "Partitioning of Heavy Metals on Suspended Sediments During Floods on the River Ystwyth, Wales." *Trace Substances in Environmental Health* 18 (1984): 548–57.

——. "Sediment Quality Related to Discharge in a Mineralized Region of Wales." In *Recent Developments in the Explanation and Prediction of Erosion and Sediment Yield*, pp. 341–50. International Association of Scientific Hydrology Publication 137. Paris: International Association of Scientific Hydrology, 1982.

Bradley, S. B., and J. J. Cox. "Heavy Metals in the Hamps and Manifold Valleys, North Staffordshire, U.K.: Distribution in Flood Plain Soils." *Science of the Total Environment* 50 (1986): 103–28.

Brookes, A. *Channelized Rivers: Perspectives for Environmental Management*. Chichester: Wiley, 1988.

Brown, D. E. "Southwestern Wetlands—Their Classification and Characteristics." In *Strategies for Protection and Management of Floodplain Wetlands and Other Riparian*

Ecosystems, edited by R. R. Johnson and J. F. McCormick, pp. 269–82. U.S. Department of Agriculture, Forest Service, General Technical Report WO-12. Washington, D.C.: U.S. Forest Service, 1978.

Brown, D. E., C. H. Lowe, and C. P. Pase. *A Digitized Systematic Classification for Ecosystems with an Illustrated Summary of the Natural Vegetation of North America*. U.S. Department of Agriculture, Forest Service General Technical Report RM-73. Fort Collins, Colo.: U.S. Forest Service, 1980.

Brown, R. B., G. F. Kling, and N. H. Cutshall. "Agricultural Erosion Indicated by Cs-137 Redistribution: II. Estimates of Erosion Rates." *Soil Science Society of America Proceedings* 45 (1981): 1191–97.

Burkham, D. E. *Channel Changes of the Gila River in Safford Valley, Arizona*. U.S. Geological Survey Professional Paper 655-G. Washington, D.C.: U.S. Geological Survey, 1972.

Burkholder, V. L. *Report of the Chief Engineer*. Albuquerque: Middle Rio Grande Conservancy District, New Mexico, 1928.

Campbell, B. L., R. J. Loughran, G. L. Elliott, and D. J. Shelly. "Mapping Drainage Basin Sediment Sources Using Caesium-137." In *Proceedings of the International Symposium on Drainage Basin Sediment Delivery, Albuquerque, New Mexico*, pp. 437–46. International Association of Hydrological Sciences Publication 159. Albuquerque: International Association of Hydrological Sciences, 1986.

Campbell, C. J., and W. A. Dick-Peddie. "Comparison of Phreatophyte Communities on the Rio Grande in New Mexico." *Ecology* 45 (1964): 492–502.

Cant, D. J., and R. G. Walker. "Fluvial Processes and Facies Sequences in the Sandy Braided South Saskatchewan River, Canada." *Sedimentology* 25 (1978): 625–48.

Carpenter, K. E. "On the Biological Factors Involved in the Destruction of River Fisheries by Pollution Due to Lead Mining." *Annals of Applied Biology* 12 (1925): 1–13.

———. "A Study of the Fauna of Rivers Polluted by Lead Mining in the Aberystwyth District of Cariganshire." *Annals of Applied Biology* 11 (1924): 1–23.

Carrigan, P. H., Jr. *Inventory of Radionuclides in Bottom Sediment of the Clinch River, Eastern Tennessee*. U.S. Geological Survey Professional Paper 433-I. Washington, D.C.: U.S. Geological Survey, 1969.

———. *Radioactive Waste Distribution in the Clinch River, Eastern Tennessee*. U.S. Geological Survey Professional Paper 433-G. Washington, D.C.: U.S. Geological Survey, 1968.

Chapin, C. E., ed. *Socorro Region II*. New Mexico Geological Society 34th Field Conference Guidebook. Socorro: New Mexico Geological Society, 1983.

Chase, G. D., and J. L. Rabinowitz. *Principles of Radioisotope Methodology*. Minneapolis: Burgess, 1967.

Chenoweth, W. L. "Uranium Occurrences of the Nacimiento–Jemez Region, Sandoval and Rio Arriba Counties, New Mexico." In *Silver Anniversary Guidebook: Ghost Ranch, Central-Northern New Mexico*, edited by C. T. Siemers, pp. 309–14. New Mexico Geological Society 25th Field Conference Guidebook. Socorro: New Mexico Geological Society, 1974.

———. "Uranium in the Santa Fe Area, New Mexico." In *Guidebook of Santa Fe Country*, edited by R. V. Ingersall, pp. 261–64. New Mexico Geological Society 30th Field Conference Guidebook. Socorro: New Mexico Geological Society, 1979.

Chorley, R. J., and P. Haggett. *Physical and Information Models in Geography*. London: Methuen, 1967.

Chow, V. T. *Open Channel Hydraulics*. New York: McGraw-Hill, 1959.

Christensen, E. M. "The Rate of Naturalization of *Tamarix* in Utah." *American Midland Naturalist* 68 (1962): 51–57.

Church, P. P. *The House at Otowi Bridge*. Albuquerque: University of New Mexico Press, 1959.

Clark, I. G. *Water in New Mexico: A History of Its Management and Use*. Albuquerque: University of New Mexico Press, 1987.

Cooperrider, C. K., and B. A. Hendricks. *Soil Erosion and Stream Flow on Range and Forest Lands of the Upper Rio Grande Watershed in Relation to Land Resources and Human Welfare*. U.S. Department of Agriculture Technical Bulletin 567. Washington, D.C.: U.S. Department of Agriculture, 1937.

Cordell, L. "Gravimetric Expression of Graben Faulting in Santa Fe Country and the Española Basin, New Mexico." In *Guidebook of Santa Fe Country*, edited by R. V. Ingersall, pp. 59–64. New Mexico Geological Society 30th Field Conference Guidebook. Socorro: New Mexico Geological Society, 1979.

Coughtrey, P. J., D. Jackson, C. H. Jones, P. Kane, and M. C. Thorne. *Radionuclide Distribution and Transport in Terrestrial and Aquatic Ecosystems*. Vol. 4: *Plutonium and Neptunium*. Rotterdam: A. A. Balkema, 1984.

Cull, R. F., K. Duggan, T. J. East, R. Martin, and A. S. Murray. "Tailings Transport and Deposition Downstream of the Northern Hercules (Moline) Mine in the Catchment of the Mary River, N.T." In *Environmental Planning and Management for Mining and Energy*, edited by P. J. Broese, R. Von Groenou, and J. R. Burton, pp. 199–216. Darwin: Northern Australia Mine Rehabilitation Workshop, 1986.

Curie, E. *Madame Curie*. Garden City, N.Y.: Doubleday, 1938.

Dahlman, R. C., C. T. Garten, Jr., and T. E. Hakonson. "Comparative Distribution of Plutonium in Contaminated Ecosystems at Oak Ridge, Tennessee, and Los Alamos, New Mexico." In *Transuranic Elements in the Environment*, edited by W. C. Hanson, pp. 71–81. U.S. Department of Energy Report DOE/TIC-22800. Washington, D.C.: U.S. Department of Energy and National Technical Information Service, 1980.

Dane, C. H., and G. O. Bachman. *Geologic Map of New Mexico*. Reston, Va.: U.S. Geological Survey, 1965.

Davies, B. E., and J. Lewin. "Chronosequences in Alluvial Soils with Special Reference to Historic Lead Pollution in Cardiganshire, Wales." *Environmental Pollution* 6 (1974): 49–57.

Davis, G. P. "Man and Wildlife in Arizona—The Pre-Settlement Era, 1823-1864." M.S. thesis, Arizona State University, 1973.

deBuys, W. *Enchantment and Exploitation: The Life and Hard Times of a New Mexico Mountain Range*. Albuquerque: University of New Mexico Press, 1985.

Dennis, J., ed. *The Nuclear Almanac: Confronting the Atom in War and Peace*. Boston: Addison-Wesley, 1984.

Dethier, D. P., and K. A. Demsey. "Erosional History and Soil Development on Quaternary Surfaces, Northwest Española Basin, New Mexico." In *Rio Grande Rift: Northern New Mexico*, edited by W. S. Baldridge, P. W. Dickerson, R. E. Reicker, and J. Zidek, pp. 227–33. New Mexico Geological Society 35th Field Conference Guidebook. Socorro: New Mexico Geological Society, 1984.

Deudney, D. "What Goes Up Must Come Down." *Bulletin of the Atomic Scientists* 40 (1984): 10–11.

Devaurs, M. A., E. P. Springer, L. J. Lane, and G. J. Langhorst. "Prediction Methodology for Contaminant Transport from Rangeland Watersheds." In American Society of Agricultural Engineers, International Symposium on Modeling Agricultural, Forest and Rangeland Hydrology, December 12–13, Chicago. Los Alamos National Laboratory Report LA-UR-88-2071. Mimeo, 1988.

De Voto, R. H., F. A. Peel, and W. H. Pierce. "Pennsylvanian and Permian Stratigraphy, Tectonism and History, Northern Sangre de Cristo Range, Colorado." In *Guidebook of the San Luis Basin, Colorado*, edited by H. L. James, pp. 141–64. New Mexico Geological Society 22nd Field Conference Guidebook. Socorro: New Mexico Geological Society, 1971.

Dewey, J. D., F. E. Roybal, and D. E. Funderburg. *Hydrologic Data on Channel Adjustments, 1970 to 1975, on the Rio Grande Downstream from Cochiti Dam, New Mexico, Before and After Closure*. U.S. Geological Survey Water Resources Investigation and National Technical Information Service Report PB-301 134. Washington, D.C.: U.S. Geological Survey, 1979.

Dungan, M. A., W. R. Muehlberger, L. Leininger, C. Peterson, N. J. McMillan, G. Gunn, M. Lindstrom, and L. Haskin. "Volcanic and Sedimentary Stratigraphy of the Rio Grande Gorge and the Late Cenozoic Geologic Evolution of the Southern San Luis Valley." In *Rio Grande Rift: Northern New Mexico*, edited by W. S. Baldridge, P. W. Dickerson, R. E. Reicker, and J. Zidek, pp. 157–70. New Mexico Geological Society 35th Field Conference Guidebook. Socorro: New Mexico Geological Society, 1984.

Dunne, T., and L. B. Leopold. *Water in Environmental Planning*. San Francisco: Freeman, 1978.

East, T. J. "Geomorphological Assessment of Sites and Impoundments for the Long Term Containment of Uranium Mill Tailings in the Alligator Rivers Region." *Australian Geographer* 16 (1986): 16–22.

East, T. J., R. J. Cull, A. S. Murray, and K. Duggan. "Fluvial Dispersion of Radioactive Mill Tailings in the Seasonally-Wet Tropics, Northern Australia." In *Fluvial Geomorphology of Australia*, edited by R. F. Warner, pp. 303–22. Sydney: Academic Press, 1988.

Echeverria, J. D., P. Barrow, and R. Roos-Collins. *Rivers at Risk: The Concerned Citizen's Guide to Hydropower*. Washington, D.C.: Island Press, 1989.

Emery, P. A., R. J. Snipes, J. M. Dumeyer, and J. M. Klein. *Water in the San Luis Valley, South-Central Colorado*. Colorado Water Conservation Board Water-Resources Circular 18. Denver: Colorado Water Conservation Board, 1973.

Environmental Surveillance Group, ed. *Environmental Surveillance at Los Alamos* [annual]. Los Alamos, N.M.: Los Alamos National Laboratory, 1971–1989.

Erel, Y., J. J. Morgan, and C. C. Patterson. "Natural Levels of Lead and Zinc in a Remote Mountain Stream." *Geochimica et Cosmochimica Acta* 55 (1991): 707–19.

Erickson, K. A., and A. W. Smith. *Atlas of Colorado*. Boulder: Colorado Associated University Press, 1985.

Everitt, B. L. "Ecology of Saltcedar—A Plan for Research." *Environmental Geology* 3 (1980): 77–84.

Fenneman, N. M. *Physiography of Western United States*. New York: McGraw-Hill, 1931.

Fergusson, H. *Rio Grande*. New York: Morrow, 1931.

Fortescue, J. A. C. *Environmental Geochemistry: A Holistic Approach*. New York: Springer-Verlag, 1980.

Foster, G. R., and T. E. Hakonson. "Erosional Losses of Fallout Plutonium." In "Proceedings of the Symposium on Environmental Research on Actinides in the Environment." U.S. Department of Energy. Mimeo, 1983.

——. "Estimating Erosional Losses of Fallout Plutonium." In *Environmental Surveillance at Los Alamos During 1982*, edited by Environmental Surveillance Group, pp. 83–84. Los Alamos, N.M.: Los Alamos National Laboratory, 1982.

——. "Predicted Erosion and Sediment Delivery of Fallout Plutonium." *Journal of Environmental Quality* 13 (1984): 595–602.

Frazier, A. H., and W. Heckler. *Embudo, New Mexico, Birthplace of Systematic Stream Gaging*. U.S. Geological Survey Professional Paper 778. Washington, D.C.: U.S. Geological Survey, 1972.

Freehling, M. D. *Riparian Woodlands of the Middle Rio Grande Valley, New Mexico: A Study of Bird Population and Vegetation with Special Reference to Russian Olive (Elaeagnus angustifolia)*. U.S. Fish and Wildlife Service Report, Region 2 Office. Albuquerque: U.S. Fish and Wildlife Service, 1982.

Frissel, M. J., A. T. Jakubick, N. van der Klugt, R. Pennders, P. Poelstra, and E. Zwemmer. *Modelling of the Transport of Radionuclides of Strontium, Caesium, and Plutonium in Soil. Experimental Verification*. Brussels: Commission of the European Communities, 1980.

Fritts, H. C. *Tree Rings and Climate*. London: Academic Press, 1976.

Gaca, J. R., and D. E. Karig. "Gravity Survey in the San Luis Area, Colorado." U.S. Geological Survey, Unnumbered Open-File Report, 1965.

Garde, R. J., and K. G. Ranga Raju. *Mechanics of Sediment Transportation and Alluvial Stream Problems*. New York: Wiley, 1977.

Garland, T. R., R. E. Wildung, and R. C. Routson. "The Chemistry of Plutonium in Soils. I. Plutonium Solubility." In *Pacific Northwest Laboratory Annual Report for 1975 to the USERDA Division of Biometical and Environmental Research*. Part 2: *Ecological Sciences*. Batelle Northwest Laboratory Report BNWL-2000. Seattle: Batelle Northwest Laboratory, 1976.

Gibbs, R. J. "Mechanisms of Trace Metal Transport in Rivers." *Science* 180 (1973): 71–73.

Gilbert, R. O., L. L. Eberhardt, E. B., Fowler, E. M. Romney, E. H. Essington, and J. E. Kinnear. "Statistical Analysis of Pu 239, 240 and Am 241 Contamination of Soil and Vegetation on NAEG Study Sites." In *The Radioecology of Plutonium and Other Transuranics in Desert Environments*, edited by M. G. White and P. B. Dunaway, pp. 242–253. U.S. Atomic Energy Commission Report NVO-153. Springfield, Va.: National Technical Information Service, 1975.

Gile, L. H., J. W. Hawley, and R. B. Grossman. *Soils and Geomorphology in the Basin and Range Area of Southern New Mexico—DGuidebook to the Desert Project*. New Mexico Bureau of Mines and Mineral Resources Memoir 39. Socorro: New Mexico Bureau of Mines and Mineral Resources, 1981.

Gjevre, J. A. *Chili Line: The Narrow Rail Trail to Santa Fe*. Española, N.M.: Rio Grande Sun Press, 1971.

Glasstone, S, and P. J. Dolan, eds. *The Effects of Nuclear Weapons*. 3rd ed. Washington, D.C.: U.S. Department of Defense and the Energy Research and Development Administration, 1977.

Glover, R. E. *Dispersion of Dissolved or Suspended Materials in Flowing Streams*. U.S. Geological Survey Professional Paper 433-B. Washington, D.C.: U.S. Geological Survey, 1964.

Godfrey, R. G., and B. J. Frederick. *Stream Dispersion at Selected Sites*. U.S. Geological Survey Professional Paper 433-K. Washington, D.C.: U.S. Geological Survey, 1970.

Goodchild, P. *J. Robert Oppenheimer: Shatterer of Worlds*. New York: Fromm International, 1985.

Goudie, A., ed. *Geomorphological Techniques*. London: Allen & Unwin, 1981.

Graf, W. L. *The Colorado River: Change and Basin Management*. Washington, D.C.: Association of American Geographers, 1985.

——. "Downstream Changes in Stream Power in the Henry Mountains, Utah." *Annals of the Association of American Geographers* 73 (1983): 373–87.

——. "Fluvial Dynamics of Thorium-230 in the Church Rock Event, Puerco River, New Mexico." *Annals of the Association of American Geographers* 80 (1990): 327–42.

——. "Fluvial Erosion and Federal Public Policy in the Navajo Nation." *Physical Geography* 7 (1986): 97–115.

——. *Fluvial Processes in Dryland Rivers*. Berlin: Springer-Verlag, 1988.

——. "Mercury Transport in Stream Sediments of the Colorado Plateau." *Annals of the Association of American Geographers* 75 (1985): 552–65.

Graf, W. L., S. L. Clark, M. T. Kammerer, T. Lehman, K. Randall, and R. Schrder. "Geomorphology of Heavy Metals in the Sediments of Queen Creek, Arizona, USA." *Catena* 18 (1991): 567–82.

Gregory, K. J., ed. *River Channel Change*. Chichester: Wiley, 1977.

Griggs, R. L. *Geology and Ground-Water Resources of the Los Alamos Area, New Mexico*. U.S. Geological Survey Water-Supply Paper 1753. Washington, D.C.: U.S. Geological Survey, 1964.

Grouef, S. *Manhattan Project: The Untold Story of the Making of the Atomic Bomb*. Boston: Little, Brown, 1967.

Groves, L. M. *Now It Can Be Told: The Story of the Manhattan Project*. New York: Harper and Brothers, 1962.

Hakonson, T. E. "Environmental Pathways of Plutonium into Terrestrial Plants and Animals." *Health Physics* 29 (1975): 583–88.

Hakonson, T. E., and K. V. Bostick. "Cesium-137 and Plutonium in Liquid Waste Discharge Areas at Los Alamos." In *Radioecology and Energy Resources*, edited by C. E. Cushing, Jr., pp. 40–48. Special Publication 1, Ecological Society of America. Stroudsburg, Pa.: Dowden, Hutchinson & Ross, 1976.

Hakonson, T. E., and L. J. Johnson. "Distribution of Environmental Plutonium in the Trinity Site Ecosystem After 27 Years." In *Proceedings of the 3rd International Congress of the Radiation Protection Association, September 9–14, 1973*, pp. 242–47. Washington, D.C.: Radiation Protection Association, 1973.

Hakonson, T. E., L. J. Johnson, and W. D. Purtymun. "The Distribution of Plutonium in Liquid Waste Disposal Areas at Los Alamos." In *Proceedings of the 3rd International Congress of the Radiation Protection Association, September 9–14, 1973*, pp. 248–53. Washington, D.C.: Radiation Protection Association, 1973.

Hakonson, T. E., and J. W. Nyhan. "Ecological Relationships of Plutonium in Southwest Ecosystems." In *Transuranic Elements in the Environment*, edited by W. C. Hanson, pp. 82–91. U.S. Department of Energy Report DOE/TIC-22800. Washington, D.C.: U.S. Department of Energy and National Technical Information Service, 1980.

Hakonson, T. E., J. W. Nyhan, and W. D. Purtymun. "Accumulation and Transport of Soil Plutonium in Liquid Waste Discharge Areas at Los Alamos." In *Transuranic Nuclides in the Environment*, pp. 175–89. International Atomic Energy Agency Report 1AEA-SM-199/3. Vienna: International Atomic Energy Agency, 1976.

Hakonson, T. E., R. L. Watters, and W. C. Hanson. "The Transport of Plutonium in Terrestrial Ecosystems." *Health Physics* 40 (1981): 63–69.

Hamilton, L. D. "Fallout and Countermeasures." *Bulletin of the Atomic Scientists* 19 (1963): 36–40.

Hammad, H. Y. "River Bed Degradation After Closure of Dams." *Proceedings of the American Society of Civil Engineers, Journal of the Hydraulics Division* 98, HY9 (1976): 591–607.

Happ, S. A. "Sedimentation in the Middle Rio Grande Valley, New Mexico." *Geological Society of America Bulletin* 59 (1948): 1191–1215.

Harley, E. P., P. W. Krey, and H. L. Volchok. "Global Inventory and Distribution of Fallout Plutonium." *Nature* 241 (1973): 444.

Harper, A. G., A. R. Cordova, and K. Oberg. *Man and Resources in the Middle Rio Grande Valley*. Albuquerque: University of New Mexico Press, 1943.

Harrington, C. D., and M. J. Aldrich, Jr. "Development and Deformation of Quaternary Surfaces on the Northeastern Flank of the Jemez Mountains." In *Rio Grande Rift: Northern New Mexico*, edited by W. S. Baldridge, P. W. Dickerson, R. E. Reicker, and J. Zidek, pp. 235–39. New Mexico Geological Society 35th Field Conference Guidebook. Socorro: New Mexico Geological Society, 1984.

Hastings, J. R. "Vegetation Change and Arroyo Cutting in Southeastern Arizona." *Journal of the Arizona Academy of Science* 1 (1959): 60–67.

Hawkins, D. *Manhattan District History, Project Y, The Los Alamos Project*. Los Alamos, N.M.: Los Alamos Scientific Laboratory, 1947.

Hawley, J. W., ed. *Guidebook to Rio Grande Rift in New Mexico and Colorado*. New Mexico Bureau of Mines and Mineral Technology Circular 163. Socorro: New Mexico Bureau of Mines and Mineral Technology, 1978.

Hawley, J. W., D. W. Love, and S. G. Wells. "Summary of the Hydrology, Sedimentology, and Stratigraphy of the Rio Puerco Valley." In *Chaco Canyon Country*, edited by S. G. Wells, D. W. Love, and T. W. Gardner, pp. 33–36. Albuquerque: American Geomorphological Field Group, 1983.

Hearne, G. A., and J. D. Dewey. *Hydrologic Analysis of the Rio Grande Basin North of Embudo, New Mexico; Colorado and New Mexico*. U.S. Geological Survey Water-Resources Investigations Report 86-4113. Washington, D.C.: U.S. Geological Survey, 1988.

Herford, R. "Climate Change and Ephemeral-Stream Processes: Twentieth Century Geomorpholgoy and Alluvial Stratigraphy of the Little Colorado River, Arizona." *Geological Society of America Bulletin* 95 (1984): 654–68.

——. "Modern Alluvial History of the Paria River Drainage Basin, Southern Utah." *Quaternary Research* 25 (1986): 293–311.

Heward, A. P. "Alluvial Fan Sequence and Megasequence Models with Examples from Westphalian D2D-Stephanian B Coalfields, Northern Spain." In *Fluvial Sedimentology*, edited by A. D. Miall, pp. 669–702. Calgary: Canadian Society of Petroleum Geologists. 1978.

Hewlett, R. G., and O. E. Anderson. *The New World, 1939–1946*. University Park: Pennsylvania State University Press, 1962.

Hey, R. D. "Dynamic Process-Response Model of River Channel Development." *Earth Surface Processes* 4 (1979): 59–72.

Hink, V. C., and Ohmart, R. D. *Middle Rio Grande Biological Survey: Final Report*. U.S. Army Corps of Engineers Contract Report DACW47-81-C-0015. Albuquerque: U.S. Army Corps of Engineers, 1984.

Hirschboeck, K. K. "Flood Hydroclimatology." In *Flood Geomorphology*, edited by V. R. Baker, R. C. Kochel, and P. C. Patton, pp. 27–49. New York: Wiley, 1988.

Holton, G. *Thematic Origins of Scientific Thought*. Cambridge, Mass.: Harvard University Press, 1973.

Hunt, C. B. *Natural Regions of the United States and Canada*. San Francisco: Freeman, 1974.

Hunt, C. B., P. Averitt, and R. L. Miller. *Geology and Geography of the Henry Mountains Region, Utah*. U.S. Geological Survey Professional Paper 228. Washington, D.C.: U.S. Geological Survey, 1953.

Huntley, D. "Ground-Water Recharge to the Aquifers of Northern San Luis Valley, Colorado–Summary." *Geological Society of America Bulletin* 90 (1976): 707–9.

Hydraulic Engineering Center. *HEC-6: Scour and Deposition in Rivers and Reservoirs*. U.S. Army Corps of Engineers User's Manual 723-G2-L2470. Davis, Calif.: U.S. Army Corps of Engineers Hydraulic Engineering Center, 1976.

——. *HEC-2: Water Surface Profiles*. U.S. Army Corps of Engineers User's Manual 723-X6-L202A. Davis, Calif.: U.S. Army Corps of Engineers Hydraulic Engineering Center, 1981.

Hyvarinen, C. E. "Channel Stabilization Practices on Middle Rio Grande in New Mexico." In *Symposium on Channel Stabilization Problems*, vol. 1, chap. 5. Technical Report 1. Vicksburg, Miss.: U.S. Army Corps of Engineers, 1963.

Imhoff, K. R., P. Koppe, and F. Deitz. "Heavy Metals in the Ruhr River and Their Budget in the Catchment Area." *Progress in Water Technology* 12 (1980): 735–49.

International Commission on Large Dams. *World Register of Dams*. Paris: International Commission on Large Dams, 1976.

International Commission on Radiological Protection. *The Metabolism of Compounds of Plutonium and Other Actinides*. International Commission on Radiological Protection Publication 19. New York: Pergamon Press, 1972.

——. *Report of Committee IV on Evaluation of Radiation Doses to Body Tissues from Internal Contamination Due to Occupational Exposures*. International Commission on Radiological Protection Publication 10. Oxford: Pergamon Press, 1959.

Jacobson, L., and P. Overstreet. "The Uptake by Plants of Plutonium and Some Productions of Nuclear Fission Adsorbed on Soil Colloids." *Soil Science* 65 (1948): 129–34.

Jakubick, A. T. "In Situ Plutonium Transport in Geomedia." In *Migration of Long-Lived Radionuclides in the Geosphere*, edited by Organization for Economic Cooperation and Development, pp. 201–23. Paris: OECD, 1979.

——. "Migration of Plutonium in Natural Soils." In *Transuranic Nuclides in the Environment*, pp. 216–27. International Atomic Energy Agency Report IAEA-SM-199/3. Vienna: International Atomic Energy Agency, 1976.

James, H. L., ed. *Guidebook of the San Luis Basin, Colorado*. New Mexico Geological Society 22nd Field Conference Guidebook. Socorro: New Mexico Geological Society, 1971.

Jarvis, P. J. *Heavy Metal Pollution: An Annotated Bibliography: 1976–1980*. East Anglia, Geobooks, 1983.

Jenne, E. A., and J. S. Wahlberg. *Role of Certain Stream-Sediment Components in Radioion Sorption*. U.S. Geological Survey Professional Paper 433-E. Washington, D.C.: U.S. Geological Survey, 1968.

Joesting, H. R., J. E. Case, and L. E. Cordell. "The Rio Grande Trough Near Albuquerque, New Mexico." In *Guidebook of the Albuquerque Country*, edited by S. A. Northrop, pp. 148–52. New Mexico Geological Society 12th Field Conference Guidebook. Sororro: New Mexico Geological Society, 1961.

Jones, J. R. E. "A Further Study of the Zinc-polluted River Ystwyth." *Journal of Animal Ecology* 27 (1958): 1–14.

——. "A Study of the Zinc-polluted River Ystwyth in Northern Cardiganshire, Wales." *Annals of Applied Biology* 27 (1940): 363–78.

Kachanoski, R. G. "Comparison of Measured Soil Cesium-137 Losses and Erosion Rates." *Canadian Journal of Soil Science* 67 (1987): 199–203.

Karl, T. R., and R. W. Knight. *Atlas of Monthly Palmer Hydrological Drought Indices for the Contiguous United States*. Asheville, N.C.: National Climatic Data Center, 1985.

Karlinger, M. R., T. R. Eschner, R. F. Hadley, and J. E. Kirchner. *Relation of Channel-Width Maintenance to Sediment Transport and River Morphology, Platte River, Nebraska*. U.S. Geological Survey Professional Paper 1277-E. Washington, D.C.: U.S. Geological Survey, 1983.

Kellerhals, R., C. R. Meil, and D. I. Bray. "Classification and Analysis of River Processes." *American Society of Civil Engineers, Journal of the Hydraulics Division* 107 (1976): 813–29.

Kelly, V. C. "General Geology and Tectonics of the Western San Juan Mountains, Colorado." In *Guidebook of Southwestern San Juan Mountains, Colorado*, edited by B. Baldwin, pp. 154–62. New Mexico Geological Society 8th Field Conference Guidebook. Socorro: New Mexico Geological Society, 1957.

——. *Geology of the Albuquerque Basin, New Mexico*. New Mexico Bureau of Mines and Mineral Resources Memoir 33. Socorro: New Mexico Bureau of Mines and Mineral Resources, 1977.

——. *Geology of the Española Basin, New Mexico*. New Mexico Bureau of Mines and Mineral Resources Geologic Map 48. Socorro: New Mexico Bureau of Mines and Mineral Resources, 1978.

——. "Geomorphology of Española Basin." In *Guidebook of Santa Fe Country*, edited by R. V. Ingersall, pp. 281–88. New Mexico Geological Society 30th Field Conference Guidebook. Socorro: New Mexico Geological Society, 1979.

Kelly, V. C., and S. A. Northrop. *Geology of Sandia Mountains and Vicinity, New Mexico*. New Mexico Bureau of Mines and Mineral Resources Memoir 29. Socorro: New Mexico Bureau of Mines and Mineral Resources, 1975.

Kennedy, V. C. *Mineralogy and Cation-Exchange Capacity of Sediments from Selected Streams*. U.S. Geological Survey Professional Paper 433-D. Washington, D.C.: U.S. Geological Survey, 1965.

Kingsley, W. H. *Survey of Los Alamos and Pueblo Canyon for Radioactive Contamination and Radioassay Test Run on Sewer-Water Samples and Water and Soil Samples Taken from Los Alamos and Pueblo Canyons*. Los Alamos National Laboratory Report LAMS-516. Los Alamos, N.M.: Los Alamos National Laboratory, 1947.

Knox, J. C. "Historical Valley Floor Sedimentation in the Upper Mississippi Valley." *Annals of the Association of American Geographers* 77 (1987): 224–44.

Komar, P. D. "Sediment Transport by Floods." In *Flood Geomorphology*, edited by V. R. Baker, R. C. Kochel, and P. C. Patton, pp. 97–112. New York: Wiley, 1988.

Kunettka, J. W. *City of Fire: Los Alamos and the Atomic Age, 1943–1945*. Albuquerque: University of New Mexico Press, 1979.

Kuzin, A. M., and A. A. Peredelskiy. "Nature Conservancy and Some Aspects of Radioactive–Ecological Relations" (in Russian). *Okhr. Prir. Zap. SSSR* 1 (1956): 65–70.

Lagasse, P. F. *An Assessment of the Response of the Rio Grande to Dam Construction—Cochiti to Isleta Reach*. 7 vols. U.S. Army Corps of Engineers Albuquerque District Office Technical Report. Albuquerque: U.S. Army Corps of Engineers, 1980.

Lambert, W. "Notes on the Late Cenozoic Geology of the Taos–Questa Area, New Mexico." In *Guidebook of Taos–Raton–Spanish Peaks Country, New Mexico and Colorado*, edited by S. A. Northrop and C. B. Read, pp. 43–50. New Mexico Geological Society 17th Field Conference Guidebook. Socorro: New Mexico Geological Society, 1966.

Lance, J. C., S. C. McIntyre, J. W. Naney, and S. S. Rousseva. "Measuring Sediment Movement at Low Erosion Rates Using Cesium-137." *Soil Science Society of America Proceedings* 50 (1986): 1303–9.

Lane, L. J., and T. E. Hakonson. "Influence of Particle Sorting in Transport of Sediment Associated Contaminants." In *Proceedings of the Waste Management Symposium, Tuscon, Arizona, 1982*, pp. 543–57. Tucson: University of Arizona Press, 1982.

Lane, L. J., W. D. Purtymun, and N. M. Becker. *New Estimating Procedures for Surface Runoff, Sediment Yield, and Contaminant Transport in Los Alamos County, New Mexico*. Los Alamos National Laboratory Report LA-10335-MS, UC-11. Los Alamos, N.M.: Los Alamos National Laboratory, 1985.

Langbein, W. B. "Geometry of River Channels." *Journal of the Hydraulics Division, American Society of Civil Engineers* 90 (1964): 301–12.

Lapp, R. E., and H. L. Andrews. *Nuclear Radiation Physics*. Englewood Cliffs, N.J.: Prentice-Hall, 1972.

Larsen, E. S., and W. Cross. *Geology and Petrology of the San Juan Region, Southwestern Colorado*. U.S. Geological Survey Professional Paper 258. Washington, D.C.: U.S. Geological Survey, 1956.

Larson, K. H., J. L. Leitch, W. F. Dunn, J. W. Neel, J. H. Olafson, E. E. Held, J. Taylor, W. J. Cross, and A. W. Bellamy. *Los Alamos*. University of California Report UCLA-108. Berkeley, Calif.: University of California, 1951.

Leeder, M. R. *Sedimentology: Process and Product*. Boston: Allen & Unwin, 1982.

Leenaers, H. *The Dispersal of Metal Mining Wastes in the Catchement of the River Geul, Belgium – The Netherlands*. Utrecht: Geografisch Instituut, Rijksuniversiteit, 1989.

Leopold, L. B., W. W. Emmett, and R. M. Myrick. *Channel and Hillslope Processes in a Semiarid Area, New Mexico*. U.S. Geological Survey Professional Paper 352-G. Washington, D.C.: U.S. Geological Survey, 1966.

Levinson, A. A. *Introduction to Exploration Geochemistry*. Wilmette, Ill.: Applied Publishing, 1980.

Lewin, J., B. E. Davies, and P. Wolfenden. "Interactions Between Channel Change and Historic Mining Sediments." In *River Channel Change*, edited by K. J. Gregory, pp. 353–67. Chichester: Wiley, 1977.

Lindolm, R. C. *A Practical Approach to Sedimentology*. London: Allen & Unwin, 1987.

Little, C. A. "Plutonium in a Grassland Ecosystem." In *Transuranic Elements in the Environment*, edited by W. C. Hanson, pp. 62–68. U.S. Department of Energy Report DOE/TIC-22800. Washington, D.C.: U.S. Department of Energy and National Technical Information Service, 1980.

Los Alamos Historical Society. *Los Alamos: Beginning of an Era, 1943–1945*. Los Alamos, N.M.: Los Alamos Historical Society, 1986.

Los Alamos Scientific Laboratory. "Minutes of the Laboratory Advisory Board, August 19, 1944." Los Alamos National Laboratory. Mimeo, 1944.

Loughran, R. J., B. L. Campbell, and D. E. Walling. "Soil Erosion and Sedimentation Indicated by Caesium 137: Jackmoor Brook Catchment, Devon, England." *Catena* 14 (1987): 201–12.

Love, D. W., and J. D. Young. "Progress Report on the Late Cenozoic Geologic Evolution of the Lower Rio Puerco." In *Socorro Region II*, edited by C. E. Chapin, pp. 277–84. New Mexico Geological Society 34th Field Conference Guidebook. Socorro: New Mexico Geological Society, 1983.

Lusby, G. C., V. H. Reid, and O. D. Knipe. *Effects of Grazing on the Hydrology and Biology of the Badger Wash Basin in Western Colorado, 1953–1966*. U.S. Geological Survey Water-Supply Paper 1532-D. Washington, D.C.: U.S. Geological Survey, 1971.

Macklin, M. G. "Flood-Plain Sedimentation in the Upper Axe Valley, Mendip, England." *Transactions of the Institute of British Geographers* 10 (1985): 235–44.

Macklin, M. G., and R. B. Dowsett. "The Chemical and Physical Speciation of Trace Metals in Fine Grained Overbank Flood Sediments in the Tyne Basin, North-east England." *Catena* 16 (1989): 135–51.

Macklin, M. G., and D. G. Passmore. "Late Quaternary Sedimentation in the Lower Tyne Valley, North East England." Seminar Paper 55. University of Newcastle upon Tyne, Department of Geography. Mimeo, 1988.

Marcus, W. A. "Copper Dispersion in Ephemeral Stream Sediments." *Earth Surface Processes and Landforms* 12 (1987): 117–28.

Mark, J. C. *A Short Account of Los Alamos Theoretical Work on Thermonuclear Weapons, 1946–1950*. Los Alamos Report LA-5647-MS. Los Alamos, N.M.: Los Alamos National Laboratory, 1974.

Marron, D. C. "Physical and Chemical Characteristics of a Metal Contaminated Overbank Deposit, West Central South Dakota, USA." *Earth Surface Processes and Landforms* 14 (1989): 419–32.

Martin, J. M., and M. Maybeck. "Elemental Mass Balance of Material Carried by Major World Rivers." *Marine Chemistry* 7 (1979): 173–206.

Martz, L. W., and E. de Jong. "Using Cesium-137 to Assess the Variability of Net Soil Erosion and Its Association with Topography in a Canadian Prairie Landscape." *Catena* 14 (1987): 439–51.

May, S. J. "Neogene Stratigraphy and Structure of the Ojo Caliente–Rio Chama Area, Española Basin, New Mexico." In *Guidebook of Santa Fe Country, New Mexico*, edited by R. V. Ingersall, pp. 83–88. Geological Society 30th Field Conference Guidebook. Socorro: New Mexico Geological Society, 1979.

McHenry, J. R., and J. C. Ritchie. "Dating Recent Sediment in Impoundments." In *Proceedings of the Symposium on Surface Water Impoundments, American Society of Civil Engineers, Minneapolis, June 2–5, 1980*, pp. 1279–D89. Washington, D.C.: American Society of Civil Engineers, 1980.

McIntyre, S. C., J. C. Lance, B. L. Campbell, and R. L. Miller. "Using Cesium-137 to Estimate Soil Erosion on a Clearcut Hillside." *Journal of Soil and Water Conservation* 42 (1987): 117–20.

McLemore, V. T., and R. M. North. "Occurrences of Precious Metals and Uranium Along the Rio Grande Rift in Northern New Mexico." In *Rio Grande Rift: Northern New Mexico*, edited by W. S. Baldridge, P. W. Dickerson, R. E. Riecker, and J. Zidek, pp. 205–12. New Mexico Geological Society 35th Annual Field Conference Guidebook. Socorro: New Mexico Geological Society, 1984.

Means, J. L., D. A. Crerar, M. P. Borcsik, and J. O. Duquid. "Adsorption of Co and Selected Actinides by Mn and Fe Oxides in Soils and Sediments." *Geochimica et Cosmochimica Acta* 42 (1978): 1763–73.

Meine, C. *Aldo Leopold: His Life and Work*. Madison: University of Wisconsin Press. 1988.

Menzel, R. G. "Enrichment Ratios for Water Quality Modeling." In *CREAMS: A Field Scale Model for Chemicals, Runoff, and Erosion from Agricultural Management Systems*, edited by W. G. Knisel, pp. 486–92. U.S. Department of Agriculture Conservation Research Report 26, III, 12. Washington, D.C.: U.S. Department of Agriculture, 1980.

Mermel, T. W. *Register of Dams in the United States: Completed, Under Construction and Proposed*. New York: McGraw-Hill, 1958.

Meyer, M. C. *Water in the Hispanic Southwest: A Social and Legal History*. Tucson: University of Arizona Press, 1984.

Miller, K. L. *Under the Cloud: The Decades of Nuclear Testing*. New York: Free Press, 1986.

Miser, H. D. *The San Juan Canyon: Southeastern Utah*. U.S. Geological Survey Water-Supply Paper 538. Washington, D.C.: U.S. Geological Survey, 1924.

Moore, J. C. "Uranium Deposits in the Galisteo Formation of the Hagan Basin, Sandoval County, New Mexico." In *Guidebook of Santa Fe Country, New Mexico*, edited by R. V. Ingersall, pp. 265–67. Geological Society 30th Field Conference Guidebook. Socorro: New Mexico Geological Society, 1979.

Morain, S. A., T. K. Budge, and M. E. White. *Vegetation and Land Use in New Mexico*. New Mexico Bureau of Mines and Mineral Technology Resource Map RM-8 [Scale 1: 1 million]. Socorro: New Mexico Bureau of Mines and Mineral Technology, 1977.

Morrison, E. E. *Turmoil and Tradition: A Biography of Henry L. Stimson*. Boston: Houghton Mifflin, 1960.

Muehlberger, W. R., G. E. Adams, T. E. Longgood, Jr., and B. E. St. John. "Stratigraphy of the Chama Quadrangle, Northern Rio Arriba County, New Mexico." In *Guidebook of Rio Chama Country*, edited by E. C. Beaumont and C. B. Read, pp. 103–9. New Mexico Geological Society 11th Field Conference Guidebook. Socorro: New Mexico Geological Society, 1960.

Muller, G., and U. Förstner. "Heavy Metals in Sediments of the Rhine and Elbe Estuaries, Mobilization or Mixing Effect?" *Environmental Geology* 1 (1977): 33–39.

Nash, R. *The Rights of Nature*. Madison: University of Wisconsin Press, 1989.

National Resources Council. *Regional Planning, Part V—Upper Rio Grande*. Washington, D.C.: U.S. Government Printing Office, 1938.

Nelson, D. M., and M. B. Lovett. "Measurement of the Oxidation State and Concentration of Plutonium in Interstitial Waters of the Irish Sea." In *Impacts of Radionuclides Released into the Marine Environment*, pp. 338–92. International Commission on Atomic Energy Report STI/PUB/5651. Vienna: International Commission on Atomic Energy, 1981.

New Mexico Geological Society, ed. *Guidebook of Southwestern San Juan Mountains, Colorado*. New Mexico Geological Society 8th Field Conference Guidebook. Socorro: New Mexico Geological Society, 1957.

Newton, L. "Pollution of the Rivers of West Wales by Lead and Zinc Mine Effluent." *Annals of Applied Biology* 31 (1944): 1–11.

Nishita, H., and M. Hamilton. "The Influence of Several Soil Components and Their Interaction on Plutonium Extractability from a Calcareous Soil." *Soil Science* 131 (1981): 56–59.

Nishita, H., and R. M. Haug. "The Effect of Fulvic and Humic Acids and Inorganic Phase of Soil on the Sorption and Extractability of Pu(IV) 239 from Several Clay Minerals." *Soil Science* 128 (1979): 291–96.

Nordin, C. F., Jr. *A Preliminary Study of Sediment Transport Parameters, Rio Puerco Near Bernardo, New Mexico*. U.S. Geological Survey Professional Paper 462-C. Washington, D.C.: U.S. Geological Survey, 1963.

Northrop, S. A., ed. *Guidebook of the Albuquerque Country*. New Mexico Geological Society 12th Field Conference Guidebook. Socorro: New Mexico Geological Society, 1961.

Noshkin, V. E. "Ecological Aspects of Plutonium Dissemination in Aquatic Environments." *Health Physics* 22 (1972): 537.

Nyhan, J. W., B. J. Drennon, W. V. Abeele, M. L. Wheeler, W. D. Purtymun, G. Trujillo, W. J. Herrera, and J. W. Booth. "Distribution of Plutonium and Americium Beneath a

33-Year-Old Liquid Waste Disposal Site." *Journal of Environmental Quality* 14 (1985): 501–9.

——. *Distribution of Radionuclides and Water in Bandelier Tuff Beneath a Former Los Alamos Liquid Waste Disposal Site After 33 Years*. Los Alamos National Laboratory Report LA-10159-LLWM. Los Alamos, N.M.: Los Alamos National Laboratory, 1984.

Nyhan, J. W., F. R. Mera, and R. E. Neher. "Distribution of Plutonium in Trinity Soils After 28 Years." *Journal of Environmental Quality* 5 (1976): 431–37.

Nyhan, J. W., F. R. Mera, and R. J. Peters. "The Distribution of Plutonium and Caesium in Alluvial Soils of the Los Alamos Environs." In *Radioecology and Energy Resources*, edited by C. E. Cushing, Jr., pp. 49–57. Special Publication 1, Ecological Society of America. Stroudsburg, Pa.: Dowden, Hutchinson & Ross, 1976.

Nyhan, J. W., T. G. Schofield, G. C. White, and G. Trujillo. *Sampling Soils for Cs-137 Using Various Field Sampling Volumes*. Los Alamos National Laboratory Report LA-8951-MS, UC-11. Los Alamos, N.M.: Los Alamos National Laboratory, 1981.

Nyhan, J. W., G. C. White, and G. Trujillo. "Soil Plutonium and Caesium in Stream Channels and Banks of Los Alamos Liquid Effluent Receiving Areas." *Health Physics* 43 (1982): 531–41.

Odum, E. P. "Consideration of the Total Environment in Power Reactor Waste Disposal." *Proceedings of the International Conference on Peaceful Uses of Atomic Energy* 13 (1956): 350–58.

Olafson, J. H., and K. H. Larson. "Plutonium, Its Biology and Environmental Persistence." In *Radioecology*, edited by V. Schultz and A. W. Klement, Jr., pp. 161–69. New York: Reinhold, 1963.

Oliphant, M. *Rutherford*. New York: Elsevier, 1972.

Pase, C. P., and E. F. Layser. "Classification of Riparian Habitat in the Southwest." In *Importance, Preservation, and Management of Riparian Habitat: A Symposium*, edited by R. R. Johnson and D. A. Jones, pp. 5–9. U.S. Department of Agriculture, Forest Service, General Technical Report RM-43. Fort Collins, Colo.: U.S. Forest Service, 1977.

Payne, W. D. "The Role of Sulfides and Other Heavy Minerals in Copper Anomalous Stream Sediments." Ph.D. diss., Stanford University, 1971.

Pearce, T. M. *New Mexico Place Names*. Albuquerque: University of New Mexico Press, 1965.

Pemberton, E. L. "Channel Changes in the Colorado River Below Glen Canyon Dam." In *Proceedings of the Third Interagency Sedimentation Conference*, pp. 5/61–5/73. Denver: U.S. Government Printing Office, 1976.

Perkins, R. W., and C. W. Thomas. "Worldwide Fallout." In *Transuranic Elements in the Environment*, edited by W. C. Hanson, pp. 8–14. U.S. Department of Energy Report DOE/TIC-22800. Washington, D.C.: U.S. Department of Energy and National Technical Information Service, 1980.

Petts, G. E. *Impounded Rivers: Perspectives for Ecological Management*. New York: Wiley, 1984.

Pfeiffer, W. C., M. Fiszman, and N. Carbonell. "Fate of Chromium in a Tributary of the Irajo River, Rio de Janeiro." *Environmental Pollution* 1 (1980): 117–26.

Pickering, R. J. *Distribution of Radionuclides in Bottom Sediment of the Clinch River, Eastern Tennessee*. U.S. Geological Survey Professional Paper 433-H. Washington, D.C.: U.S. Geological Survey, 1969.

Plikarpochkin, V. V. "The Quantitative Estimation of Ore-bearing Areas from Sample Data of the Drainage System." In *Geochemical Exploration*, edited by R. W. Boyle, pp. 585–86. Toronto: Canadian Institute of Mining and Metallurgy, 1971.

Popp, C. J., J. W. Hawley, and D. W. Love. *Radionuclide and Heavy Metal Distribution in Recent Sediments of Major Streams in the Grants Mineral Belt, New Mexico*. Final Report to Office of Surface Mining. Washington, D.C.: U.S. Department of the Interior, 1983.

Potter, L. D. "Shoreline Ecology of Lake Powell." In *Program and Abstracts of the Geological Society of America Annual Meetings, Salt Lake City Meeting*. Boulder, Colo.: Geological Society of America, 1975.

Prout, W. E. "Adsorption of Radioactive Wastes on Savannah River Plant Soil." *Soil Science* 86 (1958): 13–17.

Purtymun, W. D. *Storm Runoff and Transport of Radionuclides in DP Canyon, Los Alamos County, New Mexico*. Los Alamos National Laboratory Report LA-5744. Los Alamos, N.M.: Los Alamos National Laboratory, 1974.

Purtymun, W. D., G. L. Johnson, and E. C. John. *Distribution of Radioactivity in the Alluvium of a Disposal Area at Los Alamos, New Mexico*. U.S. Geological Survey Professional Paper 550-D. Washington, D.C.: U.S. Geological Survey, 1966.

Purtymun, W. D., R. J. Peters, T. E. Buhl, M. N. Maes, and F. H. Brown. *Background Concentrations of Radionuclides in Soils and River Sediments in Northern New Mexico, 1974–1986*. Los Alamos National Laboratory Report LA-11134-MS, UC-11. Los Alamos, N.M.: Los Alamos National Laboratory, 1987.

Purtymun, W. D., R. J. Peters, and M. N. Maes. 1990. *Transport of Plutonium in Snowmelt Runoff*. Los Alamos National Laboratory Report LA-11795-MS, UC-902. Los Alamos, N.M.: Los Alamos National Laboratory, 1990.

Rai, D., R. J. Serne, and J. L. Swanson. "Solution Species of Plutonium-239(V) in the Environment." In *Waste Isolation Safety Assessment Program, Task 4, Second Information Meeting*, edited by R. J. Serne, pp. 101–9. U.S. Department of Energy Document PNL-SA-7352. Washington, D.C.: U.S. Department of Energy, 1978.

Rhoads, R. *The Making of the Atomic Bomb*. New York: Simon and Schuster, 1986.

Ritchie, J. C., and J. R. McHenry. "Fallout Cs-137: A Tool in Conservation Research." *Journal of Soil and Water Conservation* 30 (1975): 283–85.

Ritchie, J. C., J. R. McHenry, and A. C. Gill. "Dating Recent Reservoir Sediments." *Limnology and Oceanography* 18 (1973): 254–63.

Robinson, G. O. *The Oak Ridge Story*. Nashville, Tenn.: Southern Press, 1950.

Robinson, T. W. *Introduction, Spread, and Areal Effect of Saltcedar* (Tamarix) *in the Western States*. U.S. Geological Survey Professional Paper 491-A. Washington, D.C.: U.S. Geological Survey, 1965.

——. *Phreatophytes*. U.S. Geological Survey Water-Supply Paper 1423. Washington, D.C.: U.S. Geological Survey, 1958.

Rogowski, A. S., and T. Tamura. "Movement of Cesium-137 by Runoff, Erosion, and Infiltration on the Alluvial Captina Silt Loam." *Health Physics* 11 (1965): 1333–40.

Romer, A., ed. *Radiochemistry and the Discovery of Isotopes*. New York: Dover, 1970.

Romney, E. M., and J. J. Davis. "Ecological Aspects of Plutonium Dissemination in Terrestrial Environments." *Health Physics* 22 (1972): 551–57.

Rozental, S., ed. *Niels Bohr*. London: North-Holland, 1967.

Ruhe, R. V. *Geomorphic Surfaces and Surficial Deposits in Southern New Mexico*. New Mexico Bureau of Mines and Mineral Resources Memoir 18. Socorro: New Mexico Bureau of Mines and Mineral Resources, 1967.

Rupp, C. R. "Stream-Grade Variation in Riparian Forest Ecology Along Passage Creek, Virginia." *Bulletin Torr. Botanical Club* 109 (1982): 488–99.

Sayre, W. W., and F. W. Chang. *A Laboratory Investigation of Open-Channel Dispersion Processes for Dissolved, Suspended, and Floating Dispersants*. U.S. Geological Survey Professional Paper 433-E. Washington, D.C.: U.S. Geological Survey, 1968.

Sayre, W. W., H. P. Guy, and A. R. Chamberlain. *Uptake and Transport of Radionuclides by Stream Sediments*. U.S. Geological Survey Professional Paper 433-A. Washington, D.C.: U.S. Geological Survey, 1963.

Sayre, W. W., and D. W. Hubbell. *Transport and Dispersion of Labeled Bed Material, North Loup River, Nebraska*. U.S. Geological Survey Professional Paper 433-C. Washington, D.C.: U.S. Geological Survey, 1965.

Schulz, R. K. "Root Uptake of Transuranic Elements." In *Transuranics in Natural Environments*, edited by M. G. White and P. B. Dunaway, pp. 92–99. U.S. Energy Research and Development Administration Report NVO-178. Washington, D.C.: U.S. Energy Research and Development Administration, 1977.

Schulz, R. K., R. Overstreet, and I. Barshad. "On the Soil Chemistry of Cesium-137." *Soil Science* 89 (1960): 16–27.

Schumm, S. A., and H. R. Kahn. "Experimental Study of Channel Patterns." *Geological Society of America Bulletin* 83 (1972): 1755–70.

Scott, M. R. "Thorium and Uranium Concentrations and Isotopic Ratios in River Sediments." *Earth and Planetary Science Letters* 4 (1968): 245–52.

Seaborg, G. T. "Early History of Heavy Isotope Production at Berkeley." Lawrence Berkeley Laboratory Report. Mimeo, 1976.

——. *The Transuranic Elements*. New Haven, Conn.: Yale University Press, 1958.

Seaborg, G. T., and W. D. Loveland. *The Elements Beyond Uranium*. New York: Wiley, 1990.

Segre, E. *Enrico Fermi, Physicist*. Chicago: University of Chicago Press, 1970.

——. *From X-Rays to Quarks: Modern Physicists and Their Discoveries*. San Francisco: Freeman, 1980.

Seibenthal, C. E. *Geology and Water Resources of the San Luis Valley, Colorado*. U.S. Geological Survey Water-Supply Paper 240. Washington, D.C.: U.S. Geological Survey, 1910.

Shaw, E. M. *Hydrology in Practice*. Berkshire: Van Nostrand Reinhold, 1983.

Siemers, C. T., ed. *Silver Anniversary Guidebook: Ghost Ranch, Central-Northern New Mexico*. New Mexico Geological Society 25th Field Conference Guidebook. Socorro: New Mexico Geological Society, 1974.

Simmons, V. M. *The San Luis Valley*. Boulder, Colo.: Pruett, 1979.

Simons, D. B., R-M. Li, L-Y. Li, and M. J. Ballantine. *Erosion and Sedimentation Analysis of Rio Puerco and Rio Salado Watersheds*. Technical Report, U.S. Army Corps of Engineers, Albuquerque District Office. Fort Collins, Colo.: Simons, Li & Associates, 1981.

Simons, D. B., and F. Senturk. *Sediment Transport Technology*. Fort Collins, Colo.: Water Resources Publications, 1977.

Smith, A. K. "Manhattan Project: The Atomic Bomb." In *The Nuclear Almanac: Confronting the Atom in War and Peace*, edited by J. Dennis, pp. 20–52. Reading, Mass.: Addion-Wesley, 1984.

Solomons, W., and U. Förstner. *Metals in the Hydrocycle*. Berlin: Springer-Verlag, 1984.

Spiegel, Z. "Late Cenozoic Sediments of the Lower Jemez River Region." In *Guidebook of the Albuquerque Country*, edited by S. A. Northrop, pp. 132–38. New Mexico Geological Society 12th Field Conference Guidebook. Socorro: New Mexico Geological Society, 1961.

Stewart, P. "Watershed Characteristics, Part B: Decreases in Flow in the Rio Grande." U.S. Forest Service Report, Carson National Forest, Taos, N.M. Mimeo, 1981.

Stockholm International Peace Research Institute. *SIPRI Annual Yearbook, World Armaments and Disarmament*. Cambridge, Mass.: MIT Press, 1987.

Stoker, A., A. J. Ahlquist, D. L. Mayfield, W. R. Hansen, A. D. Talley, and W. D. Purtymun. *Radiological Survey of the Site of a Former Radioactive Liquid Waste Treatment Plant (TA-45) and the Effluent Receiving Areas of Acid, Pueblo, and Los Alamos Canyons, Los Alamos, New Mexico*. Los Alamos National Laboratory and U.S. Department of Energy Report LA-8890-ENV, UC-70. Los Alamos, N.M.: Los Alamos National Laboratory, 1981.

Stone, R. S. "Health Protection Activities of the Plutonium Project." *Proceedings of the American Philosophical Society* 90 (1946): 11–19.

Stuewer, R. H., ed. *Nuclear Physics in Retrospect*. Minneapolis: University of Minnesota Press, 1979.

Szaz, F. M. *The Day the Sun Rose Twice: The Story of the Trinity Site Nuclear Explosion, July 16, 1945*. Albuquerque: University of New Mexico Press, 1984.

Tedford, R. H. "Neogene Stratigraphy of the Northwestern Albuquerque Basin." In *Albuquerque Country II*, edited by J. A. Grambling and S. G. Wells, pp. 273–78. New Mexico Geological Society 33rd Field Conference Guidebook. Socorro: New Mexico Geological Society, 1982.

Templeton, W. L., R. E. Nakatani, and E. Held. In *Radioactivity in the Marine Environment*, edited by A. H. Seymour. Washington, D.C.: National Academy Press, 1971.

Thompson, C. B. "Importance of Phreatophytes in Water Supply." *American Society of Civil Engineers, Irrigation and Drainage Division, Proceedings* 84(IRI) (1958): 1502/1–1502/17.

Thompson, J. C. "Channel Stabilization, Middle Rio Grande Project." In *Symposium on Channel Stabilization Problems*, vol. 3, chap. 1. Technical Report 1. Vicksburg, Miss.: U.S. Army Corps of Engineers, 1965.

Trimble, S. W., and S. W. Lund. *Soil Conservation and the Reduction of Erosion and Sedimentation in the Coon Creek Basin, Wisconsin*. U.S. Geological Survey Professional Paper 1234. Washington, D.C.: U.S. Geological Survey, 1982.

Tweto, O. *Geologic Map of Colorado* [Scale 1: 500,000]. Reston, Va.: U.S. Geological Survey, 1979.

Upson, J. E. "Physiographic Subdivisions of the San Luis Valley, Southern Colorado." *Journal of Geology* 47 (1939): 721–36.

——. "Physiographic Subdivisions of the San Luis Valley, Southern Colorado (Reprinted with Additional Notes)." In *Guidebook of the San Luis Basin, Colorado*, edited by H. L. James, pp. 113–22. New Mexico Geological Society 22nd Field Conference Guidebook. Socorro: New Mexico Geological Society, 1971.

U.S. Bureau of Radiological Health and Training Institute. *Radiological Health Handbook*. U.S. Public Health Service Publication 2016. Washington, D.C.: U.S. Department of Health, Education, and Welfare, 1972.

U.S. Bureau of Reclamation. *Annual Report, 1979*. Washington, D.C.: U.S. Bureau of Reclamation, 1979.

U.S. Congress. House of Representatives. *Rio Grande and Tributaries, New Mexico*. 81st Cong., 1st sess., H. Doc. 243. Washington, D.C.: U.S. Government Printing Office, 1948.

U.S. Department of Energy. *U.S. Department of Energy Order 5400.1, General Environmental Protection Program*. Washington, D.C.: U.S. Department of Energy, 1988.

——. *U.S. Department of Energy Order 5484.1*. Washington, D.C.: U.S. Department of Energy, 1981.

Van Cleave, M. "Vegetation Changes in the Middle Rio Grande Conservancy District." M.S. thesis, University of New Mexico, 1935.

Warner, R. F., ed. *Fluvial Geomorphology of Australia*. Sydney: Academic Press, 1988.

Waters, F. *The Woman at Otowi Crossing*. Athens, Ohio: Swallow Press, 1966.

Watson, J. R. "Plant Geography of North Central New Mexico." *Contributions from the Hull Botanical Laboratory* 160 (1913): 194–217. [Reprinted from *Botanical Gazette* 54 (1912): 194–217]

Watters, R. L. "Ecological Transport." In *Actinides in Man and Animals*, edited by M. E. Wrenn, pp. 503–7. Cleveland: R. D. Press, 1981.

Watters, R. L., D. N. Edgington, T. E. Hakonson, W. C. Hanson, and M. H. Smith. "Synthesis of the Research Literature." In *Transuranic Elements in the Environment*, edited by W. C. Hanson, pp. 1–10. U.S. Department of Energy Report DOE/TIC-22800. Washington, D.C.: U.S. Department of Energy and National Technical Information Service, 1980.

Watters, R. L., T. E. Hakonson, and L. J. Lane. "The Behavior of Actinides in the Environments." *Radiochimica Acta* 32 (1983): 89–103.

Weast, R. C. *CRC Handbook of Chemistry and Physics*. Boca Raton, Fla.: CRC Press, 1988.

Weimer, W. C., R. R. Kinnison, and J. H. Reeves. *Survey of Radionuclide Distributions Resulting from the Church Rock, New Mexico, Uranium Mill Tailings Pond Dam Failure*. U.S. Nuclear Regulatory Commission Report NUREG/DR-2449, PNL-4122. Richland, Wash.: Battelle Pacific Northwest Laboratory, 1981.

Welsh, M. E. *U.S. Army Corps of Engineers: Albuquerque District, 1935–1985*. Albuquerque: University of New Mexico Press, 1987.

Wendorf, F. "A Reconstruction of Northern Rio Grande Prehistory." *American Anthropologist* 56 (1954): 200–221.

Wertz, J. B. "Logarithmic Pattern in River Placer Deposits." *Economic Geology* 44 (1949): 193–209.

Whicker, F. W., C. A. Little, and T. F. Winsor. *Symposium on Environmental Surveillance Around Nuclear Installations, 5–9 November 1973*. Warsaw: International Atomic Energy Commission, 1973.

Whicker, F. W., and C. M. Loveless. "Relationships of Physiography and Microclimate to Fallout Deposition." *Ecology* 49 (1968): 363–68.

Whicker, F. W., and V. Schultz. *Radioecology: Nuclear Energy and the Environment*. 2 vols. Boca Raton, Fla.: CRC Press, 1982.

White, G. C., and T. E. Hakonson. "Statistical Considerations and Survey of Plutonium Concentration Variability in Some Terrestrial Ecosystem Components." *Journal of Environmental Quality* 8 (1979): 176–82.

White, P. S. "Pattern, Process and Natural Disturbance in Vegetation." *Botanical Review* 45 (1979): 229–99.

Wiard, L. *Floods in New Mexico, Magnitude and Frequency*. U.S. Geological Survey Circular 464. Washington, D.C.: U.S. Geological Survey, 1962.

Wilding, R. E., and T. R. Garland. "The Relationship of Microbial Processes to the Fate and Behavior of Transuranic Elements in Soils, Plants, and Animals." In *Transuranic Elements in the Environment*, edited by W. C. Hanson, pp. 31–37. U.S. Department of Energy Report DOE/TIC-22800. Washington, D.C.: U.S. Department of Energy and National Technical Information Service, 1980.

Williams, G. P., and M. G. Wolman. *Downstream Effects of Dams on Alluvial Rivers*. U.S. Geological Survey Professional Paper 1286. Washington, D.C.: U.S. Geological Survey, 1984.

Williams, J. L., ed. *New Mexico in Maps*. 2nd ed. Albuquerque: University of New Mexico Press, 1986.

Wilson, D. *Rutherford*. Cambridge, Mass.: MIT Press, 1983.

Wilson, L. *Future Water Issues, Taos County, New Mexico*. Santa Fe, N.M.: Lee Wilson and Associates, 1980.

Wise, S. M. "Cesium-137 and Lead-210: A Review of the Techniques and Some Applications in Geomorphology." In *Timescales in Geomorphology*, edited by R. A. Cullingford, D. A. Davidson, and J. Lewin, pp. 110–27. Chichester: Wiley, 1980.

Wolfenden, P. and J. Lewin. "Distribution of Metal Pollutants in Active Stream Sediments." *Catena* 5 (1978): 67–78.

——. "Distribution of Metal Pollutants in Floodplain Sediments." *Catena* 4 (1977): 309–17.

Wolman, M. G., and L. B. Leopold. *River Flood Plains: Some Observations on Their Formation*. U.S. Geological Survey Professional Paper 282-C. Washington, D.C.: U.S. Geological Survey, 1957.

Woodson, R. C. "Stabilization of the Middle Rio Grande in New Mexico." *American Society of Civil Engineers, Journal of the Waterways and Harbor Division* 87 (1961): 1–15.

Woodward, L. A. "Tectonics of Central-Northern New Mexico." In *Silver Anniversary Guidebook: Ghost Ranch, Central-Northern New Mexico*, edited by C. T. Siemers, pp. 123–30. New Mexico Geological Society 25th Field Conference Guidebook. Socorro: New Mexico Geological Society, 1974.

Yamamoto, M., T. Tanii, and M. Sakanoue. "Characteristics of Fallout Plutonium in Soil." *Journal of Radiation Research* 22 (1981): 134–42.

Index

Numbers in boldface refer to pages on which maps appear; numbers in italics refer to pages on which photos appear.

Middlebury College
0 00 02 0651365 7